BIRDS AND FROGS

Selected Papers, 1990–2014

BIRDS AND FROGS

SELECTED PAPERS, 1990–2014

Freeman J. Dyson

Institute for Advanced Study, Princeton

 **World Scientific**

NEW JERSEY · LONDON · SINGAPORE · BEIJING · SHANGHAI · HONG KONG · TAIPEI · CHENNAI

Published by

World Scientific Publishing Co. Pte. Ltd.

5 Toh Tuck Link, Singapore 596224

USA office: 27 Warren Street, Suite 401-402, Hackensack, NJ 07601

UK office: 57 Shelton Street, Covent Garden, London WC2H 9HE

Library of Congress Cataloging-in-Publication Data

Dyson, Freeman J.

   [Works. Selections. 2015]

   Birds and frogs : selected papers, 1990–2014 / Freeman J Dyson, IAS [Institute for Advanced Studies], Princeton.

     pages cm

   Follows author's Selected papers of Freeman Dyson with commentary (Providence, R.I. : American Mathematical Society, 1996).

   Includes bibliographical references and index.

   ISBN 978-9814602853 (hardcover : alk. paper) -- ISBN 981460285X (hardcover : alk. paper) --

   ISBN 978-9814602860 (pbk. : alk. paper) -- ISBN 9814602868 (pbk. : alk. paper)

   1. Field theory (Physics) 2. Quantum electrodynamics. 3. Gravitation. 4. Number theory. 5. Biology.

I. Dyson, Freeman J. Selected papers of Freeman Dyson with commentary. II. Title.

   QC173.7.D973 2015

   530.14--dc23

                                      2014042306

British Library Cataloguing-in-Publication Data

A catalogue record for this book is available from the British Library.

Printed in Singapore

CONTENTS

Section 1

Introduction and Commentary

CHAPTER 1

INTRODUCTION

I am grateful to Dr. K. K. Phua at World Scientific Publishing Company for publishing this collection of papers from my later years. It is a sequel to two earlier volumes, "Selected Papers of Freeman Dyson with Commentary", published by the American Mathematical Society in 1996, and "From Eros to Gaia", published by Pantheon Books in 1992. The two earlier collections contained technical and non-technical papers separately. This collection contains both non-technical and technical papers. For the convenience of readers, the non-technical papers are divided into popular science (Section 2), memoirs of people I have known (Section 3), and public affairs (Section 4), and the technical papers are put together at the end (Section 5). During the last 24 years, roughly half of my working hours have been devoted to scientific problems and half to human problems. I spend half my time calculating and half writing.

In this commentary I discuss the papers in chronological order. The starting-point is the year 1990, when I was more concerned than usual with human problems. I gave some public lectures on various ethical problems that were emerging from the successes and failures of modern science. I was invited to summarize these lectures in an essay, "Science in Trouble", published in the American Scholar, a literary magazine edited by Joseph Epstein. The essay is item (4.1) in this collection. It marks the start of my career as a social critic. It begins with an aggressive statement: "Trouble comes to science on three levels: personal, local and global". On each level I found ethical problems threatening the future of the scientific enterprise. The essay was widely read and widely resented. Never in my life as a writer have I received such a barrage of angry letters from readers. I had successfully gored a number of sacred cows. I wrote in response to thank my critics, telling them that the purpose of criticism is not to compel agreement but to provoke argument.

Returning to my roots in England, I contributed item (3.1) to the bicentennial celebration of the self-taught mathematician George Green at the University of

Nottingham. Green was the son of a miller in Nottingham, and as a child he had to spend long hours in a revolving room at the top of the windmill, adjusting the pitch of the blades to the strength and direction of the wind. Somehow he got hold of the works of the French mathematicians Laplace and Lagrange, and studied them in the windmill when the wind was quiet. He invented Green's Functions, which are useful devices for solving dynamical equations. He introduced French analytical methods to England, where nothing much had been done in mathematics since the death of Newton. The windmill still stands, now converted into a science museum and educational center for the local schoolchildren. Julian Schwinger and I were invited to talk at the celebration, since we had both made intensive use of Green's Functions in our development of quantum electrodynamics. We spoke about the revolution in physics that happened in the 1940s when Green's Functions were applied to quantum theory. The final act of the celebration was the dedication of a memorial tablet to Green in Westminster Abbey, with a picture of the windmill.

Item (3.2) is a short piece published by the American Philosophical Society to celebrate the astronomer James Bradley. I call him the inventor of modern science because he was the first person who ever measured anything to six-figure accuracy. He trained his assistant Charles Mason to achieve this level of accuracy, and Mason put his skill to good use when he surveyed the Mason–Dixon line separating Maryland from Pennsylvania.

The years 1990–2006 were the final years of a fifty-year friendship and collaboration with the Indian–French mathematical physicist Madan Lal Mehta. Mehta's brilliant work on the theory of random matrices attracted me to work on that subject in 1960. He was born in India in 1932, moved to France in 1958 and became a French citizen in 1971. I invited him to come as a member to the Institute for Advanced Study for the year 1962–1963, and we collaborated intensively from that time forward. The constant exchange of messages and ideas with Mehta was one of the main joys of my life as a scientist.

One of Mehta's elegant contributions to random matrix theory was a paper, "A non-linear differential equation and a Fredholm determinant", published in the Journal de Physique in 1992. That paper stimulated my response, item (5.1). It was appropriately published in the volume, "Chen Ning Yang, a Great Physicist of the Twentieth Century", edited by Shing-Tung Yao, celebrating the seventieth birthday of Yang. At the beginning of my paper, I explain why the dedication to Yang is appropriate. The paper imitates the style of the classic calculation of the long-range order of a two-dimensional Ising ferromagnet, published by Yang in 1952. I remarked that imitation is the sincerest form of flattery, even if it comes forty years late.

In my paper (5.1), I used Mehta's system of nonlinear equations to find the asymptotic behavior of the eigenvalues of a random matrix when the number of levels is large and the distortion of the series by external forces is also large. The crystalline long-range order of the eigenvalues is exactly described by Jacobian elliptic functions instead of by the sines and cosines that appear in previously known examples of random-matrix eigenvalue distributions. I had found a new asymptotic form for a function known as the Fifth Painlevé Trancendent, defined in a classic paper by the French mathematician P. Painlevé in 1902. My paper is an extreme example of mathematical baroque style applied to an unimportant physical problem. It was my farewell tribute to the beauty of random matrix theory and to the unique personality of Mehta.

Mehta loved to travel around the world with little money and few worldly possessions, adapting easily to alien cultures. In 1993, already over sixty years old and not in the best of health, he made a classic journey overland from Beijing through Tibet to India, traveling like a native on buses and trucks, and when necessary on foot. One section of the trip was a bus-ride from Golmud, in China north of Tibet, to Lhasa, crossing the central Tibetan plateau. That was before the modern railway over the plateau was built. He told us with pride that the Tibetans and Chinese on the bus were mostly suffering from altitude sickness, but he was not. After retiring from his position in France, he returned to his home in India and died there in 2006.

For the academic year 1995–1996, the French particle physicist Thibault Damour was a visiting member at the Institute for Advanced Study, and item (5.2) is the result of our collaboration. We analyzed the evidence from isotope ratios of rare-earth nuclei in the ancient fission reactors at Oklo in West Africa, looking for effects of any possible variation of the constants of nature during the two billion years since the reactors were operating. Alexandr Shlyakhter had the idea of looking for such evidence when he was a student in Leningrad in 1976. He showed that the isotope ratios give far stronger bounds on the variation of constants than any other evidence. We confirmed Shlyakhter's results and made them more precise. Shlyakhter emigrated from the Soviet Union and came to MIT in 1989. I spoke to him a few times by telephone but never met him. In the year 1998, I heard the devastating news that he was struggling with a brain tumor, and in 2000 he died.

In 1994 I retired from my position as active Professor and became Professor Emeritus at the Institute for Advanced Study. This meant that I remained a member of the Institute community but was free to travel and take other jobs. I spent the Fall of 1994 as visiting professor at Dartmouth College, invited by my historian friend Martin Sherwin. Martin and I taught a course together called "The Nuclear Age". He was then already hard at work on the book, "American Prometheus", a biography of Robert Oppenheimer finally published in 2005. The term at Dartmouth strengthened my interest in history and gave me a chance to interact strongly with undergraduates. In 1999, I spent a term teaching at Gustavus Adolphus College in

Minnesota. There I taught a "Science and Society" course with Professor Larry Potts, focussed on ethical problems of biotechnology rather than on nuclear weapons. My reception at Gustavus Adolphus was especially friendly, because the campus had been half destroyed by a huge tornado one year earlier. We were living and working in half-destroyed buildings, sharing the hardships and helping each other to survive.

In 1998, I was actively engaged in revising my book, "Origins of Life", originally published in 1985, for a second edition to be published in 1999. Biology had moved ahead rapidly in the intervening years, but I found that the origin of life remained an unsolved problem, as mysterious as ever. The main theme of my book was the hypothesis that life began twice, first with garbage-bag creatures lacking exact replication, and second with parasitic replicators using the garbage-bag creatures for life support. This theme did not need to be changed when the book was revised. It remains today a plausible hypothesis, without evidence either to contradict or to confirm it. Item (2.1) is a public lecture given at the Jet Propulsion Laboratory, summarizing the thoughts that went into the new edition of the book.

Item (3.3) is a banquet speech, celebrating my friend Yang Chen Ning on the occasion of his retirement from Stony Brook in 1999. I called attention to three outstanding qualities of Yang that are rarely combined. First, a marvelous mathematical skill, enabling him to solve technical problems. Second, a deep understanding of nature, enabling him to ask important questions. Third, a community spirit, enabling him to play a major role in the rebirth of Chinese civilization. Together, these three qualities make him what he is, a conservative revolutionary who values the past and leads the way to the future.

Item (4.2) is the acceptance speech given at the Washington National Cathedral when I won the Templeton Prize for Progress in Religion. In the speech, I asked the question why I won the prize, but I did not answer it. I still do not know the answer. I am one of a large number of scientists who find no conflict between science and religion, provided that each side treats the other with respect. Unfortunately, public attention is concentrated on a minority of militant fundamentalists and militant atheists who enjoy attacking each other. They distract attention from the majority of believers and unbelievers who are glad to live together in peace. In Washington, I was speaking for the majority.

In 2000, I spent three weeks as visiting lecturer at the University of Canterbury at Christchurch, New Zealand. I gave six public lectures of which item (4.3) was one. It arose from an essay by Michael Armstrong, published in "Outlook" in 1983, describing in detail the way Tolstoy taught the village children on his estate. "Outlook" was a little magazine devoted to the education of young children. It was founded in 1970 by David Hawkins, the philosopher who served as Oppenheimer's personal assistant at Los Alamos. I subscribed to it from the beginning until it died in 1986. The Armstrong essay, with the title, "Tolstoy on education: The pedagogy of freedom", gave me the idea for a talk comparing Tolstoy with Napoleon. Before starting his village school, Tolstoy traveled around Europe examining the systems of public education established by Napoleon. Tolstoy was horrified by what he saw, and resolved to make his school as different as possible from the Napoleonic model. The clash between Tolstoy and Napoleon is sharp, not only in the education of children but also in our views of history, science and ethics. In all these areas, Napoleon works from the top down, Tolstoy from the bottom up. I am on the side of Tolstoy.

Item (5.3) is a little piece of recreational mathematics and needs no explanation. Item (4.4) is a personal meditation about childhood, spoken on a television program in Amsterdam directed by Wim Kayzer. Kayzer had the program published in a book, "Het Boek van de Schoonheid en de Troost", with the English contributions translated into Dutch. He intended to publish an English version, "The Book of Beauty and Consolation", with the Dutch contributions translated into English. The English version never appeared, and this is the first publication of my remarks in English. Of all the items here collected, this is my favorite.

Item (5.4) is a serious discussion of the foundations of quantum mechanics, written for a symposium in 2002 honoring John Wheeler. I describe four thought-experiments that lead me to the conclusion that quantum mechanics cannot be a complete description of nature. In my opinion, a complete description of nature must contain a classical world and a quantum world. Roughly speaking, the classical world describes the past and deals with facts, while the quantum world describes the future and deals with probabilities. The simplest example of this double description is a uranium atom. The wave-function of an out-going alpha-particle tells us the probability that the atom will decay tomorrow. But the wave-function cannot state the fact that the atom decayed yesterday. Only a dualistic picture of the universe can describe my four thought-experiments consistently. Of course, I do not expect these arguments to convert the modern quantum-experts to my point of view. We all agree about the practical consequences of quantum theory while continuing to disagree about the philosophical interpretation.

Item (5.5), written in collaboration with Jeremy Bernstein, is a piece of work done 44 years earlier when we worked together on Project Orion, a spaceship propelled by nuclear bombs. The basic feasibility of the project depends on the opacity of the bomb debris. The debris is supposed to transfer momentum to the ship without transferring too much energy. This is possible only if the debris is highly opaque to its own radiation. We studied the question, how large can the opacity of matter be, without violating the laws of physics, at a given density and temperature. The answer is simple. The maximum opacity is independent of density and inversely proportional to temperature. It turned out that for our spaceship, a mixture of light elements would provide an opacity close to the maximum. We published our work 44 years later because it is relevant to radiation transport in highly evolved stars.

The next three items are autobiographical. Item (2.2) is a description of my childhood and education, published as a chapter in a book, "When we were kids: How a child becomes a scientist", edited by John Brockman. Item (3.4) describes a meeting with Enrico Fermi in 1953, when he demolished in twenty minutes my theoretical explanation of his experimental results. He showed his greatness as a physicist by seeing intuitively that my theory of meson-proton scattering could not be right. He did me an enormous favor by saving me and my students from years of fruitless calculations.

Item (4.5) is an account of the failure of the Operational Research Section of the Royal Air Force Bomber Command, where I worked for the last two years of World War II. I call it a failure because we never solved the two main problems that faced the command. The first problem was bombing accuracy. To destroy important military targets such as oil-refineries required accuracy of a few hundred meters, while the actual accuracy until the last few months of the war was several kilometers. This problem was finally solved, not by the Operational Research Section, but by innovative tactics invented by the aircrew of 5 Group. The second problem was bomber losses. The losses were of the order of five per cent for operations over Germany. The campaign was enormously costly in human lives as well as in industrial resources. The main cause of losses in the later years of the war was fighter aircraft with guns firing vertically upward. A fighter directly below could destroy a bomber without being seen. Our Research Section never guessed that this was our main problem and only learned about it after the war was over. These two failures are typical of failures of intelligence systems in more recent wars. Intelligence systems tend to fail because they are designed to convey information from war-fighters to analysts far from the fighting, and they convey very little information back to the war-fighters who might use it effectively.

Item (3.5) is a memoir of Edward Teller written for the National Academy of Sciences. I enjoyed writing it because I had enjoyed working with Teller and remained his friend for fifty years. I tried as best I could to present a balanced picture of his faults and virtues, to counteract the efforts of his political enemies to demonize him. His three closest friends, Ernest Lawrence, Enrico Fermi and John von Neumann, great scientists who could have written memoirs with more authority than mine, all died young.

Item (4.6) is a talk that I gave to the Medina Seminar, an annual event at Princeton University. Each year, a group of distinguished judges comes for two days of intellectual refreshment. The judges are a highly intelligent and critical audience. I spoke about the proper balance between individual rights and group rights. My thoughts on this subject were stimulated by Caroline Humphrey, an anthropologist who is fluent in Russian. She had given a talk to the American Philosophical Society about the difficulty of translating the English word "Freedom" into Russian. There are three Russian words that are used to translate "Freedom", and none of them has the right meaning. The problem arises from the fact that the ideas of freedom in English and Russian culture are basically different. In English, freedom means the right of an individual to live without coercion by the community. In Russian, freedom means the right of a community to manage its own affairs. This is why many people in Russia believe that they had more freedom in the days of Stalin than they have today. Starting from this linguistic problem, I discussed the delicate relation between individual and group rights in law, and between individual selection and group selection in evolutionary biology. American laws and customs are exceptional in giving precedence to the individual, while the majority of laws and customs in other societies give precedence to the community. If we are to live together in peace, we must understand our differences and be prepared to make compromises.

Item (2.3) gave its title to the book. I wrote an essay with this title for the American Mathematical Society. It was supposed to be a lecture in honor of Albert Einstein, to be given at the meeting of the Society in Vancouver. I got sick and never gave the lecture, but it was published in the American Mathematical Society Notices, and later, revised and translated into Russian, in the Uspekhi Fizicheskikh Nauk. The version published here is the English text used for the Russian translation. The birds and frogs in the essay are the two kinds of mathematician that I have encountered during my working life. But not all birds and frogs are mathematicians. They can be people in all walks of life. Birds are those who fly high and survey the landscape over a wide area. Frogs are those who stay on the ground and solve one problem at a time. Not only in science but in other human enterprises such as literature and politics, birds and frogs play leading roles and both are needed. That is why the title is appropriate for the book as a whole.

Items (2.4) and (4.7) were two talks written at the same time for different audiences, a scientific talk for an unscientific audience and an unscientific talk for a scientific audience. Item (2.4) is another talk to the judges at the Medina Seminar in Princeton. After my talk about legal problems the year before, some of the judges said that they would like to hear me talk about my own subject, not about theirs. So I spoke about the history of physics, painting a picture with a broad brush, starting with Oswald Spengler's vision of the Decline of the West, and ending with the triumph of the Large Hadron Collider in Geneva. Item (4.7) is a talk given a month later about human and political problems to an audience of scientists in California. I spoke about the two practical issues that concern me most deeply, the abolition of nuclear weapons and the development of genetic engineering. I see these two issues as overwhelmingly more important for the future of humanity than the other problems arising out of science and technology. By the abolition of nuclear weapons I mean legal abolition, not physical abolition. Legal abolition means that we sign an international agreement to outlaw nuclear weapons and reliably get rid of our own weapons. We will thereafter be living in an uncertain world with illegal weapons possibly hidden away in various places. I maintain that this uncertain world is vastly less dangerous than the world we now inhabit. The development of genetic engineering will transform every aspect of economic life, and has the promise of equalizing the distribution of wealth over our planet. The creative use of genetic engineering could bring wealth to any region with access to water, soil, sunshine and modern communications.

Items (3.6)–(3.8) are memoirs written for three great scientsts that I had the good luck to know, John Wheeler for the American Philosophical Society, Subrahmanian Chandrasekhar for the American Institute of Physics, and Nicholas Kemmer for the Royal Society of London. All three were warm and generous human beings as well as deep thinkers. Unfortunately the American Philosophical Society allowed me only four small pages, far too small a space for an adequate memorial to a long-lived and many-sided genius. The Royal Society, on the contrary, maintains a tradition of publishing biographical memoirs of its fellows without any restriction on length. That is why the memoir of Wheeler is the shortest and the memoir of Kemmer the longest. All three of them deserve a more extensive biography. A good biography of Chandrasekhar was published by Kameshwar Wali. But Wheeler and Kemmer were pre-eminent as teachers. Their star students, Richard Feynman and Peter Higgs, are now more famous than their teachers.

Item (5.6) is a technical contribution to the mathematical theory of games. It is of interest to a wider public because the game of Prisoner's Dilemma is often used by biologists as a model for the evolution of cooperation in a population of selfish individuals. William Press, who is an expert on game theory, discovered a remarkable new strategy for Prisoner's Dilemma which he calls the Extortion Strategy. Press had the idea and worked out numerical examples to prove that it worked. My contribution to the collaboration was only to translate Press's idea into old-fashioned mathematical equations. I found that the strategy works if a certain determinant is equal to zero. We published our paper in the Proceedings of the National Academy of Sciences, where it attracted considerable attention from biologists. It turned out that the extortion strategy is not a good way to evolve cooperation. On the contrary, it leads to domination of one player by the other, and can be defeated by an opponent who is determined not to cooperate. One year later, the situation was clarified in a paper by Alexander Stewart and Joshua Plotkin with the title, "From extortion to generosity: The evolution of zero-determinant strategies in the prisoner's dilemma", also published in the Proceedings of the National Academy of Sciences. Alexander and Plotkin use Press's idea and turn it upside-down. Instead of dominating the opponent by an unfair division of spoils, the generous version of the Press strategy guides the opponent to cooperate by a division that is unfair in the opposite direction. The generous strategy uses a short-range advantage for the opponent to achieve a long-range advantage for both players. The Stewart–Plotkin strategy happily confirms what moral sentiments suggest: generosity wins. Of course the Prisoner's Dilemma is a grossly over-simplified model, and there is no reason to believe that the behavior of the model resembles the behavior of real animals or humans evolving in the real world.

Item (5.7) is a talk given at the 125th birthday celebration of the Indian mathematical genius Svinivasa Ramanujan at the University of Florida. It is a piece of recreational mathematics, demonstrating that many beautiful questions raised by Ramanujan are still unanswered.

Item (2.5) is a talk written for a conference at Brown University to celebrate the life and work of John von Neumann. In 1987, I had given a talk at a similar conference at the University of Illinois with the title, "A Walk Through Ramanujan's Garden", using the metaphor of a garden to give shape and structure to Ramanujan's discoveries. In 2010, I used the same metaphor to give coherence to von Neumann's diverse achievements. I emphasized the theme of mathematical logic which was central to von Neumann's thinking, whether he was dealing with the foundations of mathematics as a student or with practical problems of applied mathematics at the end of his life. I had not intended the talk to be published, and left the text in an unfinished state. Then, two years later, I received a friendly letter from Marina Whitman, the daughter of von Neumann, who had heard the talk at Brown University. She said the talk had given her for the first time a clear picture of her father's style of thinking, and urged me to publish it. After that, I polished the manuscript and sent it to the American Mathematical Society.

The last three items, (2.6), (4.8) and (5.8), are talks given within the last year to different audiences about different subjects. All of them are technical in substance, though only the last one is technical in style. They all deal with unsolved problems at the edge of science and technology. In all three talks, I am guessing what the future may bring with new ideas and new tools that have not yet been invented. Item (2.6) is a talk given at the Rockefeller University to an audience of experts, mostly biologists, asking the question whether the human brain works as an analog computer comparing images or as a digital computer comparing numbers. I examine various lines of evidence that lead me to conclude that the correct answer is analog. This is a tentative conclusion that could easily turn out to be wrong. The strongest evidence is the dismal failure of efforts to build digital computers with anything approaching the intellectual capabilities of a three-year-old human child. Seventy years of promises to achieve artificial intelligence with digital machinery never came close to fulfilment. I am suggesting that we shall in the end achieve artificial intelligence, when we understand how the brain works as an analog computer and learn to copy its architecture.

Item (4.8) is a talk given to a mixed audience of space-scientists and science-fiction writers in San Diego. The occasion gave me a chance to let imagination run riot, describing my hopes for the long-range future in concrete form. I talked about two objects that embody the possible destiny of life in the universe. The Noah's Ark Egg is a thing of beauty, weighing a few kilograms and looking like an ordinary ostrich egg. Instead of growing into a single bird, it grows into a complete ecosystem with millions of species of microbes and plants and animals, ready to grow and evolve into a living community wherever it finds a suitable world to colonize. The Viviparous Plant is the most essential offspring of the Noah's Ark Egg. The Viviparous Plant is a plant that grows a transparent greenhouse around itself, and then gives birth to new plants which are protected by the warmth of the greenhouse until they are ready to break out into the cold world outside. The Noah's Ark Egg, with its ecology based on viviparous plants using greenhouses to keep warm, allows life to flourish on cold and small celestial bodies where it has access to ice and minerals and starlight. Such objects provide far more habitable area than planets.

Item (5.8) is a technical talk given at Nanyang Technological University in Singapore, to an audience of scientists and students at my ninetieth birthday conference. I spoke about a question that I have often thought about but never been able to answer. Is it in principle possible to observe the effects of individual gravitons? A slightly different version of the question asks whether we can design a thought-experiment to test the theory of quantum gravity. The talk describes three different kinds of graviton-detectors that might answer the question positively if any of them can be demonstrated to work. The three kinds are: first, a LIGO-style detector, measuring the change in the distance between two objects caused by a low-energy graviton passing by; second, a gravito-electric detector, observing an electron or other particle kicked out of an atom by a medium-energy graviton; third, a coherent photon detector, observing the coherent mixing of electromagnetic and gravitational fields at high energies. The only result that I am able to demonstrate mathematically is a negative one. The LIGO-style detector fails to detect single gravitons as a consequence of the laws of quantum mechanics and general relativity. This is an interesting and possibly important result, but it does not answer the question whether graviton-detectors of other kinds must fail. I end my commentary by thanking Dr. Phua for organizing the Singapore conference, and for giving me a chance to meet with a crowd of bright young colleagues from China and other parts of Asia.

SECTION 2

TALKS ABOUT SCIENCE

THE ORIGINS OF LIFE IN THE UNIVERSE*

1. Gravity is Cool

Most of the people in this room know far more about the details of space exploration and the search for life on Mars and Europa than I do. I will not try to tell you how to do your jobs. I am speaking here as an ignorant outsider, and so I will talk about things that lie outside your immediate concerns. I will talk about a general framework of ideas into which the search for life in the universe can fit. I will talk about the origins of life, a subject about which we are all equally ignorant.

To begin with, I will talk about the astronomical universe. We know that the universe is friendly to the evolution of life. Otherwise we would not be here to discuss it. But we don't know why the universe is so friendly. To make it possible for life to evolve either on the earth or anywhere else, two things were essential. There had to be violent events with very short time-scales, supernova explosions to bring the chemical elements essential to life out of the insides of stars where they are made, collisions between planetesimals crashing together to form planets. And there had to be long quiet periods in which slowly evolving life was sheltered from violence. Time in our universe has always had two faces, the quick violent face and the slow gentle face, the face of the destroyer and the face of the preserver. How did these two faces of time come to coexist in a single universe?

In the old days, people who thought about the natural history of the earth were divided into catastrophists and uniformitarians. Catastrophists believed that time molds the earth by means of sudden catastrophes, the most famous example being Noah's flood. Uniformitarians believed that time works slowly, molding the earth by long-continued action of the same processes that we observe at work today. We now know that both sides in the debate were right. Time loves to sit quietly for

---

*Lecture given at Jet Propulsion Laboratory, Pasadena, California, August 13, 1998.

millions of years, and then to pounce suddenly in a single hour of fury. Nothing is permanent, but the illusion of permanence can last for a long time. We see in the channeled scablands of Washington State the traces of a flood more violent than Noah's, when a huge lake formed by the melting Canadian ice-cap broke through an ice-dam and destroyed everything in its path as it roared down to the sea. And we see, a few hundred miles to the north in Canada, the Burgess shale, where delicate fossils of our ancient ancestors have been marvelously preserved in rocks that are five hundred million years old.

When we move from the earth to the heavens, the contrast between the quiet and the stormy faces of time becomes even starker. In our quiet little corner of the universe, the sun, moon and planets serenely ride in their orbits, the stars shine steadily for billions of years. Elsewhere, in places remote from us in space or time, sudden cataclysms fill the sky with violence as heavenly bodies are born or die. The birth of our moon was such a cataclysm, more than four billion years ago, when the earth was young. Another planet, smaller than the earth but larger than the moon, collided obliquely with the earth. In a few minutes the incoming planet was disrupted, half of it plunged into the earth, the other half formed a ring of orbiting debris which condensed to make our moon. Whatever structures existed on the earth before the collision were totally obliterated. The heat of impact made the earth white-hot, with an atmosphere of vaporized rock. Afterwards the earth cooled down and became the planet that we know. No object of comparable size has collided with it in more recent times. Later impacts were smaller and did only superficial damage. But we see, far away across the universe, cataclysms vastly more violent than the collisions of planets. We see supernova explosions in which the core of a massive star collapses in a fraction of a second and the envelope of the star is ejected into space. We know that such an explosion must have occurred in our neighborhood shortly before the earth and the sun were born, to produce the mixture of heavy atoms that we see on earth today. And we see other cataclysms even more abrupt and energetic than supernova explosions. We call these events gamma-ray bursts, but nobody knows how or why they happen. The universe is full of cataclysms. Luckily for us, they are separated by immense distances and by eons of time. During the intervals between cataclysms, time shows her quiet face and life can survive and prosper.

The existence of life depends crucially on the fact that time has these two faces, the quiet and the violent, cleanly separated from each other. The violent face created the stuff that we are made of. The quiet face sustains us and allows us to evolve. We need to understand the reasons why these faces exist. There are two main reasons, one superficial and one fundamental. The superficial reason is that the universe is extravagantly large. Distances are so great that accidental collisions hardly ever happen. This is why the sun with its family of planets can run like a perfect clock, with orbits undisturbed by close encounters with alien stars, for billions of years. We are protected from cataclysms by the sheer size of the interstellar spaces.

The more fundamental reason for the two faces of time is the fact that the universe is dominated by the force of gravitation. Gravitational energy is quantitatively the largest reserve of energy, and qualitatively the least disordered. It carries zero entropy. Because of its superior quality, gravitational energy can change easily and irreversibly into other forms of energy. Other forms of energy are associated with disorder and heat, but gravitational energy is cool. Gravity is cool in both senses of the word. This is why gravitational energy can drive turbines in a hydroelectric power-station with almost a hundred percent efficiency, while other kinds of power-station, powered by coal, oil, natural gas or uranium, struggle to reach fifty percent. The gravitational energy of the ice in the high Canadian ice-cap during the last ice age waited quietly for a hundred thousand years, until the thaw and the break in the ice-dam released it. As soon as it was released, it changed into the turbulent energy of the flood that excavated a trillion tons of rock and created the scablands. Similarly, the gravitational energy of the planet that collided with the earth stayed cool for millions of years, and then suddenly changed into heat at the moment of impact when the moon was born. All over the universe, when conditions are right for gravitational energy to be released, it can change instantly into heat and radiation, and a cataclysm results. The two faces of time are a consequence of the two faces of gravitation. Gravitation is the ordering principle that holds our earth together as a stage for us to walk on, and gravitation is the ultimate reservoir of energy that can smash our world to pieces.

The best popular account of the science that explains how the universe can be friendly to life is a book, "Creation of the Universe", by the Chinese astronomers Fang Li Zhi and Li Shu Xian. The book was translated into English by T. Kiang and published by World Scientific Publishing Company of Singapore in 1989. Fang Li Zhi is the famous dissident astronomer now living in exile in the United States. I particularly recommend Chapter 6, with the title "How Order was born of Chaos". This tells the same story that I am telling you today, but with more detail and more depth.

2. RNA and Garbage

Now I come to the second question, how life might have actually begun, either here or on Mars or anywhere else. There are two strongly divergent views of the state of affairs when life began. One view is the RNA world, which says that life began with molecules of RNA, ribonucleic acid, replicating themselves and organizing the replication of other molecules. The other view is the garbage-bag world, which says that life began with little bags of garbage, membranes made of oily scum or other common chemicals that like to form membranes, enclosing volumes of dirty water containing miscellaneous garbage. The RNA world is the orthodox theory, the party line of the molecular biology community, accepted by the majority of expert biochemists and geneticists. The experts love RNA because it is marvelous stuff to do experiments with. You can put RNA into a test-tube and watch it replicate and evolve. The experts believe in the RNA world even more strongly since Thomas Cech discovered in 1986 that RNA can not only replicate itself but can also act as an enzyme to catalyze reactions between other molecules. RNA enzymes are called ribozymes and are an important part of the genetic machinery in all modern cells. The RNA world is a beautiful scene, with busy little ribozymes cooperating to organize the beginnings of life.

The garbage-bag world is not so elegant and not so widely accepted. It was originally proposed by the Russian biologist Oparin in the 1920s. Unfortunately Oparin later fell into ill repute, because he was a prominent academician at the end of his life during the bad times when Lysenko was destroying Russian genetics. The prevalent opinion among western biologists was that Oparin was a bad guy, and therefore his theory must have been bad too. Fortunately, bad guys sometimes produce good theories. One of the main proponents of the garbage-bag theory today is Doron Lancet, a chemist at the Weizmann Institute in Israel. I myself prefer it, partly because I like to be in the minority, and partly because I find it chemically more plausible. The idea of the garbage-bag world is that a random collection of molecules in a bag may occasionally contain catalysts that cause synthesis of other molecules that act as catalysts to synthesize other molecules, and so on. Very rarely a collection of molecules may arise that contains enough catalysts to reproduce the whole population as time goes on. The reproduction does not need to be precise. It is enough if the catalysts are maintained in a rough statistical fashion. The population of molecules in the bag is reproducing itself without any exact replication. While this is happening, the bag may be growing by accretion of fresh garbage from the outside, and the bag may occasionally be broken into two bags when it is thrown around by turbulent motions. The critical question is then, what is the probability that a daughter bag produced from the splitting of a bag with a self-reproducing population of molecules will itself contain a self-reproducing population? When this probability is greater than one half, a parent produces on the average more than one functional daughter, a divergent chain reaction can occur, the bags containing self-reproducing populations will multiply, and life of a sort has begun.

The life that begins in this way is the garbage-bag world. It is a world of little proto-cells that only metabolize and reproduce themselves statistically. The molecules that they contain do not replicate themselves exactly. Reproduction is not the same thing as replication. Cells can reproduce but only molecules can replicate. The Darwinian process of evolution by natural selection does not require exact replication to be effective. Darwin had never heard of DNA or exact replication when he developed his theory of evolution. Statistical reproduction is a good enough basis for natural selection. As soon as the garbage-bag world begins with crudely reproducing proto-cells, natural selection will operate to improve the quality of the catalysts and the accuracy of the reproduction. It would not be surprising if a million years of selection would produce proto-cells with many of the chemical refinements that we see in modern cells.

Life is not one thing but two, metabolism and replication, and the two things are logically separable. There are accordingly two logical possibilities for life's origins. Either life began only once, with the functions of replication and metabolism already present and linked together from the beginning, or life began twice, with two separate kinds of creatures, one kind arising first, capable of metabolism without exact replication, the other kind coming much later, capable of replication without metabolism. If life began once, the beginning was something like the RNA world. If life began twice, the first beginning was the garbage-bag world, with creatures containing all kinds of molecules, only a few of them being proteins. These garbage-bag creatures might have existed independently for a long time, perhaps as long as one or two billion years, eating, growing and gradually evolving a more and more efficient metabolic apparatus. The second beginning might have been with replicating parasites, arriving later and preying upon the garbage-bag creatures. The parasites could use the products of the garbage-bag metabolism to achieve their own replication. We see a hint of the way this might have happened in the ATP molecule, adenosine triphosphate, which is perhaps a relic of the original linkage between metabolism and replication. The garbage-bag creatures might have developed ATP as a key molecule allowing efficient storage of energy for metabolic processes. As a by-product of their use of ATP, they might have accumulated the closely related molecule AMP, adenosine monophosphate, which happens to be one of the four nucleotides that make up RNA. In a cell with a high concentration of AMP, the AMP might have polymerized into RNA, and in this way the replicating parasites first got their feet in the door. Other nucleotides such as GMP, guanine monophosphate, could have entered the game in a similar way. Here is a possible route for the evolution of replication, using the already evolved garbage-bag creatures to provide life-support to the replicators while they are evolving.

Another feature of the universe that has been essential to the evolution of life is symbiosis. Symbiosis is the reattachment of two structures, after they have been detached from each other and have evolved along separate paths for a long time, so as to form a combined structure with behavior not seen in the separate components. Symbiosis is a familiar concept both in biology and astronomy. In biology, almost all higher plants and animals make use of symbiotic bacteria to perform many of their metabolic functions. Nitrogen-fixing bacteria in the roots of soybean plants and cellulose-digesting bacteria in the stomachs of cows are two well-known examples. Lynn Margulis collected the evidence to prove that symbiosis played an even more fundamental role in the evolution of eucaryotic cells from procaryotes. She proved that the mitochondria and chloroplasts that are essential components of modern cells were once independent free-living creatures. They first invaded the ancestral eucaryotic cell from the outside and then became adapted to living inside. The combined cell then learned to coordinate the activities of its component parts, so that it acquired a complexity of structure and function that neither component could have evolved separately. In this way symbiosis allows evolution to proceed in giant steps. A symbiotic creature can jump from simple to complicated structures much more rapidly than a creature evolving by the normal processes of genetic mutation. I am suggesting that the jump from garbage-bag creatures to cells with a modern genetic apparatus also came about by symbiosis. The nucleic acid creatures, originating as parasites within the garbage-bag creatures, gradually learned to cooperate with their hosts. The garbage-bag creatures learned to tolerate the parasites and to exploit their capacity for exact replication. Together, the two components of the symbiosis created a modern cell that was so efficient, both in metabolism and in replication, that it wiped out all earlier forms of life.

Symbiosis is also a dominant factor in the evolution of the non-living universe. Symbiosis is as frequent in the sky as it is in biology. Astronomers are accustomed to talking about symbiotic stars. The basic reason why symbiosis is important in astronomy is the double mode of action of gravitational forces. When gravity acts upon a uniform distribution of matter occupying a large volume of space, the first effect of gravity is to concentrate the matter into lumps separated by voids. The separated lumps then differentiate and evolve separately along different evolutionary histories. They become distinct types of object. But then, after a period of separate existence, gravity acts in a second way to bring lumps together and bind them into pairs. The binding into pairs is a sporadic process depending on chance encounters. It usually takes a long time for a given lump to be bound into a pair. But the universe has plenty of time. After a few billion years, a large fraction of objects of all sizes become bound in symbiotic systems, either in pairs or in clusters. Once they are bound together by gravity, dissipative processes of various kinds tend to bring them closer together. As they come closer together, they interact with one another more strongly and the effects of symbiosis become more striking.

Examples of astronomical evolution caused by symbiosis are to be seen wherever one looks in the sky. On the largest scale, symbiotic pairs and clusters of galaxies are very common. When galaxies come into close contact, their internal evolution is often profoundly modified. A common sign of symbiotic activity is an active galactic nucleus. An active nucleus is seen in the sky as an intensely bright source of light at the center of a galaxy. The various varieties of active galactic nuclei commonly arise from the symbiotic effects of other galaxies nearby. The probable cause of the intense light is gas falling into a black hole at the center of one galaxy as a result of the gravitational perturbations caused by another galaxy. It happens frequently that big galaxies swallow small galaxies. Nuclei of swallowed galaxies are observed inside the swallower, like mouse-bones in the stomach of a snake. This form of symbiosis is known as galactic cannibalism.

On the scale of stars, we can distinguish even more different types of symbiosis, because there are many types of star and many stages of evolution for each of the stars in a symbiotic pair. The most spectacular symbiotic pairs have one component that is highly condensed, either a white dwarf or a neutron star or a black hole, and the other component a normal star. If the two stars are orbiting around each other at a small distance, gas will spill over from the normal star into the deep gravitational field of the condensed star. The gas falling into the deep gravitational field will become intensely hot and will produce a variety of unusual effects such as recurrent nova outbursts, intense bursts of X-rays and rapidly flickering light-variations. The more common and less spectacular symbiotic pairs consist of two normal stars orbiting around each other close enough so that mass is exchanged between them. These pairs are seen as optically variable stars, often surrounded by rings of gas and dust. They are important to the long-term evolution of our galaxy, because they recycle matter from the stars back into the interstellar gas and dust out of which the stars were formed.

But this talk is supposed to be about life, and I must not digress further into astronomy. Let me come back to the subject of life. From the point of view of life, the most important example of astronomical symbiosis is the symbiosis of the earth and the sun. The whole system of sun and planets and satellites, the system which we call the Solar System, is a typical example of astronomical symbiosis. At the beginning, when the Solar System was formed, the sun and the earth were born with totally different chemical compositions and physical properties. The sun was made mainly of hydrogen and helium, the earth was made of heavier elements such as oxygen and silicon and iron. The sun was physically simple, a sphere of gas heated at the center by the burning of hydrogen into helium and shining steadily for billions of years. The earth was physically complicated, partly liquid and partly solid, its surface frequently transformed by phase-transitions of many kinds. The symbiosis of these two contrasting worlds made life possible. The earth provided chemical and environmental diversity for life to explore. The sun provided physical stability, a steady input of energy on which life could rely. The combination of the earth's variability with the sun's constancy provided the conditions in which life could evolve and prosper.

In addition to the sun and the planets and their satellites, the solar system also contains a large number of asteroids and comets, smaller objects gravitationally bound to the sun but not sharing in the orderly motions of the planets. The asteroids and comets are an important part of the symbiosis that binds the system together. Since they have disordered motions, they occasionally collide with planets and produce catastrophic disturbances of the local environment. Traces of these impacts are visible on the surface of the earth, and even more visible on the moon. Impacts large enough to affect the whole earth and cause extinctions of species on a global scale occur about once in a hundred million years. The random obliteration of ecologies by major impacts has been a part of the history of life on earth since the beginning. It is likely that these catastrophes drove evolution forward by destroying species that were too well adapted to static environments, making room for species that were adaptable to harsher and more rapidly changing conditions. Without the occasional impact catastrophe to reward adaptability, it is unlikely that our own species would have emerged. We are among the most adaptable of species, offspring of a symbiosis in which sun, planets, asteroids and comets all played an essential part.

The universality of the modern genetic code proves that all existing cells have a common ancestor, but does not provide an absolute date for the epoch of the latest common ancestor. Genetic evidence gives us good relative dating of the different branches of the evolutionary tree, but no absolute dating. The interval of time between the beginning of life and the latest common ancestor may have been very long. Some new geological evidence has upset preconceived notions and raised new questions about the date of life's origin. The new evidence comes from the most ancient of all known terrestrial rocks in Greenland. The Greenland rocks are reliably dated and are at least 3.8 billion years old. They contain tiny carbonaceous inclusions which must be at least as old as the rocks. Careful analysis with microprobes of the abundances of carbon isotopes in the inclusions shows that the carbon 13 isotope is depleted to a degree characteristic of biologically processed carbon. This suggests that life existed on earth very soon after the time of heavy bombardment when the lunar highlands became densely cratered. Nobody expected life on earth to be established so early, but the evidence in the Greenland rocks has to be taken seriously. The orthodox view until recently was that life originated on earth some time between 3.5 and 3.8 billion years ago. The new evidence from Greenland suggests that life may be more ancient than we supposed. It appears to have been spread widely over the earth even before the era of heavy bombardment ended.

The most famous experiment exploring the origin of life was done by Stanley Miller in Berkeley in 1953. Miller filled a flask with a reducing atmosphere composed of methane, ammonia, hydrogen and water, passed electric sparks through it and collected the reaction products. He found a mixture of organic compounds containing a remarkably high fraction of amino-acids. He also tried the experiment in an oxidizing atmosphere, and in a neutral atmosphere composed of nitrogen, carbon dioxide and water. Other people have repeated the experiment with many variations. The results are always the same. With a reducing atmosphere you get plenty of amino-acids. With an oxidizing or neutral atmosphere you do not.

We now know that the earth's reducing atmosphere, if it ever existed, had disappeared by the time the heavy meteoritic bombardment of the earth ceased, about 3.8 billion years ago. Sedimentary rocks laid down on the ancient earth, including carbonates and various oxidized forms of iron, have been reliably dated with ages going all the way back to 3.8 billion years. These rocks could not have formed under reducing conditions. Their composition proves that the atmosphere was neutral from 3.8 until about 2 billion years ago when molecular oxygen first appeared. The atmosphere became oxidizing after life was well established and photosynthetic organisms began to produce free oxygen in large quantities.

So Miller's beguiling picture of the origin of life, in a pond full of dissolved amino-acids under a reducing atmosphere, has been discredited. Recently, a new beguiling picture has come to take its place. The new picture has life originating in a hot, deep, dark little hole on the ocean floor. Four experimental discoveries came in rapid succession to make the new picture seem plausible. First, the discovery of abundant life existing today around vents on the mid-ocean ridges, several kilometers below the surface, where hot water emerging from deep below is discharged into the ocean. The water entering the ocean is saturated with hydrogen sulphide and metallic sulphides, so that it provides a reducing environment independent of the atmosphere above. Second, the discovery that bacterial life exists today in strata of rock deep underground, in places where contact with surface life is unlikely. In some cases, the deep underground bacteria do not belong to any previously known species. Third, the discovery of strikingly life-like phenomena observed in the laboratory, when hot water saturated with soluble iron sulphides is discharged into a cold water environment. The sulphides precipitate as membranes and form gelatinous bubbles. The bubbles look like possible precursors of living cells. The membrane surfaces adsorb organic molecules from solution, and the metal sulphide complexes catalyze a variety of chemical reactions on the surfaces. Fourth, the discovery that many ancient lineages of bacteria are thermophilic, that is to say, specialized to live and grow in hot environments. The ancient lineages were identified by compiling sequences of bases in ribosomal RNA of many species, and using the observed similarities and differences of the sequences to construct a phylogenetic tree. The phylogenetic tree has a root which represents the hypothetical RNA sequence of the ribosomes in the latest common ancestor of all life. The most ancient lineages are those that branch off closest to the root of the tree. Many of them are found today in hot springs, often in places where the water temperature is close to boiling.

These four lines of evidence, from ocean ridges, deep oil-well drilling, laboratory experiments and genetic analysis, combine to make the picture of life originating in a hot deep environment credible. Since we know almost nothing about the origin of life, we have no basis for declaring any possible habitat for life to be likely or unlikely until we have explored it. The picture of life beginning in a deep hot crevice in the earth is purely speculative and in no sense proved. It has an important corollary. If it is true, it implies that the origin of life was largely independent of conditions on the surface of the planet. And this in turn implies that life might have originated as easily on Mars as on the earth. Thomas Gold postulates a deep, hot biosphere still existing in the crust of the earth. He presents evidence that the deep biosphere may contain as much biomass as the surface biosphere with which we are familiar. He remarks, "If in fact such life originated at depth in the earth, there are at least ten other planetary bodies in our solar system that would have had a similar chance for originating microbial life". I don't know which ten objects Gold had in mind. Certainly Mars, Europa, Titan and Triton would be on the list.

A new version of the garbage-bag theory with some experimental evidence to support it was recently proposed by Wächtershäuser, a German patent attorney who does chemical research as a hobby. Wächtershäuser makes garbage-bags out of metal sulphides. Metal sulphides in general, and iron sulphides in particular, are good candidates for pre-biotic chemistry and are known to be abundant in hydrothermal vents. Bubbles made from iron sulphide membranes might have been the original garbage-bags. The Wächtershäuser theory fits well with the idea that life originated in a deep, hot environment. At the present time there is no compelling reason to accept or to reject any of the theories. Any of them, or none of them, could turn out to be right. We do not yet know how to design experiments which might decide between them. I happen to prefer the garbage-bag theory, not because I think it is necessarily right, but because it is unfashionable.

There is a possible analogy between the origin of life and the origin of elaborate body-plans in higher organisms. Half a billion years ago, after life had existed for about three billion years, there was a sudden efflorescence of elaborate body-plans. The efflorescence is known as the "Cambrian explosion", and produced in a geologically short time all the major body-plans from which modern higher organisms evolved. Something must have happened shortly before the Cambrian epoch to make the genetic programming of elaborate body-plans possible. What might have happened was the invention of "indirect development", the system by which an embryo sets aside a package of cells that are destined to grow into an adult, the body-plan of the adult having no connection with the body-plan of the embryo. The advantage of this system is that the embryo provides life-support to the adult during the vulnerable stages of its early growth, while the adult is free to evolve elaborate and fine-tuned structures unconstrained by existing structures of the embryo. Three California paleontologists, Davidson, Peterson and Cameron, have collected evidence that the great majority of existing body-plans arose from indirect development. This fact was overlooked until recently because the two best-known body-plans, the vertebrate and the arthropod, are exceptions to the rule. The vertebrates and arthropods, the two most successful phyla of animals, probably began like the others with indirect development, but later evolved a short-cut system of direct development with the adult body-plan growing directly from the embryo. If the system of indirect development came first, it means that multicellular organisms evolved by a two-step process. The first step was the evolution of embryonic forms of limited complexity, lacking the genetic machinery to program specialized structures. The second step was the evolution of adult forms with the modern armamentarium of genetic controls, and with life-support provided by the embryo. I am proposing that the early evolution of life followed the same two-step pattern as the evolution of higher organisms. First came the embryonic stage of life, cells with functioning metabolism but without any genetic apparatus, unable to evolve beyond a primitive level. Second came the adult stage, cells with genetic machinery allowing the evolution of far more fine-tuned metabolic pathways, and again with life-support provided by the first stage while the second stage evolved.

To me, one of the most attractive features of the garbage-bag theory of the origin of life is that it shows life following the same pattern, a major step in evolution divided into two separate jumps, at three crucial periods of its history. First, the period of origins about four billion years ago, when the two jumps were metabolism and replication. Second, the evolution of eucaryotic cells according to Margulis, about two billion years ago, when the two jumps were parasitic invasion and symbiosis. And, third, the evolution of higher organisms, about half a billion years ago, when the two jumps were the embryo and the package of cells that grew into an adult. In each of the three revolutions, the first stage relied on crude and simple modes of inheritance, and the second stage jumped to new levels of sophistication in the translation of genetic language into anatomical structure.

About a hundred and fifty million years after the Cambrian explosion, life made one of its greatest jumps when it moved from the ocean onto land. To survive on land it had to invent lungs, and weight-carrying bones, and a dry impermeable skin that could prevent loss of water. The first animals that came ashore did not yet have impermeable skins. They were the amphibians, animals like frogs that hatch from eggs in the water and live only part of their lives on land. They desiccate and die if they sit too long in sunshine. It took another fifty million years for the descendants of the amphibians to become reptiles fully adapted to living on land. The reptiles with their impermeable skins spread all over the earth and made it their home. The liberation of life from the ocean made possible all the later inventions that make the land beautiful, fur and feathers and forests and flowers.

3. Looking for Life

For the last part of this talk, I move from the past to the future. When I look at things from the point of view of an astronomer, the four hundred million years that elapsed since life escaped from the ocean until now is a short time. That was perhaps only the first jump in another revolution that needs two jumps to be complete. The first jump was from the ocean onto the land. The second jump will be from the land into space. The revolution will only be complete when life has escaped from this planet and made the universe its home. We are beginning the second jump now with our exploring of the planets and our quick trips to the moon. But for a long time, so long as we depend on spacecraft and spacesuits to stay alive in space, we shall be amphibians. We can survive in space for a limited time, and we must return to our home planet to breathe air under an open sky. This amphibian phase may last for a few hundreds of years, while the life that we carry with us away from earth is still confined to artificial habitats.

After that, perhaps sooner than most people expect, we shall breed plants and animals that do not need to be confined but are adapted to living wild in space. The jump from breathing air to living in a vacuum is no greater than the jump from breathing water to breathing air. Plants and animals will need considerable genetic engineering to be at home in a vacuum. Plants will need new organs of photosynthesis that produce liquid or solid peroxides instead of oxygen gas. Animals will need new organs of respiration to take in oxygen in the form of peroxides instead of from air. Instead of lungs, animals would have an organ like a liver that dissociates peroxides slowly into molecular oxygen and feeds the oxygen directly into the blood. Both plants and animals will need stronger skin to hold internal pressure and prevent their blood from boiling. The vapor pressure of water at blood temperature is quite small, so the skin will not need to be thick to hold it. In cold places far from the sun, animals will need thicker layers of fur and plants will need thicker layers of bark to provide thermal insulation. This will be a challenge for plant and animal breeders, but with a mastery of the techniques of genetic engineering they should be able to do it. When they have done it, life will have moved again from the age of amphibians to the age of reptiles. The second jump of the new revolution will be complete, and life will be on its way to the next phase of evolution. Life adapted to vacuum will evolve to create new ecologies on all the worlds where sunlight and the chemical elements essential to life are to be found.

I am not an astronomer myself, but I like to talk with astronomers. I see the universe through the eyes of the astronomers. Everywhere I look, I see beauty. But everything we see, except our own planet, is dead. And one planet with life is more beautiful than a whole universe dead. The universe would be far more beautiful and meaningful if it were full of life, if life were spread out over those millions of worlds. Somehow life must find a way to spread and make a home for itself in every corner of the universe, just as it made a home for itself in every corner of this planet. Perhaps our destiny is to be the midwives, to help the living universe to be born. I believe in space-travel, not as an end in itself, but as a means to bring the universe to life.

Human beings did not have to exist for life to escape from the earth. There might have been other midwives, other species that developed intelligence and tools to give a start to space-travel. It might have been possible for life to spread over the universe even without intelligent midwives. Pieces of Mars are occasionally blasted away from Mars by a comet or asteroid impact and arrive on Earth without being destroyed. If Mars had possessed living inhabitants, some Martian creatures might have survived the voyage and made their home on Earth. Perhaps we are their descendants. Organisms from Earth might occasionally arrive in the same way on Mars. No place in the universe is totally isolated from its neighbors. The universe has plenty of time. The universe could have waited long enough to see life moving out from Earth by the natural processes of comet impacts and meteorite infall. If we take responsibility for spreading life through the universe, we are only speeding up the natural processes.

Today the rockets that we use to get from earth into space are absurdly expensive. The public has learned that space is a playground where only rich governments and rich corporations can play. The public believes that space-travel will always be too expensive for ordinary people. But this need not always be so. Space-travel need not always be a spectator sport, with a small elite of stars paid for by the millions who stay on the ground and watch the show on television. To make space-travel cheap in the future, we need public highways into space. To be accessible to the general public, space-travel must be about a hundred times cheaper than it is today. The cost of launching payload into space should be reduced from ten thousand dollars a pound to a hundred dollars a pound. This sounds hard to do, but it may be possible if we develop radically new methods of propulsion, leaving the source of energy on the ground so that each spacecraft does not need to carry its own fuel. The source of energy for launching payloads from the ground into space should be like a public highway, serving anyone who arrives at the site with a spacecraft and is willing to pay for the service.

There are several possible ways of building a public highway into space. A laser beam pointing from the ground up into the sky can be a public highway. Two years ago in Princeton we saw a film of the first flight of Leik Myrabo's Lightcraft Technology Demonstrator, a little toy model of a laser-propelled spacecraft. Myrabo is a professor at Rensellaer Polytechnic Institute in Troy, New York state. His model rose three feet into the air, not so high as the first flight of the Wright brothers in December 1903. More recent Lightcraft flights flew higher, but the first flight was the decisive step. The first flight demonstrated that laser propulsion is possible, just as the 1903 flight demonstrated that human heavier-than-air flight was possible. The sound-track accompanying the film of the Lightcraft flight made a noise like a machine-gun, as the laser fired high-intensity pulses at a rate of ten per second. The vehicle has a little fish-shaped body with a blunt nose and a shiny reflecting dish around its waist. It weighs two ounces and has a diameter of six inches. Each laser pulse is focussed by the dish and heats the air at the focus to a high temperature, causing a shock-wave that propels the model upward. The average power in the beam is ten kilowatts. Myrabo borrowed the laser from the United States Air Force at White Sands, New Mexico. Last year Myrabo was back in Princeton with a film of the Lightcraft flying 75 feet up a laser beam. It flies stably up to 75 feet. To fly higher he will need a better laser. A simple calculation shows that if it takes a ten kilowatt laser to lift a two-ounce space-craft off the ground, it will take a one gigawatt laser to lift a five-ton spacecraft. Using laser-heated water as propellant, a spacecraft with a take-off weight of five tons should be able to carry two tons of propellant and a useful payload of at least a ton, enough for a couple of passengers and their baggage. A one gigawatt laser will be expensive, but it is not absurd. We already have hundred-megawatt lasers. The laser is the easy part of a laser-propulsion system. The hard part is the engine. We don't yet know how to design the engine for a full-scale system.

It took seventy years to go from the Wright brothers' Flyer One of 1903 to the modern air-transport system with huge numbers of commercial aircraft flying routinely all over the world. Perhaps it will take another seventy years to go from Myrabo's Lightcraft Technology Demonstrator to a laser-propelled public highway system with huge numbers of spacecraft traveling routinely to destinations in geostationary orbit and on the moon and beyond. At each destination there must be a massive infrastructure comparable with a major airport and including the associated industries and hotels. All this will take a long time, measured on a human time-scale, but a short time, measured on an astronomical timescale.

Building public highways is difficult because they become cheap only when they are constantly used. To keep a public highway into space busy, we need a hundred thousand launches per year, roughly one every five minutes. The first users of the highway will be big corporations and governments who can afford to pay for its initial cost. After it has been built, it will be available for ordinary people, for private groups who go into space at their own expense, like the Pilgrim Fathers who rented a second-hand boat and sailed from England to America in 1620. The ordinary people will not only take themselves and their families into space. They will take plants and animals too. Before that, we will have plants and animals genetically engineered to live wild in strange places. People will spread life with them wherever they go.

Cheap space-travel will sooner or later be developed. There is no law of physics that decrees that space-travel must always be expensive. So far as the laws of physics are concerned, if one measures the cost of moving into space by the amount of energy required, the cost of launching a person from Earth into space should be no greater than the cost of a commercial flight from New York to Tokyo. But to bring the cost of space-launches down to the cost of commercial air travel requires a huge volume of traffic. Space-travel will only be cheap when millions of people can use it. And millions of people can only use it when there are cities in the sky to which travelers can go. The growth of space-habitats and the decline of costs will be a slow process. Much can be done in a hundred years, but not enough to have a major impact on the social problems of Earth. The expansion of life into space will not come in time to solve the problems of our grandchildren. When it finally comes, it will be a mixed blessing for humanity. Only one thing is certain. Once life escapes from this little planet into the universe, there will be no stopping it. It will keep on moving and changing. It will go on its way, with or without our help. Like good midwives, we should step aside and watch it grow.

Finally, this talk has a punch-line, and here it is. I come now at last to the third of my three questions. Where should we be looking for evidence of life? Dreams of a possible future have practical consequences for the present. Things that for us are in the speculative future may have happened in the past somewhere else. When we begin a serious search for life existing elsewhere in the universe, we should keep in mind the possibility that life elsewhere is already adapted to living in vacuum. For life adapted to living in vacuum, planets are not the most likely habitat. Life would be more likely to be found on small bodies such as asteroids or comets, places where gravity is weaker and moving from one world to another is easier. For life adapted to vacuum, a planet would be a death-trap. A planet for them would be like a deep well full of water for a human child. A swarm of small objects like the Kuiper Belt, the ring of icy worlds that we see just outside the orbit of Neptune, provides far more habitable surface area than all the planets together. It would be a friendlier place for vacuum-life to flourish. We see many stars like Beta Pictoris that have more real estate in their Kuiper Belts than we have in ours. When we are searching for evidence of alien life, we should not look only at planets. Planets may be the most likely places for life to begin, but they are not the most likely places for life to be found in the sky. In searching for evidence of life, as in all branches of astronomy, you have the best chance of making important discoveries if you look where other people are not looking. I am not saying that we should give up looking for evidence of life on Mars, only that we should look in other places too, Europa and the Kuiper Belt and giant molecular clouds. Wherever there is water, carbon, nitrogen and starlight, life might already be teeming. I am willing to bet even money that when the first alien life is found it will not be on a planet. This is a bet which I will be happy either to win or to lose.

CHAPTER 2.2

WHEN WE WERE KIDS*

My strong suit was always mathematics. I was not driven to become a scientist by any craving to understand the mysteries of nature. I never sat and thought deep thoughts. I never had any ambition to discover new elements or cure diseases. I just enjoyed calculating and fell in love with numbers. Science was exciting because it was full of things I could calculate.

One episode I remember vividly. I do not know how old I was. I only know that I was young enough to be put down for an afternoon nap in my crib. The crib had solid mahogany side-pieces so that I couldn't climb out. I didn't feel like sleeping, so I spent the time calculating. I added one plus a half plus a quarter plus an eighth plus a sixteenth and so on, and I discovered that if you go on adding like this forever you end up with two. Then I tried adding one plus a third plus a ninth and so on, and discovered that if you go on adding like this forever you end up with one and a half. Then I tried one plus a quarter and so on, and ended up with one and a third. So I had discovered infinite series. I don't remember talking about this to anybody at the time. It was just a game that I enjoyed playing.

Another episode is the total eclipse of the sun in the summer of 1927. I was then three and a half. My father said the eclipse would be total at Giggleswick in Yorkshire. I lived with my family at Winchester about 200 miles south of Giggleswick. My sister and I saw the partial eclipse at Winchester, looking at it through bits of glass which were made dark by holding them over a smoky candle flame. We saw the sun slowly shrinking until it became a narrow crescent. But I was furiously angry at my father because he refused to take us to Giggleswick. I asked my father when the next total eclipse visible in England would be. He said 1999. I calculated that I would have to live to seventy five to see the next eclipse. That made me even angrier.

---

*A Chapter in *Curious Minds*: *How a child becomes a scientist*, edited by John Brockman.

After my mother's death I found among her papers some relics that she had preserved of my childhood. One of them is a paper with the heading ASTRONIMY, with a sentence for each of the planets, for example: "You can hadly ever see Murcery becose the Sun is nearly allways in frount of it". At the end in my mother's hand: FJD aged five and a half. This scrap of paper is evidence of two things. First, I had a mother who cared and encouraged me to learn. Second, the fact that the statement about Mercury is wrong shows that I did not copy it from grown-ups. I must have made it up.

My father was a musician and my mother was a lawyer, neither of them expert in science. But they read the current popular science books by Whitehead, Eddington, Jeans, Hogben and Haldane, and left them on the shelves for me to browse. Among them was Eddington's "Space, Time and Gravitation", an excellent introduction to relativity. I still have my father's copy with his signature showing that he bought it when it was published in 1920. On page 49 there is a diagram of space-time with space plotted horizontally and time vertically, the light-cones appearing as two perpendicular lines running diagonally and dividing space-time into four regions. The text explains that the region above the diagonals is the absolute future and the region below the diagonals is the absolute past, while the two regions at the sides between the diagonals are "elsewhere". When I was seven years old, the magazine Punch published a cartoon of a nanny talking to a small boy who is lying on the ground with a book. The boy was me. The book was Eddington. The nanny says, "Do you know where your sister is?" and the boy answers, "Somewhere in the absolute elsewhere". I remember being puzzled when my father said he was sending that conversation to Punch to be published. I did not see the joke. What I said was true, not funny.

In 1931, when I was seven, the asteroid Eros came unusually close to the earth. This is the same asteroid that the space-craft NEAR (Near Earth Asteroid Rendezvous) explored and landed on in 2001. It is the biggest of the asteroids that regularly come close to us. In 1931, there was much talk of the terrible damage that Eros could do if it should collide with the earth. I listened to my parents talking about this at the breakfast table. My father told me that the Astronomer Royal Sir Frank Dyson was leading an international effort to observe Eros and calculate its orbit precisely. This was important because it would give us a more accurate measure of the distance between the earth and the sun. Sir Frank was not related to us, but he came from the same part of Yorkshire as my father, and my father knew him. I liked the idea of calculating orbits precisely, and I thought maybe I will one day be Astronomer Royal and calculate orbits too. Among the papers that my mother preserved there is a fragmentary story that I wrote at the age of nine, entitled "Sir Phillip Roberts's Erolunar Collision", about an astronomer who calculates the orbit of Eros and discovers that it is heading for a collision with the moon. He predicts that the collision will happen in ten years' time, enough time for him to organize an expedition to the moon to observe the collision from close by. At that point the story stops. Reading it again after seventy years, I find it interesting that Sir Phillip makes his great discovery by calculating and not by observing.

At the age of eight I was, like other middle-class English kids at that time, sent away to boarding-school. The school was a Dickensian horror, but it had one redeeming feature, a library where I could escape from sadistic boys and sadistic headmaster. In the library was the Book of Knowledge, a popular children's encyclopedia, and the science-fiction novels of Jules Verne. I read "Hector Servadac" and other Verne stories. I found the Verne puzzling at first because I did not know it was fiction. I thought Hector Servadac had really taken a ride on a comet and visited a new planet which he called "Gallia" in honor of his native country. I was puzzled because Gallia did not appear among the planets in the Book of Knowledge. When I found out that the Verne stories were fiction, it was a big disappointment. I liked the Book of Knowledge better because I could trust it. I read about the Yosemite valley and pronounced it to rhyme with Nose-bite. I read about matter being made up of electrons and protons. Then I read a long piece about electrons and electricity and electric motors, but there was nothing about protons. I wondered why there was no knowledge about protons. Why didn't we have proticity and protic motors? I asked some of the boys and some of the teachers, but nobody knew. The school taught mostly mathematics and Latin. No science was taught. That was probably a good thing, as it made science more attractive to misfits like me. With a small group of friends I started a Science Society, circulating books and holding occasional meetings to discuss them.

At age twelve I moved to Winchester College, a high-class private school where my father was the head music teacher. There I had the enormous luck to find three kindred spirits of my own age, James Lighthill and the brothers Christopher and Michael Longuet-Higgins. All four of us later became fellows of the Royal Society. James had a brilliant career in fluid dynamics, Christopher in theoretical chemistry, Michael in oceanography. James and I were both in love with mathematics and rapidly devoured the books in the school library. Together we worked through the three volumes of Jordan's "Cours D'Analyse", the famous textbook of nineteenth-century mathematics as it was taught to students at the élite École Polytechnique in Paris a hundred years ago. We could not have had a better introduction to serious mathematics. It went far beyond anything our Winchester teachers knew or cared about. We did not ask how it happened that this treasure was in our school library. Later we suspected that it must have been put there by the famous mathematician G.H. Hardy who had been a boy in the same school forty years earlier. The teachers at Winchester wisely left us alone to educate ourselves. They gave us wide freedom and were confident that we would use it responsibly. In our last year at Winchester we spent only seven hours a week in class.

At age fourteen we read the book, "Men of Mathematics" by Eric Temple Bell, a collection of romanticized biographies of great mathematicians. Bell was a professor of mathematics at the California Institute of Technology and also a gifted writer. He wrote with authority about mathematics and knew how to pull at the heart-strings of susceptible teenagers. His book seduced a whole generation of young people to become mathematics addicts. Although many of the details in the book are historically inaccurate, the important things are true. He portrays mathematicians as real people with real faults and weaknesses. He portrays mathematics as a magic kingdom that people of many different kinds can share. The message for the young reader is, if they could do it, why not you?

At age seventeen I met my first real mathematician, a young man called Daniel Pedoe who was a junior lecturer at Southampton University. In the summer of 1941, the Winchester school authorities hired him to come once a week and give me private lessons. The sessions with Pedoe were a revelation. He had been a research student in Rome, and a member at the Institute for Advanced Study in Princeton, before World War II. He knew personally the legendary figures of mathematics, and he knew the latest fashionable stuff that they had been doing. His own field was geometry. He gave me a German translation of Severi's Algebraic Geometry to study. Pedoe and I became life-long friends, and Severi's book became one of my prized possessions. I did not become a geometer, but I acquired from Pedoe a taste for the geometric style that makes mathematics an art rather than a science. Pedoe later became a professor at the University of Minnesota and a leading spokesman for geometry in the international mathematical community. He published a book, "Japanese Temple Geometry Problems", with his friend Hidetosi Fukagawa, about an elegant indigenous version of geometry that flourished in Japan during the centuries of isolation from Western influences.

A few months after I met Pedoe, I went as a student to Trinity College at Cambridge University. There I listened to lectures by the famous English mathematicians, Hardy, Littlewood, Hodge and Besicovitch. My favorite was Besicovitch, an emigre Russian who worked on the boundary between geometry and set-theory. At that time in the middle of the war there were very few students, and I had Besicovitch all to myself. He taught me Russian as well as mathematics. We went regularly for long walks on which only Russian was spoken. He had a billiard-table in his living-room and we played billiards when the weather was too wet for walking. He gave me problems to work on which were impossibly difficult but taught me how to think. His great work on the geometry of plane sets of points became a model for my own later work in physics.

After two years as a student and two years working as a statistician for the Royal Air Force, at age twenty one I had my first academic job as instructor in mathematics at London University. I attached myself as an unofficial student to Harold Davenport, a number-theorist who was a professor at University College London. So I became a number-theorist. Davenport was like Besicovitch, a friend as well as a mentor. Davenport gave me problems to work on that were easier than Besicovitch problems. Davenport problems were difficult but not impossible. I solved two of them and published the solutions in mathematical journals. Davenport was one of those rare people who could gauge the capacity of a student and supply a problem just hard enough but not too hard for the student to solve. I owe him my start as a professional mathematician.

While I was working with Davenport, I was also thinking seriously about switching from mathematics to physics. I read the Smyth Report, "Atomic Energy: A general account of the development of methods of using atomic energy for military purposes under the auspices of the United States government", which was published in the fall of 1945 and described the achievements of the physicists who had built nuclear reactors and bombs during the war. The report gives a vivid and detailed picture, both of the science and the scientists. I felt strongly tempted to join the club. The mathematics that I was doing would never be of interest to anybody outside the small community of number-theorists. It was not even modern mathematics. It belonged to the nineteenth rather than to the twentieth century. If I wanted to become a modern mathematician, I would have to go back to school and learn the modern stuff. Why not learn physics instead? Physics had two great advantages. First, I would be doing something important rather than solving esoteric puzzles. Second, physics was well suited to my skills, since the mathematics needed for physics was nineteenth-century rather than twentieth-century mathematics. For me the switch to physics was not difficult. It only meant that I would have to find a physicist as helpful as Davenport, to supply me with problems to solve.

As a result of my work with Davenport, Trinity College invited me to come back to Cambridge as a research fellow. That meant I was free to do anything I wanted. What I wanted was to leave England and see the world. After six years of war, either fighting or cooped up in England, everyone wanted to travel. I applied for a Harkness Fellowship to spend a year in America and was one of the lucky winners. At the Cavendish Laboratory in Cambridge, I happened to meet Sir Geoffrey Taylor, an expert in fluid dynamics who had been at Los Alamos during the war. I did not know Taylor but I ventured to ask him where I should go in America. He said without any hesitation, "Oh, you should go to Cornell, that is where all the brightest people from Los Alamos went after the war". He said Hans Bethe, who had been head of the theoretical division at Los Alamos, was the person I should work with. He knew Bethe well and would put in a good word for me. I knew almost nothing about Cornell, but I took Taylor's advice. I went to Cornell to work with Bethe. And Bethe turned out to be even better than Davenport. He gave me a difficult but not impossible problem to solve, and I published the solution in a physics journal. That gave me confidence. Now I really belonged to the club.

My final stroke of luck was meeting Richard Feynman. Feynman was a young professor at Cornell, not yet famous. I had never heard of him before I came to America. He was rebuilding the whole of physics from the bottom up, using a geometrical language with diagrams that nobody except Feynman understood. I recognized that Feynman was a genius and my job was to understand his language and explain it to the world. So that is what I did. I spent as much time as I could with Feynman. I watched him drawing diagrams on the blackboard and listened to him talking. Like Besicovitch, he used to go for long walks and talk about everything under the sun. After a year at Cornell, I understood Feynman's way of thinking and translated it into the old-fashioned mathematics that I had learned in England. I published two papers explaining why Feynman's methods worked. My papers were bestsellers, and Feynman's language became the standard language of particle physicists all over the world. At the age of twenty-five I was a famous physicist. At a meeting of the American Physical Society where I was one of the main speakers, Feynman said to me, "Well, Doc, you're in". Childhood was over, and I was free to spend the rest of my life finding problems in various areas of science where a tablespoonful of elegant mathematics could make a big difference.

CHAPTER 2.3

BIRDS AND FROGS*

1. Francis Bacon and René Descartes

Some mathematicians are birds, others are frogs. Birds fly high in the air and survey broad vistas of mathematics out to the far horizon. They delight in concepts that unify our thinking and bring together diverse problems from different parts of the landscape. Frogs live in the mud below and see only the flowers that grow nearby. They delight in the details of particular objects, and they solve problems one at a time. I happen to be a frog, but many of my best friends are birds. The main theme of my talk tonight is this. Mathematics needs both birds and frogs. Mathematics is rich and beautiful because birds give it broad visions and frogs give it intricate details. Mathematics is both great art and important science, because it combines generality of concepts with depth of structures. It is stupid to claim that birds are better than frogs because they see farther, or that frogs are better than birds because they see deeper. The world of mathematics is both broad and deep, and we need birds and frogs working together to explore it.

This talk is called the Einstein lecture, and I am grateful to the American Mathematical Society for inviting me to do honor to Albert Einstein. Einstein was not a mathematician, but a physicist who had mixed feelings about mathematics. On the one hand, he had enormous respect for the power of mathematics to describe the workings of nature, and he had an instinct for mathematical beauty which led him onto the right track to find nature's laws. On the other hand, he had no interest in pure mathematics, and he had no technical skill as a mathematician. In his later years he hired younger colleagues with the title of assistants to do mathematical calculations for him. His way of thinking was physical rather than mathematical.

---

*Einstein Lecture prepared for the American Mathematical Society meeting, Vancouver, Canada, November 4, 2008. Revised for Russian translation, June 2009.

He was supreme among physicists as a bird who saw further than others. I will not talk about Einstein since I have nothing new to say.

At the beginning of the seventeenth century, two great philosophers, Francis Bacon in England and René Descartes in France, proclaimed the birth of modern science. Descartes was a bird and Bacon was a frog. Each of them described his vision of the future. Their visions were very different. Bacon said, "All depends on keeping the eye steadily fixed on the facts of nature". Descartes said, "I think, therefore I am". According to Bacon, scientists should travel over the earth collecting facts, until the accumulated facts reveal how Nature works. The scientists will then induce from the facts the laws that Nature obeys. According to Descartes, scientists should stay at home and deduce the laws of Nature by pure thought. In order to deduce the laws correctly, the scientists will need only the rules of logic and knowledge of the existence of God. For four hundred years since Bacon and Descartes led the way, science has raced ahead by following both paths simultaneously. Neither Baconian empiricism nor Cartesian dogmatism has the power to elucidate Nature's secrets by itself, but both together have been amazingly successful. For four hundred years, English scientists have tended to be Baconian and French scientists Cartesian. Faraday, Darwin and Rutherford were Baconians: Pascal, Laplace and Poincaré were Cartesians. Science was greatly enriched by the cross-fertilization of the two contrasting national cultures. Both cultures were always at work in both countries. Newton was at heart a Cartesian, using pure thought as Descartes intended, and using it to demolish the Cartesian dogma of vortices. Marie Curie was at heart a Baconian, boiling tons of crude uranium ore to demolish the dogma of the indestructibility of atoms.

In the history of twentieth-century mathematics, there were two decisive events, one belonging to the Baconian tradition and the other to the Cartesian tradition. The first was the International Congress of Mathematicians in Paris in 1900, at which Hilbert gave the keynote address, charting the course of mathematics for the coming century by propounding his famous list of twenty three outstanding unsolved problems. Hilbert himself was a bird, flying high over the whole territory of mathematics, but he addressed his problems to the frogs who would solve them one at a time. The second decisive event was the formation of the Bourbaki group of mathematicians in France in the 1930s, dedicated to publishing a series of textbooks that would establish a unifying framework for all of mathematics. The Hilbert problems were enormously successful in guiding mathematical research into fruitful directions. Some of them were solved and some remain unsolved, but almost all of them stimulated the growth of new ideas and new fields of mathematics. The Bourbaki project was equally influential. It changed the style of mathematics for the next fifty years, imposing a logical coherence that did not exist before, and moving the emphasis from concrete examples to abstract generalities. In the Bourbaki scheme of things, mathematics is the abstract structure included in the Bourbaki textbooks. What is not in the textbooks is not mathematics. Concrete examples, since they do not appear in the textbooks, are not mathematics. The Bourbaki program was the extreme expression of the Cartesian style of mathematics. It narrowed the scope of mathematics by excluding the beautiful flowers that Baconian travelers might collect by the wayside.

2. Jokes of Nature

For me, as a Baconian, the main thing missing in the Bourbaki program is the element of surprise. The Bourbaki program tried to make mathematics logical. When I look at the history of mathematics, I see a succession of illogical jumps, improbable coincidences, jokes of nature. One of the most profound jokes of nature is the square-root of minus one that the physicist Erwin Schrödinger put into his wave equation when he invented wave mechanics in 1926. Schrödinger was a bird who started from the idea of unifying mechanics with optics. A hundred years earlier, Hamilton had unified classical mechanics with ray optics, using the same mathematics to describe optical rays and classical particle trajectories. Schrödinger's idea was to extend this unification to wave optics and wave mechanics. Wave optics already existed, but wave mechanics did not. Schrödinger had to invent wave mechanics to complete the unification. Starting from wave optics as a model, he wrote down a differential equation for a mechanical particle, but the equation made no sense. The equation looked like the equation of conduction of heat in a continuous medium. Heat conduction has no visible relevance to particle mechanics. Schrödinger's idea seemed to be going nowhere. But then came the surprise. Schrödinger put the square root of minus one into the equation, and suddenly it made sense. Suddenly it became a wave equation instead of a heat conduction equation. And Schrödinger found to his delight that the equation has solutions corresponding to the quantized orbits in the Bohr model of the atom.

It turns out that the Schrödinger equation describes correctly everything we know about the behavior of atoms. It is the basis of all of chemistry and most of physics. And that square-root of minus one means that nature works with complex numbers and not with real numbers. This discovery came as a complete surprise, to Schrödinger as well as to everybody else. According to Schrödinger, his fourteen-year-old girl-friend Itha Junger said to him at the time, "Hey, you never even thought when you began that so much sensible stuff would come out of it". All through the nineteenth century, mathematicians from Abel to Riemann and Weierstrass had been creating a magnificent theory of functions of complex variables. They had discovered that the theory of functions became far deeper and more powerful when it was extended from real to complex numbers. But they always thought of complex numbers as an artificial construction, invented by human mathematicians as a useful and elegant abstraction from real life. It never entered their heads that this artificial number-system that they had invented was in fact the ground on which atoms move. They never imagined that nature had got there first.

Another joke of nature is the precise linearity of quantum mechanics, the fact that the possible states of any physical object form a linear space. Before quantum mechanics was invented, classical physics was always nonlinear, and linear models were only approximately valid. After quantum mechanics, nature itself suddenly became linear. This had profound consequences for mathematics. During the nineteenth century, Sophus Lie developed his elaborate theory of continuous groups, intended to clarify the behavior of classical dynamical systems. Lie groups were then of little interest either to mathematicians or to physicists. The nonlinear theory was too complicated for the mathematicians and too obscure for the physicists. Lie died a disappointed man. And then, fifty years later, it turned out that nature was precisely linear, and the theory of linear representations of Lie algebras was the natural language of particle physics. Lie groups and Lie algebras were reborn as one of the central themes of twentieth-century mathematics.

A third joke of nature is the existence of quasi-crystals. In the nineteenth century, the study of crystals led to a complete enumeration of possible discrete symmetry-groups in Euclidean space. Theorems were proved, establishing the fact that in three-dimensional space discrete symmetry-groups could contain only rotations of order three, four or six. Then in 1984 quasi-crystals were discovered, real solid objects growing out of liquid metal alloys, showing the symmetry of the icosahedral group which includes fivefold rotations. Meanwhile, the mathematician Roger Penrose had discovered the Penrose tilings of the plane. These are arrangements of parallelograms that cover a plane with pentagonal long-range order. The alloy quasi-crystals are three-dimensional analogs of the two-dimensional Penrose tilings. After these discoveries, mathematicians had to enlarge the theory of crystallographic groups so as to include quasi-crystals. That is a major program of research which is still in progress.

A fourth joke of nature is a similarity in behavior between quasi-crystals and the zeros of the Riemann zeta-function. The zeros of the zeta-function are exciting to mathematicians because they are found to lie on a straight line and nobody understands why. The statement that with trivial exceptions they all lie on a straight line is the famous Riemann hypothesis. To prove the Riemann hypothesis has been the dream of young mathematicians for more than a hundred years. I am now making the outrageous suggestion that we might use quasi-crystals to prove the Riemann Hypothesis. Those of you who are mathematicians may consider the suggestion frivolous. Those who are not mathematicians may consider it uninteresting. Nevertheless I am putting it forward for your serious consideration.

When the physicist Leo Szilard was young, he became dissatisfied with the ten commandments of Moses and wrote a new set of ten commandments to replace them. Szilard's second commandment says: "Let your acts be directed towards a worthy goal, but do not ask if they can reach it: they are to be models and examples, not means to an end". Szilard practised what he preached. He was the first physicist to imagine nuclear weapons and the first to campaign actively against their use. His second commandment certainly applies here. The proof of the Riemann hypothesis is a worthy goal, and it is not for us to ask whether we can reach it. I will give you some hints describing how it might be achieved. Here I will be giving voice to the mathematician that I was fifty years ago before I became a physicist. I will talk first about the Riemann Hypothesis and then about quasi-crystals.

There were until recently two supreme unsolved problems in the world of pure mathematics, the proof of Fermat's Last Theorem and the proof of the Riemann Hypothesis. Twelve years ago, my Princeton colleague Andrew Wiles polished off Fermat's Last Theorem, and only the Riemann Hypothesis remains. Wiles' proof of the Fermat Theorem was not just a technical stunt. It required the discovery and exploration of a new field of mathematical ideas, far wider and more consequential than the Fermat Theorem itself. It is likely that any proof of the Riemann Hypothesis will likewise lead to a deeper understanding of many diverse areas of mathematics and perhaps of physics too. Riemann's zeta-function, and other zeta-functions similar to it, appear ubiquitously in number theory, in the theory of dynamical systems, in geometry, in function theory and in physics. The zeta-function stands at a junction where paths lead in many directions. A proof of the hypothesis will illuminate all the connections. Like every serious student of pure mathematics, when I was young I had dreams of proving the Riemann Hypothesis. I had some vague ideas that I thought might lead to a proof. In recent years, after the discovery of quasi-crystals, my ideas became a little less vague. I offer them here for the consideration of any young mathematician who has ambitions to win a Fields Medal.

Quasi-crystals can exist in spaces of one, two or three dimensions. From the point of view of physics, the three-dimensional quasi-crystals are the most interesting, since they inhabit our three-dimensional world and can be studied experimentally. From the point of view of a mathematician, one-dimensional quasi-crystals are much more interesting than two-dimensional or three-dimensional quasi-crystals because they exist in far greater variety. The mathematical definition of a quasi-crystal is as follows. A quasi-crystal is a distribution of discrete point masses whose Fourier transform is a distribution of discrete point frequencies. Or to say it more briefly, a quasi-crystal is a pure point distribution that has a pure point spectrum. This definition includes as a special case the ordinary crystals which are periodic distributions with periodic spectra.

Excluding the ordinary crystals, quasi-crystals in three dimensions come in very limited variety, all of them being associated with the icosahedral rotation-group. The two-dimensional quasi-crystals are more numerous, roughly one distinct type associated with each regular polygon in a plane. The two-dimensional quasi-crystal with pentagonal symmetry is the famous Penrose tiling of the plane. Finally, the one-dimensional quasi-crystals have a far richer structure since they are not tied to any rotational symmetries. So far as I know, no complete enumeration of one-dimensional quasi-crystals exists. It is known that a unique quasi-crystal exists corresponding to every Pisot–Vijayaraghavan number or PV number. A PV number is a real algebraic integer, a root of a polynomial equation with integer coefficients, such that all the other roots have absolute value less than one [Bertin *et al.*, 1992]. The set of all PV numbers is infinite and has a remarkable topological structure. The set of all one-dimensional quasi-crystals has a structure at least as rich as the set of all PV numbers and probably much richer. We do not know for sure, but it is likely that a huge universe of one-dimensional quasi-crystals not associated with PV numbers is waiting to be discovered.

Here comes the connection of the one-dimensional quasi-crystals with the Riemann hypothesis. If the Riemann hypothesis is true, then the zeros of the zeta-function form a one-dimensional quasi-crystal according to the definition. They constitute a distribution of point masses on a straight line, and their Fourier transform is likewise a distribution of point masses, one at each of the logarithms of ordinary prime numbers and prime-power numbers. My friend Andrew Odlyzko has published a beautiful computer calculation of the Fourier transform of the zeta-function zeros [Odlyzko, 1990]. The calculation shows precisely the expected structure of the Fourier transform, with a sharp discontinuity at every logarithm of a prime or prime-power number and nowhere else.

My suggestion is the following. Let us pretend that we do not know that the Riemann hypothesis is true. Let us tackle the problem from the other end. Let us try to obtain a complete enumeration and classification of one-dimensional quasi-crystals. That is to say, we enumerate and classify all point distributions that have a discrete point spectrum. Collecting and classifying new species of objects is a quintessentially Baconian activity. It is an appropriate activity for mathematical frogs. We shall then find the well-known quasi-crystals associated with PV numbers, and also a whole universe of other quasi-crystals, known and unknown. Among the multitude of other quasi-crystals we search for one corresponding to the Riemann zeta-function and one corresponding to each of the other zeta-functions that resemble the Riemann zeta-function. Suppose that we find one of the quasi-crystals in our enumeration with properties that identify it with the zeros of the Riemann zeta-function. Then we have proved the Riemann hypothesis and we can wait for the telephone call announcing the award of the Fields Medal.

These are of course idle dreams. The problem of classifying one-dimensional quasi-crystals is horrendously difficult, probably at least as difficult as the problems that Andrew Wiles took seven years to explore. But if we take a Baconian point of view, the history of mathematics is a history of horrendously difficult problems being solved by young people too ignorant to know that they were impossible. The classification of quasi-crystals is a worthy goal, and might even turn out to be achievable. Problems of that degree of difficulty will not be solved by old men like me. I leave this problem as an exercise for the young frogs in the audience.

3. Abram Besicovich and Hermann Weyl

Let me now introduce you to some notable frogs and birds that I knew personally. I came to Cambridge University as a student in 1941 and had the tremendous luck to be given the Russian mathematician Abram Samoilovich Besicovich as my supervisor. Since this was in the middle of World War II, there were very few students in Cambridge, and almost no graduate students. Although I was only seventeen years old and Besicovich was already a famous professor, he gave me a great deal of his time and attention, and we became life-long friends. He set the style in which I began to work and think about mathematics. He gave wonderful lectures on measure-theory and integration, smiling amiably when we laughed at his glorious abuse of the English language. I remember only one occasion when he was annoyed by our laughter. He remained silent for a while and then said, "Gentlemen. Fifty million English speak English you speak. Hundred and fifty million Russians speak English I speak".

Besicovich was a frog, and became famous when he was young by solving a problem in elementary plane geometry known as the Kakeya problem. The Kakeya problem was the following. A line-segment of length one is allowed to move freely in a plane while rotating through an angle of 360 degrees. What is the smallest area of the plane that it can cover during its rotation? The problem was posed by the Japanese mathematician Kakeya in 1917 and remained a famous unsolved problem for ten years. George Birkhoff, the leading American mathematician at that time, publicly proclaimed that the Kakeya problem and the four-color problem were the outstanding unsolved problems of the day. It was widely believed that the minimum area was $\pi/8$, which is the area of a three-cusped hypocycloid. The three-cusped hypocycloid is a beautiful three-pointed curve. It is the curve traced out by a point on the circumference of a circle with radius one-quarter, when the circle rolls around the inside of a fixed circle with radius three-quarters. The line segment of length one can turn while always remaining tangent to the hypocycloid with its two ends also on the hypocycloid. This picture of the line turning while touching the inside of the hypocycloid at three points was so elegant that most people believed it must give the minimum area. Then Besicovich surprised everyone by proving that the area covered by the line as it turns can be less than ϵ for any positive ϵ.

Besicovich had actually solved the problem in 1920 before it became famous, not even knowing that Kakeya had proposed it. In 1920, he published the solution in Russian in the Journal of the Perm Physics and Mathematics Society, a journal that was not widely read. The University of Perm, a city 1100 kilometers east of Moscow, was briefly a refuge for many distinguished mathematicians after the Russian revolution. They published two volumes of their journal before it died amid the chaos of revolution and civil war. Outside Russia, the journal was not only unknown but unobtainable. Besicovich left Russia in 1925 and arrived at Copenhagen, where he learned about the famous Kakeya problem that he had solved five years earlier. He published the solution again, this time in English in the Mathematische Zeitschrift. The Kakeya problem as Kakeya had proposed was a typical frog problem, a concrete problem without much connection with the rest of mathematics. Besicovich gave it an elegant and deep solution, which revealed a connection with general theorems about the structure of sets of points in a plane.

The Besicovich style is seen at its finest in his three classic papers with the title, "On the fundamental geometric properties of linearly measurable plane sets of points", published in Mathematische Annalen in the years 1928, 1938 and 1939. In these papers he proved that every linearly measurable set in the plane is divisible into a regular and an irregular component, that the regular component has a tangent almost everywhere, and the irregular component has a projection of measure zero onto almost all directions. Roughly speaking, the regular component looks like a collection of continuous curves, while the irregular component looks nothing like a continuous curve. The existence and the properties of the irregular component are connected with the Besicovich solution of the Kakeya problem. One of the problems that he gave me to work on was the division of measurable sets into regular and irregular components in spaces of higher dimensions. I got nowhere with the problem, but became permanently imprinted with the Besicovich style. The Besicovich style is architectural. He builds out of simple elements a delicate and complicated architectural structure, usually with a hierarchical plan, and then, when the building is finished, the completed structure leads by simple arguments to an unexpected conclusion. Every Besicovich proof is a work of art, as carefully constructed as a Bach fugue.

A few years after my apprenticeship with Besicovitch, I came to Princeton and got to know Hermann Weyl. Weyl was a prototypical bird, just as Besicovitch was a prototypical frog. I was lucky to overlap with Weyl for one year at the Princeton Institute for Advanced Study before he retired from the Institute and moved back to his old home in Zürich. He liked me because during that year I published papers in the Annals of Mathematics about number theory and in the Physical Review about the quantum theory of radiation. He was one of the few people alive who was at home in both subjects. He welcomed me to the Institute, in the hope that I would be a bird like himself. He was disappointed. I remained obstinately a frog. Although I poked around in a variety of mud-holes, I always looked at them one at a time and did not look for connections between them. For me, number theory and quantum theory were separate worlds with separate beauties. I did not look at them as Weyl did, hoping to find clues to a grand design.

Weyl's great contribution to the quantum theory of radiation was his invention of gauge-fields. The idea of gauge-fields had a curious history. Weyl invented them in 1918 as classical fields in his unified theory of general relativity and electromagnetism [Weyl, 1918]. He called them "gauge-fields" because they were concerned with the non-integrability of measurements of length. His unified theory was promptly and publicly rejected by Einstein. After this thunderbolt from on high, Weyl did not abandon his theory but moved on to other things. The theory had no experimental consequences that could be tested. Then wave mechanics was invented by Schrödinger, and three independent publications by Fock, Klein and Gordon in 1926 proposed a relativistic wave equation for a charged particle interacting with an electromagnetic field. Only Fock noticed that the wave equation was invariant under a group of transformations which he called "gradient transformations" [Fock, 1926]. The authoritative Russian textbook on Classical Field Theory [Landau and Lifshitz, 1941] calls the invariance "gradient-invariance" and attributes its discovery to Fock. Meanwhile, F. London in 1927 and Weyl in 1928 observed that the gradient-invariance of quantum mechanics is closely related to the gauge-invariance of Weyl's version of general relativity. For a detailed account of this history see [Okun, 1998]. Weyl realized that his gauge-fields fitted far better into the quantum world than they did into the classical world [Weyl, 1929]. All that he needed to do, to change a classical gauge into a quantum gauge, was to change real numbers into complex numbers. In quantum mechanics, every quantum of electric charge carries with it a complex wave-function with a phase, and the gauge-field is concerned with the non-integrability of measurements of phase. The gauge-field could then be precisely identified with the electromagnetic potential, and the law of conversation of charge became a consequence of the local gauge-invariance of the theory.

Weyl died four years after he returned from Princeton to Zurich, and I wrote his obituary for the journal Nature [Dyson, 1956]. "Among all the mathematicians who began their working lives in the twentieth century", I wrote, "Hermann Weyl was the one who made major contributions in the greatest number of different fields. He alone could stand comparison with the last great universal mathematicians of the nineteenth century, Hilbert and Poincaré. So long as he was alive, he embodied a living contact between the main lines of advance in pure mathematics and in theoretical physics. Now he is dead, the contact is broken, and our hopes of comprehending the physical universe by a direct use of creative mathematical imagination are for the time being ended". I mourned his passing, but I had no desire to pursue his dream. I was happy to see pure mathematics and physics marching ahead in opposite directions.

The obituary ended with a sketch of Weyl as a human being: "Characteristic of Weyl was an aesthetic sense which dominated his thinking on all subjects. He once said to me, half joking, 'My work always tried to unite the true with the beautiful; but when I had to choose one or the other, I usually chose the beautiful'. This remark sums up his personality perfectly. It shows his profound faith in an ultimate harmony of Nature, in which the laws should inevitably express themselves in a mathematically beautiful form. It shows also his recognition of human frailty, and his humor, which always stopped him short of being pompous. His friends in Princeton will remember him as he was when I last saw him, at the Spring Dance of the Institute for Advanced Study last April: a big jovial man, enjoying himself splendidly, his cheerful frame and his light step giving no hint of his sixty-nine years".

The fifty years after Weyl's death were a golden age of experimental physics and observational astronomy, a golden age for Baconian travelers picking up facts, for frogs exploring small patches of the swamp in which we live. During these fifty years, the frogs accumulated a detailed knowledge of a large variety of cosmic structures and a large variety of particles and interactions. As the exploration of new territories continued, the universe became more complicated. Instead of a grand design displaying the simplicity and beauty of Weyl's mathematics, the explorers found weird objects such as quarks and gamma-ray bursts, weird concepts such as supersymmetry and multiple universes. Meanwhile, mathematics was also becoming more complicated, as exploration continued into the phenomena of chaos and many other new areas opened by electronic computers. The mathematicians discovered the central mystery of computability, the conjecture represented by the statement, P is not equal to NP. The conjecture asserts that there exist mathematical problems which can be quickly solved in individual cases but cannot be solved by a quick algorithm applicable to all cases. All the experts believe that the conjecture is true. The most famous example of such a problem is the traveling salesman problem, to find a short route for a salesman to visit a set of cities, knowing the distance between each pair. For technical reasons, we do not ask for the shortest route but for a route with length less than a given upper bound. Then the traveling salesman problem is conjectured to be NP but not P. But nobody has a glimmer of an idea how to prove it. This is a mystery that could not even have been formulated within the nineteenth-century mathematical universe of Hermann Weyl.

4. Frank Yang and Yuri Manin

The last fifty years have been a hard time for birds. Even in hard times, there is work for birds to do, and birds have appeared with the courage to tackle it. Soon after Weyl left Princeton, Frank Yang arrived from Chicago and moved into Weyl's old house. Yang took Weyl's place as the leading bird among my generation of physicists. While Weyl was still alive, Yang and his student Robert Mills discovered the Yang–Mills theory of non-Abelian gauge fields, a marvelously elegant extension of Weyl's idea of a gauge field [Yang and Mills, 1954]. Weyl's gauge-field was a classical quantity, satisfying the commutative law of multiplication. The Yang–Mills theory had a triplet of gauge-fields which did not commute. They satisfied the commutation-rules of the three components of a quantum-mechanical spin, which are generators of the simplest non-Abelian Lie algebra A1. The theory was later generalized so that the gauge-fields could be generators of any finite-dimensional Lie algebra. With this generalization, the Yang–Mills gauge-field theory provided the framework for a model of all the known particles and interactions, a model that is now known as the Standard Model of particle physics. Yang put the finishing touch to it by showing that Einstein's theory of gravitation fits into the same framework, with the Christoffel three-index symbol taking the role of gauge-field [Yang, 1974].

In an appendix to his 1918 paper, added in 1955 for the volume of selected papers published to celebrate his seventieth birthday, Weyl expressed his final thoughts about gauge-field theories (my translation) [Weyl, 1956]. "The strongest argument for my theory seemed to be this, that gauge-invariance was related to conservation of electric charge in the same way as coordinate-invariance was related to conservation of energy and momentum". Thirty years later, Yang was in Zürich for the celebration of Weyl's hundredth birthday. In his speech [Yang, 1986], Yang quoted this remark as evidence of Weyl's devotion to the idea of gauge-invariance as a unifying principle for physics. Yang then went on, "Symmetry, Lie groups and gauge invariance are now recognized, through theoretical and experimental developments, to play essential roles in determining the basic forces of the physical universe. I have called this the principle that symmetry dictates interaction". This idea, that symmetry dictates interaction, is Yang's generalization of Weyl's remark. Weyl observed that gauge-invariance is intimately connected with physical conservation laws. Weyl could not go further than this, because he knew only the gauge-invariance of commuting Abelian fields. Yang made the connection much stronger by introducing non-Abelian gauge-fields. With non-Abelian gauge-fields generating non-trivial Lie algebras, the possible forms of interaction between fields become unique, so that symmetry dictates interaction. This idea is Yang's greatest contribution to physics. It is the contribution of a bird, flying high over the rain-forest of little problems in which most of us spend our lives.

Another bird for whom I have a deep respect is the Russian mathematician Yuri Manin, who recently published a delightful book of essays with the title, "Mathematics as Metaphor" [Manin, 2007]. The book was published in Moscow in Russian, and by the American Mathematical Society in English. I wrote a preface for the English version, and I give you here a short quote from my preface. "Mathematics as Metaphor" is a good slogan for birds. It means that the deepest concepts in mathematics are those which link one world of ideas with another. In the seventeenth century, Descartes linked the disparate worlds of algebra and geometry with his concept of coordinates, and Newton linked the worlds of geometry and dynamics with his concept of fluxions, nowadays called calculus. In the nineteenth century, Boole linked the worlds of logic and algebra with his concept of symbolic logic, and Riemann linked the worlds of geometry and analysis with his concept of Riemann surfaces. Coordinates, fluxions, symbolic logic and Riemann surfaces are all metaphors, extending the meanings of words from familiar to unfamiliar contexts. Manin sees the future of mathematics as an exploration of metaphors that are already visible but not yet understood. The deepest such metaphor is the similarity in structure between number theory and physics. In both fields he sees tantalizing glimpses of parallel concepts, symmetries linking the continuous with the discrete. He looks forward to a unification which he calls the quantization of mathematics.

Manin disagrees with the Baconian story, that Hilbert set the agenda for the mathematics of the twentieth century when he presented his famous list of twenty-three unsolved problems to the International Congress of Mathematicians in Paris in 1900. According to Manin, Hilbert's problems were a distraction from the central themes of mathematics. Manin sees the important advances in mathematics coming from programs, not from problems. Problems are usually solved by applying old ideas in new ways. Programs of research are the nurseries where new ideas are born. He sees the Bourbaki program, rewriting the whole of mathematics in a more abstract language, as the source of many of the new ideas of the twentieth century. He sees the Langlands program, unifying number theory with geometry, as a promising source of new ideas for the twenty-first. People who solve famous unsolved problems may win big prizes, but people who start new programs are the real pioneers.

The Russian version of "Mathematics as Metaphor" contains ten chapters that were omitted from the English version. The American Mathematical Society decided that these chapters would not be of interest to English-language readers. The omissions are doubly unfortunate. First, readers of the English version see only a truncated view of Manin, who is perhaps unique among mathematicians in his broad range of interests extending far beyond mathematics. Second, we see a truncated view of Russian culture, which is less compartmentalized than English-language culture, and brings mathematicians into closer contact with historians, artists and poets.

5. John von Neumann

Another important figure in twentieth-century mathematics was John von Neumann. Von Neumann was a frog, applying his prodigious technical skill to solve problems in many branches of mathematics and physics. He began with the foundations of mathematics. He found the first satisfactory set of axioms for set-theory, avoiding the logical paradoxes that Cantor had encountered in his attempts to deal with infinite sets and infinite numbers. Von Neumann's axioms were used by his bird friend Kurt Gödel a few years later to prove the existence of undecidable propositions in mathematics. Gödel's thorems gave birds a new vision of mathematics. After Gödel, mathematics was no longer a single structure tied together with a unique concept of truth, but an archipelago of structures with diverse sets of axioms and diverse notions of truth. Gödel showed that mathematics is inexhaustible. No matter which set of axioms is chosen as the foundation, birds can always find questions that those axioms cannot answer.

Von Neumann went on from the foundations of mathematics to the foundations of quantum mechanics. To give quantum mechanics a firm mathematical foundation, he created a magnificent theory of rings of operators. Every observable quantity is represented by a linear operator, and the peculiarities of quantum behavior are faithfully represented by the algebra of operators. Just as Newton invented calculus to describe classical dynamics, von Neumann invented rings of operators to describe quantum dynamics.

Von Neumann made fundamental contributions to several other fields, especially to game theory and to the design of digital computers. For the last ten years of his life, he was deeply involved with computers. He was so strongly interested in computers that he decided not only to study their design but to build one with real hardware and software and use it for doing science. I have vivid memories of the early days of von Neumann's computer project at the Institute for Advanced Study in Princeton. At that time he had two main scientific interests, hydrogen bombs and meteorology. He used his computer during the night for doing hydrogen bomb calculations and during the day for meteorology. Most of the people hanging around the computer building in daytime were meteorologists. Their leader was Jule Charney. Charney was a real meteorologist, properly humble in dealing with the inscrutable mysteries of the weather, and skeptical of the ability of the computer to solve the mysteries. John von Neumann was less humble and less skeptical. I heard von Neumann give a lecture about the aims of his project. He spoke, as he always did, with great confidence. He said, "The computer will enable us to divide the atmosphere at any moment into stable regions and unstable regions. Stable regions we can predict. Unstable regions we can control". Von Neumann believed that any unstable region could be pushed by a judiciously applied small perturbation so that it would move in any desired direction. The small perturbation would be applied by a fleet of airplanes carrying smoke-generators, to absorb sunlight and raise or lower temperatures at places where the perturbation would be most effective. In particular, we could stop an incipient hurricane by identifying the position of an instability early enough, and then cooling that patch of air before it started to rise and form a vortex. Von Neumann, speaking in 1950, said it would take only ten years to build computers powerful enough to diagnose accurately the stable and unstable regions of the atmosphere. Then, once we had accurate diagnosis, it would take only a short time for us to have control. He expected that practical control of the weather would be a routine operation within the decade of the 1960s.

Von Neumann, of course, was wrong. He was wrong because he did not know about chaos. We now know that when the motion of the atmosphere is locally unstable, it is very often chaotic. The word "chaotic" means that motions that start close together diverge exponentially from each other as time goes on. When the motion is chaotic, it is unpredictable, and a small perturbation does not move it into a stable motion that can be predicted. A small perturbation will usually move it into another chaotic motion that is equally unpredictable. So von Neumann's strategy for controlling the weather fails. He was, after all, a great mathematician but a mediocre meteorologist.

Edward Lorenz discovered in 1963 that the solutions of the equations of meteorology are often chaotic. That was six years after von Neumann died. Lorenz was a meteorologist and is generally regarded as the discoverer of chaos. He discovered the phenomena of chaos in the meteorological context and gave them their modern names. But in fact I had heard the mathematician Mary Cartwright, who died recently at the age of ninety-seven, describe the same phenomena in a lecture in Cambridge in 1943, twenty years before Lorenz discovered them. She called the phenomena by different names, but they were the same phenomena. She discovered them in the solutions of the Van der Pol equation which describe the oscillations of a nonlinear amplifier [Cartwright and Littlewood, 1945]. The Van der Pol equation was important in World War II because nonlinear amplifiers fed power to the transmitters in early radar systems. The transmitters behaved erratically, and the Royal Air Force blamed the manufacturers for making defective amplifiers. Mary Cartwright was asked to look into the problem. She showed that the manufacturers were not to blame. She showed that the Van der Pol equation was to blame. The solutions of the Van der Pol equation have precisely the chaotic behavior that the Royal Air Force was complaining about. I heard all about chaos from Mary Cartwright seven years before I heard Von Neumann talk about weather control, but I was not far-sighted enough to make the connection. It never entered my head that the erratic behavior of the Van der Pol equation might have something to do with meteorology. If I had been a bird rather than a frog, I would probably have seen the connection, and I might have saved von Neumann a lot of trouble. If he had known about chaos in 1950, he would probably have thought about it deeply, and he would have had something important to say about it in 1954.

Von Neumann got into trouble at the end of his life because he was really a frog but everyone expected him to fly like a bird. In 1954, there was an International Congress of Mathematicians in Amsterdam. These congresses happen only once in four years and it is a great honor to be invited to speak at the opening session. The organizers of the Amsterdam congress invited von Neumann to give the keynote speech, expecting him to repeat the act that Hilbert had performed in Paris in 1900. Just as Hilbert had provided a list of unsolved problems to guide the development of mathematics for the first half of the twentieth century, von Neumann was invited to do the same for the second half of the century. The title of von Neumann's talk was announced in the program of the congress. It was "Unsolved Problems in Mathematics: Address by Invitation of the Organizing Committee". After the congress was over, the complete proceedings were published, with the texts of all the lectures except this one. In the proceedings, there is a blank page with von Neumann's name and the title of his talk. Underneath, it says, "No manuscript of this lecture was available".

What happened? I know what happened, because I was there in the audience, at 3 pm on Thursday September 2, 1954, in the Concertgebouw concert hall. The hall was packed with mathematicians, all expecting to hear a brilliant lecture worthy of such a historic occasion. The lecture was a huge disappointment. Von Neumann had probably agreed several years earlier to give this lecture about unsolved problems, and had then forgotten about it. Being busy with many other things, he had neglected to prepare the lecture. Then, at the last moment, when he remembered that he had to travel to Amsterdam and talk about mathematics, he pulled an old lecture from the 1930s out of a drawer and dusted it off. The lecture was about rings of operators, a subject that was new and fashionable in the 1930s. Nothing about unsolved problems. Nothing about the future. Nothing about computers, the subject that we knew was dearest to von Neumann's heart. He might at least have had something new and exciting to say about computers. The audience in the concert hall became restless. Somebody said in a voice loud enough to be heard all over the hall, "Aufgewärmte Suppe", which is German for "Warmed-up Soup". In 1954, the great majority of mathematicians knew enough German to understand the joke. Von Neumann, deeply embarrassed, brought his lecture to a quick end and left the hall without waiting for questions.

6. Weak Chaos

If von Neumann had known about chaos when he spoke in Amsterdam, one of the unsolved problems that he might have talked about was weak chaos. The problem of weak chaos is still unsolved fifty years later. The problem is to understand why chaotic motions often remain bounded and do not cause any violent instability. A good example of weak chaos is the orbital motions of the planets and satellites in the solar system. It was discovered only recently that these motions are chaotic. This was a surprising discovery, upsetting the traditional picture of the solar system as the prime example of orderly stable motion. The mathematician Laplace two hundred years ago thought he had proved that the solar system is stable. It now turns out that Laplace was wrong. Accurate numerical integrations of the orbits show clearly that neighboring orbits diverge exponentially. It seems that chaos is almost universal in the world of classical dynamics.

Chaotic behavior was never suspected in the solar system before accurate long-term integrations were done, because the chaos is weak. Weak chaos means that neighboring trajectories diverge exponentially but never diverge far. The divergence begins with exponential growth but afterwards remains bounded. Because the chaos of the planetary motions is weak, the solar system can survive for four billion years. Although the motions are chaotic, the planets never wander far from their customary places, and the system as a whole does not fly apart. In spite of the prevalence of chaos, the Laplacian view of the solar system as a perfect piece of clockwork is not far from the truth.

We see the same phenomena of weak chaos in the domain of meteorology. Although the weather in New Jersey is painfully chaotic, the chaos has firm limits. Summers and winters are unpredictably mild or severe, but we can reliably predict that the temperature will never rise to 45 degrees Celsius or fall to minus 30, extremes that are often exceeded in India or in Minnesota. There is no conservation law of physics that forbids temperatures from rising as high in New Jersey as in India, or from falling as low in New Jersey as in Minnesota. The weakness of chaos has been essential to the long-term survival of life on this planet. Weak chaos gives us a challenging variety of weather while protecting us from fluctuations so severe as to endanger our existence. Chaos remains mercifully weak for reasons that we do not understand. That is another unsolved problem for young frogs in the audience to take home. I challenge you to understand the reasons why the chaos observed in a great diversity of dynamical systems is generally weak.

The subject of chaos is characterized by an abundance of quantitative data, an unending supply of beautiful pictures, and a shortage of rigorous theorems. Rigorous theorems are the best way to give a subject intellectual depth and precision. Until you can prove rigorous theorems, you do not fully understand the meaning of your concepts. In the field of chaos I know only one rigorous theorem, proved by Tien-Yien Li and Jim Yorke in 1975 and published in a short paper with the title, "Period three implies chaos" [Li and Yorke, 1975]. The Li–Yorke paper is one of the immortal gems in the literature of mathematics. Their theorem concerns nonlinear maps of an interval onto itself. The successive positions of a point when the mapping is repeated can be considered as the orbit of a classical particle. An orbit has period N if the point returns to its original position after N mappings. An orbit is defined to be chaotic, in this context, if it diverges from all periodic orbits. The theorem says that if a single orbit with period three exists, then chaotic orbits also exist. The proof is simple and short. To my mind, this theorem and its proof throw more light than a thousand beautiful pictures on the basic nature of chaos. The theorem explains why chaos is prevalent in the world. It does not explain why chaos is so often weak. That remains a task for the future. I believe that weak chaos will not be understood in a fundamental way until we can prove rigorous theorems about it.

7. String Theorists

I would like to say a few words about string theory. Few words, because I know very little about string theory. I never took the trouble to learn the subject or to work on it myself. But when I am at home at the Institute for Advanced Study in Princeton, I am surrounded by string theorists, and I sometimes listen to their conversations. Occasionally I understand a little of what they are saying. Three things are clear. First, what they are doing is first-rate mathematics. The leading pure mathematicians, people like Michael Atiyah and Isadore Singer, love it. It has opened up a whole new branch of mathematics, with new ideas and new problems. Most remarkably, it gave the mathematicians new methods to solve old problems that were previously unsolvable. Second, the string theorists think of themselves as physicists rather than mathematicians. They believe that their theory describes something real in the physical world. And third, there is not yet any proof that the theory is relevant to physics. The theory is not yet testable by experiment. The theory remains in a world of its own, detached from the rest of physics. String theorists make strenuous efforts to deduce consequences of the theory that might be testable in the real world, so far without success.

My colleagues Ed Witten and Juan Maldacena and others who created string theory are birds, flying high and seeing grand visions of distant ranges of mountains. The thousands of humbler practitioners of string theory in universities around the world are frogs, exploring fine details of the mathematical structures that birds first saw on the horizon. My anxieties about string theory are sociological rather than scientific. It is a glorious thing to be one of the first thousand string-theorists, discovering new connections and pioneering new methods. It is not so glorious to be one of the second thousand or one of the tenth thousand. There are now about ten thousand string-theorists scattered around the world. This is a dangerous situation for the tenth thousand and perhaps also for the second thousand. It may happen unpredictably that the fashion changes and string theory becomes unfashionable. Then it could happen that nine thousand string-theorists lose their jobs. They have been trained in a narrow specialty and they may be unemployable in other fields of science.

Why are so many young people attracted to string theory? The attraction is partly intellectual. String theory is daring and mathematically elegant. But the attraction is also sociological. String theory is attractive because it offers jobs. And why are so many jobs offered in string theory? Because string theory is cheap. If you are the chairperson of a physics department in a remote place without much money, you cannot afford to build a modern laboratory to do experimental physics, but you can afford to hire a couple of string-theorists. So you offer a couple of jobs in string theory, and you have a modern physics department. The temptations are strong for the chairperson to offer such jobs and for the young people to accept them. This is a hazardous situation for the young people and also for the future of science. I am not saying that we should discourage young people from working in string theory if they find it exciting. I am saying that we should offer them alternatives, so that they are not pushed into string theory by economic necessity.

Finally, I give you my own guess for the future of string theory. My guess is probably wrong. I have no illusion that I can predict the future. I tell you my guess, just to give you something to think about. I consider it unlikely that string theory will turn out to be either totally successful or totally useless. By totally successful I mean that it is a complete theory of physics, explaining all the details of particles and their interactions. By totally useless I mean that it remains a beautiful piece of pure mathematics. My guess is that string theory will end somewhere between complete success and failure. I guess that it will be like the theory of Lie groups, which Sophus Lie created in the nineteenth century as a mathematical framework for classical physics. So long as physics remained classical, Lie groups remained a failure. They were a solution looking for a problem. But then, fifty years later, the quantum revolution transformed physics, and Lie algebras found their proper place. They became the key to understanding the central role of symmetries in the quantum world. I expect that fifty or a hundred years from now another revolution in physics will happen, introducing new concepts of which we now have no inkling, and the new concepts will give string theory a new meaning. After that, string theory will suddenly find its proper place in the universe, making testable statements about the real world. I warn you that this guess about the future is probably wrong. It has the virtue of being falsifiable, which according to Karl Popper is the hallmark of a scientific statement. It may be demolished tomorrow by some discovery coming out of the Large Hadron Collider in Geneva.

8. Manin Again

To end this talk, I come back to Yuri Manin and his book, "Mathematics as Metaphor". The book is mainly about mathematics. It may come as a surprise to Western readers that he writes with equal eloquence about other subjects such as the collective unconscious, the origin of human language, the psychology of autism, and the role of the trickster in the mythology of many cultures. To his compatriots in Russia, such many-sided interests and expertise would come as no surprise. Russian intellectuals maintain the proud tradition of the old Russian intelligentsia, with scientists, poets, artists and musicians belonging to a single community. They are still today, as we see them in the plays of Chekhov, a group of idealists bound together by their alienation from a superstitious society and a capricious government. In Russia, mathematicians, composers and film-producers talk to one another, walk together in the snow on winter nights, sit together over a bottle of wine, and share each others' thoughts.

Manin is a bird whose vision extends far beyond the territory of mathematics into the wider landscape of human culture. One of his hobbies is the theory of archetypes invented by the Swiss psychologist Carl Jung. An archetype, according to Jung, is a mental image rooted in a collective unconscious that we all share. The intense emotions that archetypes carry with them are relics of lost memories of collective joy and suffering. Manin is saying that we do not need to accept Jung's theory as true in order to find it illuminating.

More than thirty years ago, the singer Monique Morelli made a recording of songs with words by Pierre MacOrlan. One of the songs is "La Ville Morte", the dead city, with a haunting melody tuned to Morelli's deep contralto, with an accordion singing counterpoint to the voice, and with verbal images of extraordinary intensity. Printed on the page, the words are nothing special:

"En pénétrant dans la ville morte,
Je tenait Margot par le main. . .
Nous marchions de la nécropole,
Les pieds brisés et sans parole,
Devant ces portes sans cadole,
Devant ces trous indéfinis,
Devant ces portes sans parole
Et ces poubelles pleines de cris".

"As we entered the dead city, I held Margot by the hand . . . We walked from the graveyard on our bruised feet, without a word, passing by these doors without locks, these vaguely glimpsed holes, these doors without a word, these garbage-cans full of screams".

I can never listen to that song without a disproportionate intensity of feeling. I often ask myself why the simple words of the song seem to resonate with some deep level of unconscious memory, as if the souls of the departed are speaking through Morelli's music. And now unexpectedly in Manin's book I find an answer to my question. In his chapter, "The Empty City Archetype", Manin describes how the archetype of the dead city appears again and again in the creations of architecture, literature, art and film, from ancient to modern times, ever since human beings began to congregate in cities, ever since other human beings began to congregate in armies to ravage and destroy them. The character who speaks to us in MacOrlan's song is an old soldier who has long ago been part of an army of occupation. After he has walked with his wife through the dust and ashes of the dead city, he hears once more:

"Chansons de charme d'un clairon
Qui fleurissait une heure lointaine
Dans un rêve de garnison".

"The magic calls of a bugle that came to life for an hour in an old soldier's dream".

The words of MacOrlan and the voice of Morelli seem to be bringing to life a dream from our collective unconscious, a dream of an old soldier wandering through a dead city. The concept of the collective unconscious may be as mythical as the concept of the dead city. Manin's chapter describes the subtle light that these two possibly mythical concepts throw upon each other. He describes the collective unconscious as an irrational force that powerfully pulls us toward death and destruction. The archetype of the dead city is a distillation of the agonies of hundreds of real cities that have been destroyed since cities and marauding armies were invented. Our only way of escape from the insanity of the collective unconscious is a collective consciousness of sanity, based upon hope and reason. The great task that faces our contemporary civilization is to create such a collective consciousness.

References

M. J. Bertin *et al.* [1992] *Pisot and Salem Numbers* (Basel, Birkhäuser Verlag).

M. L. Cartwright and J. E. Littlewood [1945] "On non-linear differential equations of the second order, I", *J. London Math. Soc.* **20**, 180–189.

F. Dyson [1956] "Prof. Hermann Weyl, For.Mem.R.S.", *Nature* **177**, 457–458.

V. A. Fock [1926] "On the invariant form of the wave equation and of the equations of motion for a charged massive point", *Zeitschrift f. Physik* **39**, 226–232.

L. Landau and E. Lifshitz [1941] "Teoria polya", *GITTL M.-L.*, Section 16.

T.-Y. Li and J. A. Yorke [1975] "Period three implies chaos", *Am. Math. Monthly* **82**, 985–992.

Y. I. Manin [2007] *Mathematics as Metaphor: Selected Essays* (American Mathematical Society, Providence, Rhode Island). The Russian version is Manin, Yu. I., "Matematika kak Metafora", Moskva, Izdatelstvo MTsNMO, 2008.

A. M. Odlyzko [1990] "Primes, quantum chaos and computers", in *Number Theory, Proc. Symp.* (National Research Council, Washington DC), pp. 35–46.

L. B. Okun [1998] "V. A. Fock and gauge symmetry", in *Quantum Theory, in Honour of Vladimir A. Fock: Proc. VIII UNESCO Int. School of Physics, St. Petersburg, 1998*, ed. Y. Novozhilov and V. Novozhilov, Part II, pp. 13–17.

H. Weyl [1918] "Gravitation und elektrizität", *Sitz. König. Preuss. Akad. Wiss.* **26**, 465–480.

H. Weyl [1929] "Elektron und gravitation", *Zeits. Phys.* **56**, 350–352.

H. Weyl [1956] *Selecta* (Basel, Birkhäuser Verlag), p. 192.

C. N. Yang and R. L. Mills [1954] "Conservation of isotopic spin and isotopic gauge invariance", *Phys. Rev.* **96**, 191–195.

C. N. Yang [1974] "Integral formalism for gauge fields", *Phys. Rev. Lett.* **33**, 445–447.

C. N. Yang [1986] "Hermann Weyl's contribution to physics", in *Hermann Weyl, 1885–1985*, ed. K. Chandrasekharan (Springer-Verlag, Berlin), p. 19.

CHAPTER 2.4

THE CURRENT STATE OF PHYSICS*

1. Prophets of Doom

Last year and the year before, I talked to this seminar about human problems arising out of modern advances in technology. This year Judge Cotter asked me to talk about pure science, and about physics in particular, since physics is my own specialty. So I chose for my title, "The Current State of Physics". But I will not bore you with technical jargon and equations. I give you a broad historical view of physics as a part of our culture, and I include astronomy as the part of physics that has penetrated our culture most deeply. Astronomy has penetrated deepest because it is the part of physics that ordinary people can understand.

Recently I heard a talk by George Steiner, a famous literary scholar, with the title, "Is Science Nearing its Limits?" He answered the question with a resounding "Yes". His view of science can be summarized in one sentence. Science was a passing phase of human culture and is now fortunately over. He painted a picture of science as an institution in an advanced state of decline. Year after year, decade after decade, the input of money into scientific research increases while the output of important discoveries decreases. Constantly increasing numbers of researchers are pursuing problems that are of interest to diminishing numbers of people. Scientists are increasingly isolated in specialized compartments, unable to communicate with each other or with the society to which they belong. While science proliferates into esoteric specialties of little practical relevance, the society which used to look to science for guidance now relies instead on astrology and superstition. Professional astrologers now outnumber professional astronomers by five to one, and popular books on mysticism and magic outnumber popular books on science by ten to one.

_____

*David Wilkinson Lecture, Medina Seminar, given to Judiciary Leadership Development Council, Princeton University, June 15, 2009.

Steiner's picture of science in decline reminds me in many ways of Oswald Spengler's "Untergang des Abendlandes", "The Decline of the West", an apocalyptic description of the collapse of western civilization, published in Munich in July 1918 just as World War I was ending with the collapse of Germany. Spengler had been a student of physics before he became a historian. He devoted many of his pages to the state of science and technology as he saw it in 1918. He wrote with expert knowledge and awareness of the crisis in classical physics that began in 1900 with Max Planck's postulate of the quantum and came to a climax in 1916 with Einstein's publication of the theory of general relativity. "Western European physics", he wrote, "has reached the limit of its possibilities. This is the origin of the sudden and annihilating doubt that has arisen about things that even yesterday were the unchallenged foundation of physical theory, about the meaning of the energy principle, the concepts of mass, space, absolute time, and causal natural laws generally". Spengler was a master of style as well as historical erudition. His great work went through sixty editions in eight years after 1918. It was read by almost everybody in the German-speaking literary world, and also by most of the leading German-speaking scientists. Spengler's literary brilliance convinced them that classical physics had reached a dead end and the only way ahead was a radical revolution.

It turned out, as we now know, that Spengler was right. Seven years after his book was published, there was a radical revolution in physics. Heisenberg discovered the true limits of causality in atomic processes. This revolution did what Spengler had predicted. Heisenberg's quantum mechanics demolished the grand edifice of classical physics as it had existed in the nineteenth century. But after the revolution was over, things did not continue to happen as Spengler had expected. Quantum mechanics was not the end of physics. Instead of the inexorable decline that Spengler had imagined, the revolution led to a new golden age. After quantum mechanics, physics and chemistry flourished as never before, with theory and experiment racing ahead to explain and verify the finest details of atoms and molecules. The revolution was not an ending but a new beginning. The generation of physicists that grew up after the revolution did not read Spengler. That generation had lived through the decline of the West and survived it. They were still alive after living through two world wars and many other catastrophes. They were engaged in rebuilding Western civilization from the wreckage of the wars and revolutions, and they saw science still stretching ahead of them as an endless frontier. That generation happens to be my own.

2. A Hopeful Future

I am not claiming that I can predict the future. The real future will undoubtedly bring surprises and changes which neither Steiner nor I have imagined. I am only describing one possible future, a future of continued progress and achievement. I hope that you will find it more plausible than Steiner's future of failure and decline.

Here are some facts about the present state of science. First, science is now more than ever an international enterprise. It is a shining example of an international enterprise that works efficiently and informally, paying almost no attention to differences of language and nationality. Scientists are not immune from patriotic prejudice and often make exaggerated claims for the achievements of their own countries. But they also know that they must pay attention to the work of other countries if they are to stay ahead in the game. A large fraction of scientists have friends and collaborators scattered over the world. During the last twenty years, as the internet and the world wide web have become the main channels of communication between scientists, international connections have become closer. Even in remote and inaccessible parts of the world, scientists learn almost instantaneously of new discoveries or new ideas. No country can be, or should be, preeminent in all areas of science. Preeminence, when it happens, is always transient.

Another fact. The basic tools of science are growing cheaper, more powerful, more accurate and more widely accessible as time goes on. The basic tools are objects such as personal computers, digital cameras, Global Positioning System receivers, DNA sequencers, DNA synthesizers, lasers and gigabyte memories. While big and expensive scientific projects involving huge apparatus and armies of people receive most of the public attention, most of the good science is still done by a multitude of small projects involving modest apparatus and small groups of people. The future looks bright for small-scale science, as instruments become smaller and more sensitive and data-handling becomes cheaper and faster.

Another fact. The many specialized disciplines into which science is divided are not as isolated from each other as they were in the past. It is becoming easier for young people to jump from one specialty to another. Jumping across disciplines became easier for two reasons. First, the problems which scientists are trying to solve are increasingly multidisciplinary. For example, the understanding of climate change requires experts in meteorology, oceanography, glaciology, ecology, cloud physics and atmospheric chemistry to talk to one another and understand each other's jargon. Once you learn to speak another person's jargon, it is not so difficult to jump over the fence and become a member of his team. I myself am a physicist by training, but when I went to the Oak Ridge National Laboratory to study climate change, I made friends with Alan Poole who is an expert on forest ecology, and I quickly became more interested in forest ecology than in the physical aspects of climate change. I did not become a real expert in forest ecology, but I learned enough about it to know that it is crucial to any real understanding of climate.

Another reason why jumping across disciplines is becoming easier is the convergent evolution of tools. As time goes on, the tools that scientists use in different disciplines are becoming more similar and are often the same. For example, the techniques of computer simulation and data-mining are easily transferred from studies of cosmic evolution in astronomy to studies of molecular dynamics in chemistry or to studies of gene expression in molecular biology. Astronomers, chemists and biologists are all facing the problem of organising immense quantities of information in such a way that the underlying laws and patterns can be understood. The tools and skills required for organizing large data-bases can be applied equally to all three disciplines. A young astronomer, who learns her trade by studying the dynamics of clusters of galaxies in the sky, can make a career switch and apply the same skills to studying reaction pathways of genes and molecules in living cells.

To summarize these facts, we can say that modern science is changing in three ways, all of them helpful. Scientists are becoming more international, tools are becoming more accessible, and barriers between specialties are becoming lower. Science is vigorously alive, flourishing in a greater variety of places than ever before, and bringing the people of the planet together to work on the great human problems of poverty, disease, extinction of species and destruction of habitats. Science is going strong and is ready to meet new challenges.

Scientific projects can be divided roughly into two categories, small and large. Small science is done by groups of less than a hundred people, loosely organized without any hierarchical chain of command. Big science is done by large groups in the style of a military operation, with formal bureaucratic procedures and plans. Big science is expensive and gets most of the attention from the politicians and the public. It needs public attention in order to survive. For the media, small science hardly exists. Fortunately, small science does not need the media. Small science goes quietly ahead, employing more people and making more discoveries in the aggregate than big science. A healthy development of science needs both big and small. Some important problems are beyond the reach of small science. But there are plenty of important problems that small science can tackle, taking advantage of the agility and flexibility that big science lacks.

Until now I have been speaking about science in general. From now on I will speak about physics in particular. In modern physics there are many areas of lively research that are predominantly small. For example, there is low-temperature physics. Enormous progress has been made in the last twenty years, as refrigerators have been invented that extend our reach from millidegrees to nanodegrees above absolute zero temperature. A nanodegree is a million times colder than a millidegree. My experimental colleagues tell me that it is in many ways easier to handle atoms at nanodegree temperatures than at millidegree temperatures. At millidegree temperatures, everything moves around, while at nanodegree temperatures stuff stays where you put it. As you go down in temperature, refrigerators and thermometers become smaller rather than larger. The measuring instruments become more sensitive as they become smaller. Another area of exquisite small-scale experiments is high-pressure physics. The experimental tool is the diamond-anvil cell, which squeezes a tiny sample of material between two flat-faced diamonds. Since the volume of the cell is tiny, enormous pressure can be reached with modest forces. Diamond is chosen to transmit the forces because it is the strongest of all known materials. Nature helps us to do the experiments because diamond, in addition to being the strongest material, is also transparent. We can see directly what happens to the material in the cell while we squeeze it. These experiments are immensely useful for understanding the interior of the earth, because the diamond anvil, a small piece of apparatus standing on a table-top, can reach pressures all the way down to the earth's center. A third area of small-scale physics is nanotechnology, the exploration of mechanical and electrical structures with size measured in nanometers. A nanometer is about ten times as large as an atom, so nanotechnology is the study of things that are bigger than atoms but smaller than grains of sand. Nanotechnology is a rapidly expanding field of science, spreading over the boundaries between physics, chemistry and biology. No matter how far it spreads, it remains by definition small-scale.

In addition to these flourishing fields of small-scale physics that escaped the notice of George Steiner, there are two areas of physics that are inescapably big science, namely astronomy and particle physics. Big science becomes embroiled in politics because it is expensive. It is opposed by some politicians because it is a waste of taxpayers' money. It is supported by others because it is a source of jobs for industry. Such political conflicts do not improve the quality of the science. Nevertheless, I will argue that the future of both areas is bright. The future is bright because there is no shortage of great unsolved mysteries, in astronomy and in particle physics, to keep scientists busy for the next hundred years.

Two of the enticing mysteries of astronomy are the nature of the dark energy and dark matter that are now known to pervade the universe. The tools for exploring these mysteries are advancing rapidly. The traditional tools used by astronomers are telescopes, which improve slowly from year to year. But modern telescopes are only a small part of the astronomer's tool-box. The telescopes feed optical data into electronic detectors, and the detectors feed data into computers with powerful software programs to interpret and distribute the data to the astronomers who try to understand it. The detectors and the software are the cutting-edge tools of astronomy. In the last ten years, detectors and software have improved spectacularly. The improvement is likely to continue in the future, since it is driven by the continuing improvement in the performance of computers and computer networks.

3. The Ghost of Aristotle

Astronomy is a science with a long history, and it is helpful to go far back into the past to illuminate the present and the future. One of the central themes of astronomy for hundreds of years, and even up to the present day, has been to exorcize the ghost of Aristotle. Here is a quotation from Aristotle's treatise, "On the Heavens", written in the fourth century BC.

> "The heaven is a divine body, that is why it has its circular body, which by nature moves for ever in a circle ... Since, then, the heaven must move within its own boundaries, and the stars must not move forward of themselves, we may conclude that both are spherical. This will best ensure to the one its movement and to the others their immobility".

In other books he describes the sublunary world, the world in which we live, constantly changing, filled with irregularly moving objects, winds and waterfalls, animals and plants. The sublunary world is built out of the classical four elements, earth, water, air and fire. The interactions of the four elements give our world its disorder and diversity. Aristotle imagined the sublunary world and the heavens to be different in substance as well as in behavior. The heavens have no disorder and no diversity. They have no earth, water, air or fire. The heavenly bodies and the spheres on which they ride are made of a single fifth element, quintessence, which is changeless and indestructible. The heavens are the realm of perpetual harmony and perpetual peace. The sublunary world is the realm of dissonance and discord. The sublunary world is the home of animals and humans. The heavens are the home of gods. Aristotle did not write like a scientist. He wrote like a theologian.

It was an unhappy accident of history that Aristotle's writings became known in Western Europe during the cultural revolution of the Middle Ages, when universities were founded in Bologna, Paris and Oxford. There was a striking affinity between Aristotle's view of the heavens as the home of gods and the traditional Christian story of Jesus ascending into heaven to sit at the right hand of God after his resurrection. There was also an affinity between Aristotle's style of thinking and the style of Christian theology. The affinities were powerfully strengthened by Saint Thomas Aquinas, who wrote his "Summa Theologiae" in 1265, incorporating Aristotelian doctrines wholesale into a Christian context. The "Summa" became an official statement of theological belief for the Catholic church, and as a consequence Aristotle's speculations about heaven and earth became a part of Christian dogma. Throughout the later Middle Ages, the Aristotelian world-view was taught in the schools and universities of Europe and accepted by all educated and right-thinking Christians. Aristotle's ghost haunted the intellectual life of Europe for many centuries. The exorcism was a long and difficult process, beginning with Copernicus in the year 1543 when modern science was struggling to be born.

Three hundred years later, at the beginning of the twentieth century, nobody any longer believed in Aristotelian cosmology. Even in the official doctrines of the Catholic church, the Copernican world-view was accepted. Catholic theologians understood that the arrangement of planets around the sun had no theological importance. Heaven was a spiritual and not a material realm. It seemed that Aristotelian cosmology had been swept into the dustbin of history. And yet, Aristotle's ghost was still alive, powerfully distorting our view of the astronomical universe. Up to the middle of the twentieth century, the old Aristotelian view of the celestial sphere as a place of perfect peace and harmony still dominated the practice of astronomy. This view had survived the intellectual revolutions associated with the names of Copernicus, Newton and Einstein. It was still taken for granted that the universe was static. The job of an astronomer was like the job of a terrestrial map-maker, to explore the universe and make a map of an unchanging landscape. The mapping would need to be done only once. Whatever was found would not change. Einstein himself shared the general belief that the universe must be static. After he had discovered the general theory of relativity, he found a cosmological solution of his equations describing a static universe. A few years later, when Alexander Friedman found solutions describing an expanding universe, Einstein did not accept them. Einstein was still under the influence of Aristotle's ghost.

There were two turning-points in the exorcising of Aristotle's ghost. The two decisive dates were 1610 and 1931. Both turning-points were brought about by small pieces of glass used to focus light from celestial objects in novel ways. The names associated with the two events are Galileo and Zwicky. Galileo was the first astronomer to imagine a coherent universe with celestial and terrestrial objects made of the same materials and subject to the same laws. He saw Jupiter with four moons orbiting around it, like our moon orbiting around the earth, and understood that Jupiter and the earth must be in many ways alike. He also saw that our moon is not a perfect sphere made of quintessence but an earth-like ball with some areas flat and some areas covered with mountains. Zwicky was the first astronomer to imagine a violent universe. He found that violent events which he called supernovae are occurring frequently all over the universe. Galileo broke the Aristotelian separation of heaven and earth. Zwicky demolished the Aristotelian heaven of unchanging peace and harmony.

The decisive event of 1610 was Galileo making a telescope for himself out of two glass lenses and turning it to the heavens. The decisive event of 1931 was Fritz Zwicky hearing about a new kind of telescope invented by the optician Bernhard Schmidt in Germany two years earlier. Schmidt's design produced sharply focussed images over a field of view with about a hundred times the area of the field of view of other telescopes. This meant that the sky could be photographed a hundred times faster. In 1931, Zwicky was an assistant professor in the physics department at the

California Institute of Technology. He had no official credentials as an astronomer. The professional astronomers never accepted him as a colleague. They thought he was crazy and he thought they were stupid. In retrospect, we can say that both judgments were partially correct. In 1928, the Rockefeller Foundation had awarded a large sum of money to Caltech to build a major astronomical observatory. Caltech acquired Palomar Mountain as the site for the new observatory, and the professional astronomers made plans to build the biggest telescope in the world, the 200-inch Hale telescope that began to operate in 1947. Meanwhile, in 1931, Zwicky heard about the Schmidt telescope and decided to beat the professional astronomers at their own game. He quickly persuaded Caltech to buy an 18-inch Schmidt telescope, the smallest telescope that could do the job that he had in mind, and install it on Palomar Mountain. He made sure that he would have full-time use of the telescope, which was at that time the only wide-field telescope in the world at a site with good astronomical seeing. His telescope was up and running long before the bigger instruments on Palomar Mountain were started.

Zwicky had a hunch that the universe was full of violent events, unseen because nobody had been looking for them. He published papers about neutron stars, supernovae, black holes and gravitational lenses, long before these subjects became fashionable. Supernovae were new stars that occasionally shone in the sky for a few weeks with extraordinary brilliance. Zwicky understood that supernovae were events of extreme violence, probably resulting in the disruption of an entire star. He understood that to see rare, violent and short-lived events in the universe, it was necessary to photograph large areas of sky repeatedly. With his little Schmidt camera, he would have a unique opportunity to photograph the entire northern sky over and over again. Starting in 1936, with a single assistant to help him, he continued for four years to survey the northern sky repeatedly. He and the assistant discovered twenty supernovae, a large enough sample to allow him to classify them into several types and to infer their different modes of origin. Following the example set by Zwicky, the job of an astronomer today is not to make maps of an unchanging sky, but to record and interpret the processes of change.

After Zwicky's sky survey was finished, he proudly wrote: "For the construction of the 18-inch Schmidt telescope, its housing, a full-size objective prism, a small remuneration for my assistant, and the operational costs for the whole project during ten years, only about fifty thousand dollars were expended. This probably represents the highest efficiency, as measured in results achieved per dollar invested, of any telescope presently in use, and perhaps of any ever built, with the exception of Galilei's little refractor". His survey set the pattern for many later surveys done with bigger instruments and bigger investments of manpower and money. The newest sky survey, due to begin this year with the name Pan-Starrs, will follow Zwicky in emphasizing rapid and repeated coverage of the sky. It will discover a wealth of short-lived phenomena at all distances, from near-earth asteroids to optical after-glows of gamma-ray bursts in remote galaxies. The gamma-ray bursts are the newest astronomical mystery, monstrous outpourings of energy that are seen occurring all over the universe at a rate of about one per day. They are more violent than supernovae, and much briefer. They are providing further evidence, if any is needed, that Aristotle's serene and tranquil heavens were an illusion.

With modern detectors and software, a modest telescope can scan the entire sky visible from its site in a few nights, making it possible to discover transient events that traditional methods of observation would miss. Rapid scanning of the sky is giving us a new view of the universe as a dynamic arena in which violent explosions and short-lived objects play a dominant role. With the new tools spreading rapidly around the world, and observatories linked together by communication channels with high band-width, astronomy is entering a new golden age. Ten years ago, only about one percent of the volume of the universe had been explored. Now we have scanned in a preliminary way the entire universe all the way back to the beginning of time. Within the next twenty years we will be looking out at the universe with even more rapid surveys and understanding in detail its unexpected and dramatic history.

Science cannot be close to its limits so long as there are open frontiers to be explored and new tools for explorers. The space frontier is wide open in all directions. Even in the field of physics, the most mature of all branches of science, there are many open frontiers and frequent new discoveries, ranging from quantum computation to quantum condensation of gases at ultra-low temperatures. Except in string-theory, which is only a small part of physics, theory and experiment still go hand in hand. Science is inexhaustible because the complexity of structures is inexhaustible, both in the non-living world of snow-flakes and galaxies, and in the living world of birds and forests.

4. Particle Physics

I will end my talk with some speculations about particle physics. Recently my friend Frank Wilczek published a popular book, "The Lightness of Being", about particle physics. At the end of his book, a chapter entitled, "Anticipating a new golden age", describes his hopes for the future. He sees the Golden Age of particle physics starting very soon. His hopes are based on the Large Hadron Collider (LHC), the biggest and newest particle accelerator, built by the European Center for Nuclear Research (CERN) in Geneva. LHC is a splendid machine, accelerating two beams of particles in opposite directions around a circular vacuum-pipe. Particle detectors surround the beams where they collide, so that the products of the collisions are detected. The energy of each accelerated particle is more than seven times the energy of a particle in any other accelerator. Wilczek sees the advent of the LHC as a culminating moment in the history of science. "Through patchy clouds, off in the distance", he writes, "we seem to glimpse a mathematical Paradise, where the elements that build reality shed their dross. Correcting for the distortions of our everyday vision, we create in our minds a vision of what they might really be: pure and ideal, symmetric, equal, and perfect".

Wilczek, like most scientists who are actively engaged in exploring, does not pay much attention to the history of his science. He lives in the era of particle accelerators, and assumes that particle accelerators in general, and LHC in particular, will be the main source of experimental information about particles in the future. Since I am older and left the field of particle physics many years ago, I look at the field with a longer perspective. I find it useful to examine the past, to explain why I disagree with Wilczek about the future. Here is a summary of the history as I remember it.

Before World War II, particle physics did not exist. We had atomic physics, the science of atoms and electrons, and nuclear physics, the science of atomic nuclei. Beyond these well-established areas of knowledge, there was a dimly lit zone of peculiar phenomena which were called cosmic rays. Cosmic rays were a gentle rain of high-energy particles and radiation that came down onto the earth from outer space. Nobody knew what they were, where they came from, or why they existed. They appeared to come more-or-less uniformly from all directions, at all times of day or night, summer or winter. They were an enduring mystery, not yet a science.

Particle physics emerged unexpectedly in the 1940s, during the early post-war years, while the soldiers were still coming home from battle-fields and prison-camps. Particle physics started with makeshift equipment salvaged from the war to explore a new universe. It was a symbol of hope for a generation battered by war. It proved that former enemies could work together fruitfully on peaceful problems. It gave us reason to dream that friendly collaboration could spread from the world of science to the more contentious worlds of power and politics.

In 1947, Cecil Powell did a historic experiment in Bristol. He was an Englishman who worked during the war on photographic plates that could take photographs from airplanes at night. He knew how to cook photographic plates so as to make them sensitive to cosmic rays. In his plates he could see tracks of cosmic rays coming to rest. When an object comes to rest in a known place at a known time, it is no longer a vague flow of unknown stuff. It is a unique and concrete object. It is accessible to the tools of science. After Powell detected a cosmic-ray coming to rest, he knew where it was, and he could see what it did next. What it often did next was to produce a secondary particle moving close to the speed of light. When he started to study the secondary particles, the mystery of cosmic rays was transformed into the science of particle physics.

After Powell, the pioneers of particle physics continued for five years to work with cosmic rays, finding several more species of particle with passive detectors. The photographic plate was one example of a passive detector. But a new experimental tool, the high-energy accelerator, rapidly took over the field. Particle accelerators had many advantages over passive detectors. Accelerators provided particles in far greater numbers, with precisely known energies, under control of the experimenter. Accelerator experiments were more quantitative and more precise. But accelerators also had some serious disadvantages. They were more expensive than passive detectors, they required teams of engineers to keep them running, and they produced particles with a limited range of energies. Nature provided among the cosmic rays a small number of particles with energies millions of times larger than the largest accelerator could reach. If the distribution of effort between accelerators and passive detectors had been rationally planned, particle physicists would have maintained a balance between the two types of instrument, perhaps three quarters of the money for accelerators and one quarter for passive detectors. Instead, accelerators became the prevailing fashion. The era of accelerator physics had begun, and big accelerators became political status symbols for countries competing for scientific leadership. For forty years after 1955, the United States built a succession of big accelerators and only two passive detectors. The Soviet Union and the European laboratory CERN followed suit, putting almost all their efforts into accelerators. Meanwhile, serious research using passive detectors continued in Canada and Japan, countries with high scientific standards and limited resources.

So much for the history. Now I turn from the past to the future. Wilczek's expectation, that the advent of LHC will bring a Golden Age of particle physics, is widely shared among physicists and widely propagated by the public media. The public is led to believe that LHC is the only road to glory. This belief is dangerous because it promises too much. If it should happen that LHC fails, the public may decide that particle physics is no longer worth supporting. The public needs to hear some bad news and some good news. The bad news is that LHC may fail. The good news is that if LHC fails there are other ways to explore the world of particles and arrive at a Golden Age. The failure of LHC would be a serious setback, but it would not be the end of particle physics.

There are two reasons to be skeptical about the importance of LHC, one technical and one historical. The technical weakness of LHC arises from the nature of the collisions that it studies. These are proton–proton collisions, and they have the unfortunate habit of being messy. Two protons colliding at LHC energies behave rather like two sandbags, splitting open and strewing sand in all directions. A typical proton–proton collision in LHC will produce a large spray of secondary particles, and the collisions are occurring at a rate of millions per second. The machine must automatically discard the vast majority of the collisions, so that the small minority that might be scientifically important can be precisely recorded and analyzed. The criteria for discarding events must be written into the software program that controls the handling of information. The software program tells the detectors which collisions to ignore. There is a serious danger that LHC can discover only things that the programmers of the software expected. The most important discoveries may be things that nobody expected. LHC may miss the most important discoveries. Another way to go ahead with particle physics is to build large passive detectors observing natural radiation. In the last twenty years, the two most ambitious passive detectors were built in Canada and Japan. Both these detectors made important discoveries. In a well-designed passive detector deep underground, events of any kind are rare, every event is recorded in detail, and if something unexpected happens you will see it.

There are also historical reasons to be skeptical about the importance of LHC. I have made a survey of the history of important discoveries in particle physics over the last sixty years. To avoid making personal judgments about importance, I define an important discovery to be one that resulted in a Nobel Prize for the discoverers. This is an objective criterion, and it usually agrees with my subjective judgment. In my opinion, the Nobel Committee has made remarkably few mistakes in its awards. There have been sixteen important experimental discoveries between 1945 and 2008.

Each experimental discovery lies on one of three frontiers between known and unknown territories. It is on the energy frontier if it reaches a new range of energies of particles. It is on the rarity frontier if it reaches a new range of rarity of events. It is on the accuracy frontier if it reaches a new range of accuracy of measurements. I assigned each of the sixteen important discoveries to one of the three frontiers. In most cases, the assignments are unambiguous. The results of my survey are then as follows: four discoveries on the energy frontier, four on the rarity frontier, eight on the accuracy frontier. Only a quarter of the discoveries were made on the energy frontier, while half of them were made on the accuracy frontier. For making important discoveries, high accuracy was more useful than high energy. The historical record contradicts the prevailing view that LHC is the indispensable tool for new discoveries because it has the highest energy.

The majority of young particle physicists today believe in big accelerators as the essential tools of their trade. Like Napoleon, they believe that God is on the side of the big battalions. They consider passive detectors of natural radiation to be quaint relics of ancient times. When I say that passive detectors may still beat accelerators at the game of discovery, they think this is the wishful thinking of an old man in love with the past. I freely admit that I am guilty of wishful thinking. I have a sentimental attachment to passive detectors, and a dislike of machines that cost billions of dollars to build and inevitably become embroiled in politics. But I see evidence, in the recent triumphs of passive detectors and the diminishing fertility of accelerators, that nature may share my prejudices. I leave it to nature to decide whether passive detectors or LHC will prevail in the race to discover her secrets.

Fortunately, passive detectors are much cheaper than LHC. The best of the existing passive detectors were built by Canada and Japan, countries that could not afford to build giant accelerators. The race for important discoveries does not always go to the highest energy and the most expensive machine. More often than not, the race goes to the smartest brain. After all, that is why Wilczek won a Nobel Prize.

CHAPTER 2.5

A WALK THROUGH
JOHNNY VON NEUMANN'S GARDEN*

1. Foundations of Mathematics

Johnny von Neumann left behind him six massive volumes of collected works, assembled and edited by Abraham Taub [von Neumann, 1961–1963]. The collected works are his garden, containing a large and heterogeneous set of objects that he planted. Each of them grew from a seed, from an idea or a problem that came into his head. He developed the idea or solved the problem, and then wrote it down and published it. He wrote fast and published fast, so that the flowers are still fresh. For my talk this morning I decided to take a walk through the garden and see what I could find. Luckily only two of the papers are in Hungarian. He wrote mostly in German until he came to live permanently in the United States at the age of thirty, and after that in English.

Johnny was educated at the famous Lutheran High School in Budapest from age ten to eighteen. There he had excellent teachers and even more excellent school-mates. One of the school-mates was Eugene Wigner, who became an outstanding physicist and a life-long friend. But Johnny's father understood that the Lutheran High School was not giving Johnny everything he needed. Johnny had a passion for mathematics going far beyond what the school could teach. So his father hired Michael Fekete, a mathematician from the University of Budapest, to work with Johnny at home. The first flower in Johnny's garden is a paper, "On the position of zeroes of certain minimum polynomials" [von Neumann, 1922], published jointly by Fekete and von Neumann when Johnny was eighteen. The style of the paper is dry and professional, following the tradition set by Euclid two

---

*Talk given at Brown University, Providence, Rhode Island, May 4, 2010.

thousand years earlier. Almost everything that Johnny wrote as a mathematician is in the Euclidean style, stating and proving theorems one after another with no wasted words.

Although the subject of his first paper was probably suggested by Fekete, the style is already recognizable as Johnny. Johnny's unique gift as a mathematician was to transform problems in all areas of mathematics into problems of logic. He was able to see intuitively the logical essence of problems, and then to use the simple rules of logic to solve the problems. His first paper is a fine example of his style of thinking. A theorem which appears to belong to geometry, restricting the possible positions of points where some function of a complex variable is equal to zero, is transformed into a statement of pure logic. All the geometrical complications disappear and the proof of the theorem becomes short and easy. In the whole paper there are no calculations, only verbal definitions and logical deductions.

The next flower in the garden is Johnny's first solo paper, "On the introduction of transfinite numbers" [von Neumann, 1925], which he published at age nineteen. This shows where his strongest interests lay at the beginning of his career, when he was a young bird ready to leave the nest and stretch out his mathematical wings. His dominating passion, then and for the next five years, was to understand and reconstruct the logical foundations of mathematics. He was lucky to arrive on the scene at the historical moment when confusion about the foundations of mathematics was at a maximum. In the nineteenth century, Georg Cantor had greatly enlarged the scope of mathematics by creating a marvelous theory of transfinite numbers, giving precise definitions to a vast hierarchy of infinities. Then, at the beginning of the twentieth century, Bertrand Russell and other critics discovered that Cantor's theory led to logical contradictions. Russell's paradox threw doubt not only onto Cantor's creation of a new world of infinities but also onto the established concepts of classical mathematics. Johnny became aware, as soon as he began to talk with Fekete and to read the mathematical literature, that mathematics was in a state of crisis. Since Cantor's mathematical reasoning had led to logical absurdities, nobody knew how to draw the line between reliable mathematics and imaginative nonsense. Johnny decided at the age of nineteen that it was his task to resolve the crisis and to put mathematics back onto a firm logical foundation.

The first paragraph of Johnny's first solo paper consists of a single sentence, "The purpose of this work is to make the idea of Cantor's ordinal numbers unambiguous and concrete". The rest of the paper provides a new definition of ordinal numbers and demonstrates that the new definition leads to the same results as Cantor's old definition. Johnny makes no claim to have resolved the crisis that arose from Cantor's theory. He has only made the crisis more acute by giving Cantor's concepts a sharper definition. To make the crisis more acute means to understand it better, and to understand it better is the first step toward resolving it.

Johnny's second solo paper, "An axiomatization of set theory" [von Neumann, 1925], appeared two years later, when he was twenty-one years old and a student at the University of Berlin. Set theory means the theory of things and collections of things, considering only their logical relationships and forgetting about their individual qualities. From the point of view of set theory, you and I, stars and planets, words and numbers are all just things and are all treated the same way. Axiomatization means to describe set theory in the same style that Euclid used to describe geometry two thousand years ago, building the theory by logical deduction from a few basic assumptions which he called axioms. Johnny found a new set of axioms for set theory. He hoped that his new axioms could serve as a consistent logical basis for all the useful parts of mathematics while avoiding the paradoxes. But he was well aware that his consistent basis for mathematics was a hope and not a proven reality.

The essential novelty of Johnny's axioms was to introduce two species of object which he called one-things and two-things. He used these abstract names in order to avoid possibly misleading impressions that might arise from using more familiar words. To make Johnny's ideas easier to understand, I will use the names "sets" for one-things and "classes" for two-things. So Johnny had a version of set-theory with two kinds of objects, the sets which are in some sense small enough to be handled collectively by the normal rules, and the classes which are in some sense too big to be handled collectively. The axioms are constructed so that the "Class of all sets" exists as a well-defined object. It is a class but not a set. Neither the "set of all sets" nor the "class of all classes" exists in the theory. This simple trick, using different names and different rules for small and large collectives, allows Johnny to avoid the logical paradoxes. The paradoxes arose in the older versions of set-theory from using the concept "set of all sets" too freely. In Johnny's new version, this concept is forbidden, but the "class of all sets" is allowed, providing the framework for a logical construction of mathematics. The class of all sets is the universe of mathematics, the framework within which all mathematical collectives are defined.

Before writing his paper, Johnny had been talking with David Hilbert in Göttingen. Hilbert was forty years older than Johnny and was the most famous mathematician in the world. Hilbert was passionately promoting a program for resolving the crisis of mathematics by solving the Entscheidungsproblem, the decision problem. To solve the decision problem meant to find a formal method of deciding the truth or falsehood of every mathematical statement. If he could solve the decision problem, that would show that the axioms of mathematics were both consistent and categorical. To be consistent means that they can never prove both a statement and its negation. To be categorical means that for every statement the axioms prove either the statement or its negation. Hilbert proclaimed, with all his authority as spiritual father of mathematicians, that to resolve the crisis of mathematics it was necessary to find a set of axioms that were proven to be both consistent and categorical. Mathematics would only rest on a firm logical foundation if every meaningful mathematical statement could be proved true or false.

At the end of his axiomatization paper, Johnny put a brief and modest summary of his claims. He did not claim to have resolved the crisis of mathematics. He claimed only to have opened the way to a possible resolution by finding a set of axioms that is not known to be self-contradictory. He had not proved that his axioms were consistent, and he had not proved that they were categorical. He ended his paper with two sentences expressing not very diplomatically his skepticism about Hilbert's program: "Even Hilbert's approaches are here powerless, for this objection concerns the categoricity and not the consistency of set-theory. All that we can do now is to recognize that another argument against set-theory has arisen, and that we see no way ahead leading to rehabilitation".

Three years later, Johnny published two much longer papers about the foundations of mathematics. One was "On Hilbert's proof theory" [von Neumann, 1927]. The other was his Ph.D. thesis, with the title "The axiomatization of set theory" [von Neumann, 1928a], an expanded version of the 1925 paper. These two papers show that Johnny was still desperately trying to rescue mathematics by following Hilbert's program. Johnny was stuck. He had created a simple and beautiful new set of axioms, which were later shown by Kurt Gödel to be exactly what was needed for understanding the true nature of mathematics, but he did not know what to do with them. At that point, he gave up trying to rescue mathematics and devoted the rest of his life to other things.

Another three years later, in 1931, Kurt Gödel in Vienna proved two theorems that totally devastated the Hilbert program. Gödel proved that no system of axioms for mathematics could be categorical, and that no system of axioms could prove itself to be consistent. After Gödel, mathematics could never be the unique compendium of absolute truth that mathematicians from Euclid to Hilbert had imagined. After Gödel, mathematics was a free creation of the human mind, with truth and falsehood depending on human tastes and preferences. For Hilbert and many of his contemporaries, the discoveries of Gödel appeared to be a disaster. Their hopes of building a unique and solid foundation for mathematics had collapsed. But Johnny understood immediately that the new freedom created by Gödel was a gain and not a loss. Johnny said in a public lecture that Gödel was the greatest logician since Aristotle. Johnny regretted that he had not made Gödel's discoveries himself three years earlier, but he was happy to see that Gödel used his 1925 system of axioms with separate names for sets and classes. Johnny was proud to have made a substantial contribution to the foundations of the new mathematics.

2. Games and Quanta

The next flower, "Theory of party games" [von Neumann, 1928b], comes from a different corner of the garden. At the age of 24, Johnny had become a professional mathematician with a position as instructor at the University of Berlin. He enjoyed the night life of Berlin and was intrigued by the logic of games such as poker and baccarat in which the outcome depends on a mixture of luck and skill. The question, whether a logical strategy exists for a player to have the best chance to win such games, had been raised by the French mathematician Émile Borel. Borel had asked the question but was unable to answer it. Johnny found the answer, which turned out to be a deep mathematical theorem. For a game with only two players, there exists a unique strategy which gives each of them the best outcome on the average. The proof that such a strategy exists is another fine example of Johnny's style, reducing a problem of calculation to a problem of logic.

The optimum strategy usually requires a large element of randomness, so that the moves of the players are truly unpredictable. Player A must throw dice to decide how to move, so that player B cannot win by predicting what player A will do. In the game of poker, the throw of the dice will occasionally require player A to bet high on a weak hand, a move that is called bluffing. If player A never bluffs, player B can win by guessing more accurately the strength of player A's cards. At the end of his paper Johnny writes, "The agreement of the mathematical results with the empirically known rules of successful gambling, for example the necessity of bluffing in poker, can be considered as experimental confirmation of our theory".

For games with three or more players, Johnny found no such elegant solution to the problem. To have the best chance of winning a game with three players, player A must bribe or threaten player B to form a coalition against player C. The players must compete for the roles of the winners A and B and try to escape the role of the loser C. The result of the competition is decided by personal will-power or spite and not by mathematics. At the end of his discussion of the three-person game, Johnny says, "The decisive factor, which is altogether absent from the orderly and equitable two-person game, is combat".

In another corner of the garden there is a little flower all by itself, a short paper with the title, "The division of an interval into a denumerable infinity of identical parts" [von Neumann, 1928c]. This solves a problem raised by the Polish mathematician Hugo Steinhaus. I met Steinhaus after World War II in America. He was one of the few survivors of the group of brilliant mathematicians who emerged in Poland between the wars. Half of them were Jewish and half were Gentile. The chance of survival was about the same for both, since those who emigrated were mostly Jews and those who survived in Poland were all Gentiles. Johnny solved the Steinhaus problem quickly and never returned to it. The theorem that he proved is counter-intuitive, and the proof is astonishing. The theorem is about sets of points on an interval. An interval means a finite piece of a straight line. A denumerable infinity means a collection of objects that can be labeled with whole numbers, $1, 2, 3, \ldots$ all the way to infinity. The theorem says that there exists a collection of sets of points $S_1, S_2, S_3, \ldots$ with the following properties. (1) Every point on the interval belongs to exactly one S_j. (2) The sets S_j are identical in all respects except for position, each S_j being obtained from any other by displacing it bodily through a certain distance along the line.

The theorem is counter-intuitive because it is impossible to visualize the sets S_j. If you try to imagine how the points of the set S_j are arranged near to the ends, you fail. You fail because the sets are non-measurable, and nobody has ever visualized a non-measurable set of points. Non-measurable sets cannot be constructed using any of the familiar tools of geometry. Johnny's proof of the theorem is astonishing because it is totally abstract. He never even mentions the geometry of the sets S_j. He gives no clue to their appearance or their construction. He proves their existence by reducing it to a proposition in pure logic, and proves the proposition by purely logical arguments. This little paper is the most extreme manifestation of the Johnny style.

During his Berlin years, Johnny made frequent visits to Göttingen, where Heisenberg had recently invented quantum mechanics and Hilbert was the presiding mathematician. Hilbert was intensely interested in quantum mechanics and encouraged collaboration between mathematicians and physicists. From the point of view of Hilbert, quantum mechanics was a mess. Heisenberg had no use for rigorous mathematics and no wish to learn it. Dirac made free use of his famous delta-function which was defined by a mathematical absurdity, being infinite at a single point and zero everywhere else. When Hilbert remarked to Dirac that the delta-function could lead to mathematical contradictions, Dirac replied, "Did I get into mathematical contradictions?" Dirac knew that his delta-function was a good tool for calculating quantum processes, and that was all he needed. Twenty years later, Laurent Schwartz provided a rigorous basis for the delta-function and proved that Dirac was right. Meanwhile, Johnny worked with Hilbert and published a series of papers cleaning up the mess. For several years, quantum mechanics was Johnny's main interest. In 1932, he published the book, "Mathematical Foundations of Quantum Mechanics" [von Neumann, 1932a], which occupies a substantial piece of his garden.

Johnny's book was the first exposition of quantum mechanics that made the theory mathematically respectable. The concepts were rigorously defined and the consequences rigorously deduced. Much of the work was original, especially the chapters on quantum statistics and the theory of measurement. I read the book in 1946, when I was still a pure mathematician, but already intending to switch my attention to physics. I found it enormously helpful. It gave me what I needed, a mathematically precise statement of the theory, explaining the fine points that the physicists had been too sloppy to mention. From that book I learned most of what I know about quantum mechanics. But then, after I had made the transition to physics and had begun to read the current physics journals, I was surprised to discover that nobody in the physics journals ever referred to Johnny's book. So far as the physicists were concerned, Johnny did not exist. Of course, their ignorance of Johnny's work was partly a problem of language. The book was in German, and the first English translation was only published in 1955. But I think, even if the book had been available in English, the physicists of the 1940s would not have found it interesting. That was the time when the culture of physics and the culture of mathematics were most widely separated. The culture of physics was dominated by people like Oppenheimer who made friends with poets and art-historians but not with pure mathematicians. The culture of mathematics was dominated by the Bourbaki cabal which tried to expunge from mathematics everything that was not purely abstract. The gap between physics and mathematics was as wide as the gap between science and the humanities described by C. P. Snow in his famous lecture on the Two Cultures. Johnny was one of the very few people who was at home in all four cultures, in physics and mathematics, and also in science and the humanities.

The central concept in Johnny's version of quantum mechanics is the abstract Hilbert Space. Hilbert Space is the infinite-dimensional space in which quantum states are vectors and observable quantities are linear operators. Hilbert had defined and explored Hilbert Space long before quantum mechanics made it useful. The unexpected usefulness of Hilbert Space arises from the fact that the equations of quantum mechanics are exactly linear. The operators form a linear algebra, and the states can be arranged in multiplets defined by linear representations of the algebra. Johnny liked to formulate physical problems in abstract and general language, so he formulated quantum mechanics as a theory of rings of linear operators in Hilbert Space. A ring means a set of operators that can be added or subtracted or multiplied together but not divided. Any physical system obeying the rules of quantum mechanics can be described by a ring of operators. Johnny began studying rings of operators to find out how many different types of quantum system could exist.

After Johnny had published his quantum mechanics book, he continued for several years to develop the theory of rings of operators. The third volume of his collected works consists entirely of papers on rings of operators. He published seven long papers with a total of more than five hundred pages. I will not discuss these monumental papers this morning. They contain Johnny's deepest work as a pure mathematician. He proved that every ring of operators is a direct product of irreducible rings that he called factors. He discovered that there are five types of factor of which only two were previously known. Each of the types has unique and unexpected properties. Exploring the ocean of rings of operators, he found new continents that he had no time to survey in detail. He left the study of the three new types of factor unfinished. He intended one day to publish a grand synthesis of his work on rings of operators. The grand synthesis remains an unwritten masterpiece, like the eighth symphony of Sibelius.

The quantum mechanics book is the last item on my list of flowers that Johnny published in German. It was published in 1932 when he was dividing his time equally between Berlin and Princeton. In the same year he began writing papers in English. One of his first papers to appear in English was "Proof of the quasi-ergodic hypothesis" [von Neumann, 1932b], which he published in the Proceedings of the National Academy of Sciences to make sure that American mathematicians would read it. This paper solved an important problem in classical mechanics, using the same concept of Hilbert Space that he had used to solve problems in quantum mechanics. A classical dynamical system is said to be ergodic if, after we put it into an initial state and then leave it alone for an infinite time, it will come arbitrarily close to any final state with probability independent of the initial state. Johnny proved that, under certain clearly specified conditions, a system is ergodic if and only if there exist no constants of the motion. A constant of the motion means a quantity depending on the state of the system which does not change as the system moves forward in time. Johnny's theorem provides a firm mathematical basis for the assumptions that are customarily made by physicists using classical statistical mechanics. Translated into the sloppy language used by physicists, the theorem says that the time-average of any single trajectory of the system over a long time is equal to the statistical average of all trajectories. Even more sloppily, physicists say that time-averages are equal to ensemble averages, and we use the word ensemble to mean the set of all states of the system.

One of the American mathematicians who read Johnny's paper in the Proceedings of the National Academy was Garrett Birkhoff. Garrett was the son of George Birkhoff, and both father and son were famous mathematicians. Garrett and Johnny became close friends, and Garrett came to Princeton for frequent visits. After Johnny died, Garrett wrote a memoir about the work that Johnny did in the 1930s. Here is a sentence from Garrett's memoir: "Anyone wishing to get an unforgettable impression of the razor edge of von Neumann's mind, need merely try to pursue this chain of exact reasoning for himself, realizing that often five pages of it were written down before breakfast, seated at a living room writing-table in a bathrobe".

A minor offshoot of Johnny's thinking about operators in Hilbert space was his invention of continuous geometry, a new kind of geometry in which the dimension of a subspace is a continuous variable. A couple of short papers, "Continuous geometry" [von Neumann, 1936a] and "Examples of continuous geometries" [von Neumann, 1936b], are to be found in his garden. These papers were published in

1936 when Johnny was settled in Princeton. Johnny writes at the beginning, "We will give only the axioms, some commentaries on them, and then the main definitions and results. A detailed account will apppear soon in a mathematical periodical". This is a promise that was never fulfilled. From this time in his life onward, he made many such broken promises. He got into the habit of working on a problem, solving it to his own satisfaction, and then not taking the time to publish the results in detail. He gave lectures in Princeton on continuous geometry. His lecture notes were published in a book "Continuous Geometry" that appeared in 1960 after his death. The book is boring. It is probably the most boring stuff that ever appeared under Johnny's name. You can tell from the book that Johnny was already bored by continuous geometry while he was giving the lectures. He had good reasons for not publishing the notes while he was alive. He had no need to publish or perish. He was a tenured professor at the Institute for Advanced Study. After 1936 he published only stuff that he considered important and not boring. He became increasingly interested in a wide range of subjects outside pure mathematics. He had, after all, earned a degree in chemical engineering at the ETH in Zurich at the same time as he was studying mathematics in Budapest.

3. Bombs and Computers

The next flower is a report, "Theory of Detonation Waves" [von Neumann, 1942], written in 1942, presenting a scholarly and thorough analysis of what happens when chemical high explosives detonate. Johnny had seen his homeland Hungary dismembered, as a result of military defeat, in 1918. He was even more eager than other European Jews to join the fight against Hitler. He was delighted to apply his mathematical skills and his knowledge of engineering to military problems, and became a consultant to the United States Army before the United States went to war in 1941. His 1942 report was one of a series providing a theoretical basis for the improvement of military explosives. Military explosives are a delicate compromise between two conflicting requirements. They should detonate with maximum efficiency when fired in anger, and should resist detonation with maximum safety when exposed to gunfire or accidental explosions nearby. When you are trying to find the best compromise, it is a great help to have a consultant who understands the chemistry as well as the mathematics.

Johnny's report does not discuss particular weapons but supplies a mathematical theory that designers of weapons can use to optimize designs. When he began work for the military, the applications were to artillery shells and anti-submarine depth-charges. In 1943, he was invited by his friend Robert Oppenheimer to visit Los Alamos and apply his ideas to the design of nuclear weapons. His understanding of shock-waves made a big contribution to the success of the Los Alamos project.

At Los Alamos he saw monstrous numerical calculations carried out laboriously by gangs of human computers. He began to think seriously about the possibility of electronic computers that could do such calculations better and faster than humans. In 1944, he met Herman Goldstine, who was then a young army officer involved in a project to build a real electronic computer, the ENIAC, at the University of Pennsylvania. Johnny and Herman became close friends. Herman later said of Johnny, "While he was indeed a demi-god, he had made a detailed study of humans and could imitate them perfectly. Actually he had great social presence, a very warm, human personality, and a wonderful sense of humor". They worked out a plan to do something spectacular with computers as soon as the war was over. In Johnny's garden there is a paper, "On the principles of large scale computing machines" [von Neumann, 1946], describing their plans. I won't say more about this since Johnny's work on computers is covered by other speakers.

I got to know Johnny personally when I came to the Institute for Advanced Study in 1948. He was then actively engaged in building the Institute computer and learning how to use it. He understood from the beginning that two of the most important uses of the machine would be to predict weather and to model climate. He hired engineers to build the machine and meteorologists to use it. The chief engineer was Julian Bigelow and the chief meteorologist was Jule Charney. Each of them had a gang of young people to do the heavy work, persuading a totally new kind of machine to produce some real science. I enjoyed very much the young people with their rowdy conversation and irreverent behavior. There was an amusing clash of cultures between these young hooligans and the older members of the Institute. As Einstein wrote to his friend the Queen of the Belgians when he arrived at the Institute in 1933, Princeton was a quaint and ceremonious village populated by demi-gods on stilts. The culture of the older members was based on formal politeness and respect for the academic hierarchy. Johnny and I were on the side of the hooligans.

When Johnny died, the Institute quickly got rid of the computer project, and the older culture reasserted itself. No more hooligans were hired, and the breath of fresh air that they had brought to the Institute was blown away with them to UCLA and MIT. In 1980, the Institute celebrated its fiftieth birthday by publishing a volume with the title, "A Community of Scholars, 1930–1980", consisting of biographies and bibliographies of the members. Not one of the young hooligans who built the machine and predicted the weather is mentioned in the book. They were not scholarly enough to be officially recognized as belonging to the Institute. But there is a flower in Johnny's garden, a paper, "Numerical integration of the barotropic vorticity equation" [von Neumann *et al.*, 1950], by Charney, Fjortoft and von Neumann, describing their first attempts to predict weather. Since the Institute computer was not yet running, they did their calculations with the ENIAC. Using the ENIAC, the numerical simulation moved ahead in time more slowly than the weather that it was supposed to simulate, so there was no real prediction. At the end they express the hope that the Institute computer will be fast enough to keep ahead of real time. Four years later, when Johnny's machine and others like it were running, their hopes were fulfilled. Johnny then announced that a prediction of weather twenty-four hours ahead could be done in less than an hour. That was as far as he was able to go toward his dream of understanding climate. One year later, he was diagnosed with terminal cancer, and three years later he was dead.

4. Summing Up

In the last decade of his life, Johnny did not find time to write formal mathematical papers. Instead he wrote informal essays, sometimes addressed to his colleagues in the government agencies that supported his work, and sometimes to the general public. The last two flowers in my tour of his garden are addressed to the public. They are thoughtful and beautifully written. He took a lot of trouble to think clearly and write simply. The first of the two was titled, "The Mathematician" [von Neumann, 1947]. It was published in 1947 as a chapter in a book of essays, "The Works of the Mind", by a variety of authors. It is a swan-song, summarizing in simple words the conclusions that Johnny had reached at the end of his life as a pure mathematician. He had devoted the best years of his life to pure mathematics, when he was, as Newton said of his own early years, "in the prime of my life for invention". From age nineteen to age twenty-seven he had struggled to build firm logical foundations for pure mathematics, preparing the ground for Gödel's discovery that no set of foundations could be complete. After the Gödel revolution, he took advantage of the new freedom to experiment with logical foundations for quantum mechanics and for the discipline that was later given the name of computer science. His essay, "The Mathematician", describes the development of mathematics as a free creation of the human mind, with foundations either borrowed from empirical science or freely invented.

The main message of Johnny's essay is stated at the end in words that have become famous among mathematicians. "As a mathematical discipline travels far from its empirical source, or still more, if it is a second and third generation only indirectly inspired by ideas coming from reality, it is beset with very grave dangers.... At a great distance from its empirical source, or after much abstract inbreeding, a mathematical subject is in danger of degeneration. At the inception the style is usually classical. When it shows signs of becoming baroque, then the danger signal is up. It would be easy to give examples, to trace specific evolutions into the baroque and the very high baroque, but this, again, would be too technical. In any event, whenever this stage is reached, the only remedy seems to me to be the rejuvenating return to the source, the reinjection of more or less directly empirical ideas. I am convinced that this was a necessary condition to conserve the freshness and the vitality of the subject, and that this will remain equally true in the future". Johnny probably had in mind his elaborate fiddling with continuous geometry as an example of very high baroque, and his plunge into the empirical world of computer science as the rejuvenating return to the source.

After this farewell to pure mathematics, the last seven years of Johnny's life were divided between running the computer project in Princeton and advising the government in Washington. During this period he became known to the public as a military hard-liner. For a few years he publicly advocated a preventive war against the Soviet Union. He became deeply involved with high-level committees considering problems of military strategy. One of the committees, known to historians as the von Neumann committee, advocated the fateful decision to base United States strategy on a force of intercontinental ballistic missiles, combining the technologies of multistage rocketry and hydrogen bombs. This decision would make it technically possible for the United States to destroy the Soviet Union in forty minutes, with the inevitable consequence that the Soviet Union would be able to destroy the United States with a similar force of missiles a few years later.

The idea of a preventive nuclear war conveys today an impression of militarism gone mad. But to the generation that lived and suffered through the 1930s, the idea had another meaning. It was widely held, especially by liberal intellectuals, that the French and British governments had behaved in a cowardly and immoral fashion when they failed to march into Germany in 1936 to stop Hitler from remilitarizing the Rhineland. A preventive war in 1936, when Germany was still effectively disarmed and incapable of serious resistance against invading forces, might have overturned Hitler's regime in a few days and saved the fifty million human beings who were to die in World War II. We cannot know whether a preventive war in 1936 would have been either feasible or effective. We know only that the idea of preventive war as a morally acceptable option was widely accepted by the people of Johnny's generation, who looked back to 1936 as a tragically missed opportunity. To them, the idea of forestalling a terrible catastrophe by a bold preventive action was neither insane nor criminal.

Johnny argued in the 1940s that America was facing the same choice that France and Britain faced in 1936. The Soviet Union was then just beginning to build the industrial base for the mass production of nuclear weapons. Johnny saw the 1940s as the last chance for America to overthrow the Stalin regime, as 1936 had been the last chance to overthrow Hitler, without a war of annihilation. He saw a preventive war in the 1940s as preferable, not only for America but for humanity as a whole, to a war of annihilation later. I am not saying that he was right. I consider it unlikely that preventive war could have achieved its objective, either in 1936 or in the 1940s. I am only saying that to talk of Johnny's advocacy of preventive war, without mentioning the events of 1936 which dominated his perception of the moral issues, is to miss the main point of his argument.

The last flower on my tour of Johnny's garden is a paper written for the general public and published in Fortune magazine in June 1955, two months before the onset of his fatal illness. The title is "Can We Survive Technology?" [von Neumann, 1955]. Johnny is now no longer concerned with the intellectual problems of mathematicians but with the human problems of war and peace, nuclear weapons and nuclear power, global warming and climate control, computers changing the rules of economics and politics. In the last seven years of his life, as he moved into the centers of power in Washington and made friends with generals and politicians, he understood that the urgent problems of society were human rather than technical. His view of human nature was bleak. "It is just as foolish to complain that people are selfish and treacherous as it is to complain that the magnetic field does not increase unless the electric field has a curl. Both are laws of nature". His view of the future was equally bleak. "Present awful possibilities of nuclear warfare may give way to others even more awful. After global climate control becomes possible, perhaps all our present involvements will seem simple. We should not deceive ourselves. Once such possibilities become actual, they will be exploited. . . . The one solid fact is that the difficulties are due to an evolution that, while useful and constructive, is also dangerous. Can we produce the required adjustments with the necessary speed? The most hopeful answer is that the human species has been subjected to similar tests before and seems to have a congenital ability to come through, after varying amounts of trouble. To ask in advance for a complete recipe would be unreasonable. We can specify only the human qualities required: patience, flexibility, intelligence". Johnny possessed these qualities himself. They are still the qualities that we need, in order to have the best chance of survival, as we move into the world that he created.

References

John von Neumann *Collected Works*, A. H. Taub (ed.) [1961–1963] (Pergamon Press, New York).

J. von Neumann and M. Fekete [1922] "Über die lage der nullstellen gewisser minimalpolynomen", *Jahresbericht* **31**, 125–138.

J. von Neumann [1923] "Zur einführung der transfiniten zahlen", *Acta Szeged.* **1**, 199–208.

J. von Neumann [1925] "Eine axiomatisierung der mengenlehre", *J. Math.* **154**, 219–240.

J. von Neumann [1927] "Zur hilbertschen beweistheorie", *Math. Zeitschr.* **26**, 1–46.

J. von Neumann [1928a] "Die axiomatisierung der mengenlehre", *Math. Zeitschr.* **27**, 669–752.

J. von Neumann [1928b] "Zur theorie der gesellschaftsspiele", *Math. Ann.* **100**, 295–320.

J. von Neumann [1928c] "Die zerlegung eines intervalles in abzählbar viele kongruente teilmengen", *Fund. Math.* **11**, 230–238.

J. von Neumann [1932a] *Mathematische Grundlagen der Quantenmechanik* (Springer, Berlin).

J. von Neumann [1932b] "Proof of the quasi-ergodic hypothesis", *Proc. Nat. Acad. Sci.* **18**, 70–82.

J. von Neumann [1936a] "Continuous geometry", *Proc. Nat. Acad. Sci.* **22**, 92–100.

J. von Neumann [1936b] "Examples of continuous geometries", *Proc. Nat. Acad. Sci.* **22**, 101–108.

J. von Neumann [1942] "Theory of detonation waves", A progress report to April 1, 1942, Office of Scientific Research and Development Section B-1, report 549, 34 pp.

J. von Neumann and H. H. Goldstine [1946] "On the principles of large-scale computing machines", Office of Research and Inventions, Navy Department, unpublished.

J. von Neumann [1947] "The mathematician", *The Works of the Mind*, R. B. Heywood (ed.) (University of Chicago Press), pp. 180–196.

J. von Neumann, J. G. Charney and R. Fjortoft [1950] "Numerical integration of the barotropic vorticity equation", *Tellus* **2**, 237–254.

J. von Neumann [1955] "Can we survive technology?" *Fortune*.

Chapter 2.6

Are Brains Analog or Digital?*

1. Change of Title

The title of this talk was announced as "Is Life Analog or Digital?". That was a mistake. I gave a talk with that title in this place twelve years ago. I was talking about life in a totally abstract way, considering life as a system for processing information [Dyson, 1979]. I was asking the question, whether life could in principle survive for ever using a finite store of free energy. The answer was no if the life was digital, yes if the life was analog. Analog processing is in principle more economical than digital processing in its use of free energy. But this statement is an abstract mathematical proposition.

Today I will talk about real biology. We know that creatures like us have two quite separate systems for processing information, the genome and the brain. We know that the genome is digital, and we can accurately transcribe our genomes onto digital machines. We cannot transcribe our brains, and the processing of information in our brains is still a great mystery. So the title of today's talk is "Are Brains Analog or Digital?" I will be talking about real brains and real people, asking a question that will have practical consequences when we are able to answer it. Of course I am not able to answer it now. All I can do is to examine the evidence and explain why I consider it probable that the answer will be that brains are analog.

2. The Road Not Taken

Robert Frost said it. "Two roads diverged in a wood, and I..., I took the one less traveled by, And that has made all the difference". In the 1930s, there were five kinds of computers available for doing serious calculations. Two of them were digital,

*Talk at Joshua Lederberg–John von Neumann Symposium — Towards Quantitative Biology, The Rockefeller University, October 22, 2013.

the March and calculator for doing accurate arithmetic and the Hollerith card-punch machine for tabulating big masses of data. Three of them were analog, the slide-rule for doing fast arithmetic with three-figure accuracy, the Bush differential analyzer for solving differential equations, and the Lehmer photo-electric number sieve for solving equations in integers and factorizing big numbers. The Lehmer machine worked with integers but was still an analog machine. It did calculations using finite Galois fields, and identified integers by counting holes in a wheel, not by a digital code. Then in 1936, Alan Turing wrote his paper "On computable numbers" [Turing, 1936], which revealed the power and beauty of digital computing as an abstract logical construction. After that, analog computers slowly went out of fashion. Analog computers became the road not taken. I still used my slide-rule for calculating my income tax until the 1980s, when the tax collectors of the State of New Jersey demanded four-figure accuracy in the calculation of interest-rates. For four-figure accuracy, I was forced to switch to a digital calculator. Now we are so immersed in a world of digital computers that it is hard to imagine things going the other way.

If it is true that brains are analog, then we must sooner or later take the road that Turing did not take and explore what analog computers can do. There are good reasons why the world of science and commerce went digital in the twentieth century. Digital data and digital calculation have huge advantages over analog in accuracy, speed, and reliability. There are many situations in life and in business where three-figure accuracy is not good enough. There are even some situations in science where twelve-figure accuracy is not good enough. For all kinds of practical purposes, digital computers are here to stay. But the brain has big advantages in flexibility and versatility. The brain somehow manages to do a dozen different jobs, seeing, hearing, walking, talking, thinking and controlling our vital functions, all at the same time, without getting confused. If it is true that the brain is an analog machine, then we should be able to build analog machines with the same kind of flexibility and versatility. That possibility is the subject of my talk.

3. Dynamic Quantum Clustering

One of the big unsolved problems of the modern world is to convert information into understanding. This is a big problem in science, government, business administration and economics. In each of these areas we have huge amounts of information and very limited understanding. As a result of the rapid progress of digital technology, it has become far cheaper to collect information than to understand it. Recently a new method called Dynamic Quantum Clustering has been invented to extract small nuggets of understanding from large amounts of information. I am grateful to my friend Marvin Weinstein, who is one of the inventors of DQC, for telling me about it [Weinstein *et al.*, 2013]. The idea of DQC is to present the information in a large database to a human viewer in the form of a moving picture, using a mathematical algorithm borrowed from the motion of systems of particles in quantum mechanics. The quantum-mechanical motion is entirely fictitious. It does not represent any process happening in the real world. It is designed simply to make a big collection of data intelligible to a human brain. The human brain is at home in the world of moving pictures. It can pick out structures in moving pictures quickly and efficiently. After the human viewer has called attention to a structure, the DQC program can record the structure and analyze it using digital processes. The human eye and brain serve as a sub-routine in a digital program, making use of the brain's peculiar skill in extracting structure from patterns of movement in space and time.

In the Weinstein paper, the DQC method is illustrated by applying it to big databases in five different fields, nano-chemistry, condensed matter physics, biology, seismology and finance. In each field the method finds subsets of the data that contain important information. The information remains hidden if the data is analyzed by digital processes. For example, in the field of nano-chemistry, the object of study is a fragment of an ancient Roman pot, and the hidden structure is a filamentary distribution of oxidized and reduced phases of iron. Similar structures are found in the electrodes of lithium-ion batteries, in human bone and in Roman pottery. The filaments are evidence that the same kind of chemical reaction is occurring in all three materials. The chemistry of the lithium-ion electrodes was understood first, and the understanding could then be transferred to the bone and the pot, materials that are less accessible to laboratory experiments. Weinstein describes the purpose of DQC: "Look for the needle in the haystack, determine what it is, and find what this means". The reason why DQC is powerful is that you discover the needle without knowing in advance what you are looking for. You do not need to write a detailed description of the needle into the search algorithm. The human brain has the capacity to make unexpected discoveries.

I see DQC as the first step on a long road, to create islands of understanding in a sea of information. To progress further along this road, we must not only exploit the marvelous faculties of the human brain but understand how they work. Practical use of DQC for exploring big databases must go hand in hand with neurological exploring of the brain. The first big mystery to be attacked is the physical basis of memory. We know that the brain has at least two systems for recording memory, one short-term and one long-term. We have no idea how memories are encoded, either chemically or physically. After that mystery is resolved, we may go further to attack the second great mystery, how memories are accessed and retrieved. The logical structure of our memory system appears to be associative. We retrieve a memory by following a chain of thought that connects it with another memory. But we have no understanding of the language that the brain is using to encode a chain of thought. If we can ever develop a version of DQC that incorporates a chain of thought, it will become a far more powerful tool for converting information into understanding.

I am not here to talk about particle physics, but I make a brief digression to discuss the application of DQC to the understanding of particle experiments. Recently the world has been celebrating the discovery of the Higgs particle at the Large Hadron Collider (LHC), a huge and expensive particle accelerator at the European Center for Nuclear Research (CERN) at Geneva. The Higgs particle is indeed an important discovery, and Peter Higgs and Francois Englert richly deserve the Nobel Prizes that they won for predicting the existence of the particle forty years ago. I am happy that the particle is finally discovered after many years of effort. But I am unhappy with the way the particle was discovered.

The LHC is not a good machine for making discoveries. The LHC makes protons collide together at very high energies, and every collision makes a big cascade of uninteresting particles that swamp the detectors. Only a tiny fraction of rare collisions produce Higgs particles. The background of uninteresting events is so enormous that the machine cannot possibly look at them. The only way to see the rare Higgs particles is to write, into the program that controls the detector, detailed instructions for saving the interesting events and discarding the others. The Higgs particle was only discovered with this machine because we told the machine what we expected it to discover. It is an unfortunate deficiency of the LHC that it cannot make unexpected discoveries. Big steps ahead in science usually come from unexpected discoveries. It remains to be seen whether an analysis of LHC data with DQC methods might enable the LHC to make unexpected discoveries. Unfortunately, the data-stream from the LHC is so intense that it is impossible to display it without some drastic pruning. If something unexpected is hidden in the data, even if DQC is used to look at it, there is always a risk that the unexpected treasure gets chopped out in the pruning.

I think the future of particle experiments belongs to passive underground detectors. These are detectors looking for rare high-energy particles coming from outer space. Nature provides a generous supply of astronomical accelerators, distributed over the universe and accelerating particles to far higher energies than we can reach with our biggest man-made devices. Together with the rare high-energy particles come numerous low-energy particles that do not penetrate so far underground. Detectors are put deep underground so as to eliminate almost all the uninteresting background events. Deeply penetrating particles are rare enough, so that every event in the detectors can be recorded and examined in detail. If new species of particle or new kinds of interaction are present in the detectors, they will be seen. The detectors do not need to be told in advance what they are supposed to discover. Big passive detectors are expensive, but still much cheaper than LHC.

If high-energy particle experiments are to make unexpected discoveries, this is the way to go.

4. The Failure of Artificial Intelligence

Now I come back from particle physics to brains and computers. I am looking for evidence to confirm or refute the hypothesis that the brain is more analog than digital in its operation. My first piece of evidence is the failure of artificial intelligence. Sixty years ago, when the digital revolution was beginning and the pioneers of digital computer technology were dreaming of future triumphs, one of their grandest dreams was the creation of artificial intelligence. Artificial intelligence was imagined as a digital computer that could think and act and communicate like a human brain. Artificial intelligence meant a digital machine that would be as smart as a human. Many brilliant and ambitious people devoted their lives to this dream.

Twenty years later, my friend Sir James Lighthill was asked by the British government to advise the Science Research Council about the support of research in Artificial Intelligence. Lighthill wrote a famous report with the title, "Artificial Intelligence: A General Survey", which made him a lot of enemies. Lighthill divided the artificial intelligence activities into three categories, A, B and C. Category A was Advanced Automation, the development of automatic control machinery for practical purposes. Category C was Central Nervous System research, the use of computers for scientific study of neurology and psychology. Category B was supposed to be the Bridge between A and C. Category B was the Building of Robots not for utilitarian jobs but for imitating the behavior of human beings.

Lighthill discussed the progress that had been made by the year 1972 in these three areas and then delivered his verdict. He concluded that areas A and C were legitimate programs of research, A belonging to engineering and C to science. The work in A and C was moving ahead more slowly than had been expected. Two of the most important projects in area A were automatic speech-recognition and automatic translation, both of which had failed miserably. Three important objectives in area C were the understanding of memory and the understanding of learning and forgetting, none of which had been achieved. But work in areas A and C still had great promise for the future and deserved continued support. On the other hand, area B had no scientific substance. The bridge between A and C did not exist. A and C should be supported as separate enterprises with separate tools and separate objectives. Area B did not deserve support.

Here is Lighthill's sketch of area B. "Most robots are designed to operate in a world like the conventional child's world as seen by a man: they play games, they do puzzles, they build towers of bricks, they recognise pictures in drawing-books ('bear on rug with ball'); although the rich emotional character of the child's world is totally absent. A relationship which may be called pseudomaternal comes into play between a Robot and its Builder".

Now another forty years have gone by and Lighthill's verdict on Artificial Intelligence still stands. A lot has happened in these forty years in area A. In area A, after forty years of blood, sweat and tears, the Dragon programs of automatic speech-recognition are finally working well, and automatic translation programs are producing results that are imperfect but useful. In area C there has been less progress, and in area B none at all. Progess in understanding the human brain has come from new physical tools, functional MRI and genome-sequencing, rather than from new ideas about neural circuitry. Progress in building autonomous robots has come from imitating insects rather than from imitating humans. Artificial Intelligence, after sixty years of strenuous efforts in many countries, is still a failure.

I am saying that the failure of artificial intelligence was no accident. It failed because the goal was to imitate an analog device, the human brain, with a digital device, the electronic computer. The successes in area A with speech recognition and translation were possible because speech and written language are handled by special areas of our brains which have evolved to process digital information. If I am right, the successes of artificial intelligence using digital machines will be limited to these areas.

5. Maps and Feelings

Another striking fact that we know about brains is that they use maps to process information. Information from the retina goes to several areas of the brain where the picture seen by the eye is converted into maps of various kinds. Information from sensory nerves in the skin goes to areas where the information is converted into distorted maps of the body. The brain is full of maps, and a big part of its activity is transferring information from one map to another. As we know from our own use of maps, mapping from one picture to another can be done either by digital or analog processing. Because digital cameras are now cheap, and film cameras are old-fashioned and rapidly becoming obsolete, many people assume that the process of mapping images in the brain must be digital. But the brain has evolved over millions of years and does not follow our ephemeral fashions. A map is in its essence an analog device, using a picture to represent another picture, and the imaging in the brain may be done by direct comparison of pictures rather than by translation of pictures into digital data.

Introspection tells us that our brains are spectacularly quick in performing two tasks essential to our survival in a natural environment, the recognition of images in space and the recognition of patterns of sound in time. We recognize a human face or a snake in the grass in a fraction of a second. We recognize the sound of a voice or a footstep equally fast. The process of recognition requires the comparison of a perceived image with an enormous database of remembered images. How this is done in a quarter of a second with no conscious effort, we have no idea. It seems likely that the scanning of images in associative memory is done by direct comparison of analog data rather than by digitization.

Another fact that we know for sure is that our processing of sensory information is strongly coupled with emotions. Seeing and hearing and smelling are inextricably mixed up with feeling. A big part of human enterprise and creativity is devoted to the arts, music to explore the emotional dimensions of our hearing, painting and architecture to explore the emotional dimensions of seeing. Our perception of sensory information is concerned with quality more than with quantity. Subjective impressions may be misleading, but our emotional response to music or to a beautiful landscape suggests that we are processing sounds and images as shapes rather than as digits. The glory of a desert sunrise or of a Rembrandt painting belongs to the whole scene and not to the individual pixels.

As a general rule, our perception of sensory information appears to be continuous rather than discrete. There is one important exception to this rule. The exception is our perception of language. Spoken language is digital, using a finite set of discrete phonemes to convey meaning. Our auditory system is trained to convert a continuous flow of sound into a discrete sequence of phonemes, and then to convert a sequence of phonemes into words and sentences. There are specialized areas of the brain, mainly in the left hemisphere, where this digital processing of information is done. Written language is also digital, and there are other areas of the brain where visual images are translated into digital sequences of letters or ideograms. The invention of Braille allows blind people to transfer the process of digital translation from visual to tactile information. These capabilities give us proof that certain parts of the brain are digital, processing phonemes or written symbols as discrete objects. But our understanding of language is also associated with strong feelings and emotions that go far beyond the processing of sounds and symbols. Language is not perceived by us as a string of phonemes. A big part of our culture is concerned with art-forms arising from language, poetry, drama, preaching and story-telling. Beyond the digital words and phrases, there is a whole world of literature. We perceive literature as having style and sparkle, depth and resonance. Our perception of literature belongs to the overall shape of the language, not to the individual words.

The quality of a poem such as Homer's "Odyssey" or Eliot's "Waste Land" is like the quality of a human personality. A large part of our brain is concerned with social interactions, getting to know other people and learning how to live in social groups. The observed correlation between size of brain and size of social groups in primate species makes it likely that large brains evolved primarily to deal with social problems. Our ability to see others as analogs to ourselves is basic to our existence as social animals.

The computer engineer Danny Hillis published twenty-five years ago a delightful speculation with the title, "Intelligence as an Emergent Behavior; or, the Songs of Eden" [Hillis, 1988]. He tells a story to describe how it might have happened in the evolution of humans, that music came first and speech second. "Once upon a time, about two and a half million years ago, there lived a race of apes that walked upright. The young apes, like many young apes today, had a tendency to mimic the actions of others. In particular, they had a tendency to imitate sounds. Some sequences of sounds were more likely to be repeated than others. I will call these songs. The songs survived, bred, competed with one another, and evolved according to their own criterion of fitness. One successful strategy for competition was for a song to specialize, to find a particular niche where it would be likely to be repeated.

"Once the songs began to specialize, it became advantageous for an ape to pay attention to the songs of others and to differentiate between them. By listening to songs, a clever ape could gain useful information. Once the apes began to take advantage of the songs, a mutually beneficial symbiosis developed. Songs enhanced their survival by conveying useful information. Apes enhanced their survival by improving their capacity to remember, replicate, and understand songs. Evolution created a partnership between the songs and the apes that thrived on the basis of mutual self-interest. Eventually this partnership evolved into one of the world's most successful symbionts: us".

This fable of Danny Hillis may not be true, but it explains two great mysteries of human evolution that are otherwise hard to understand. Why did we evolve people like Mozart and Beethoven? And why did we evolve people like Sophocles and Shakespeare? It seems that we have far greater capacity for composing and appreciating music, and far greater capacity for elaborating and enjoying speech, than would be required for survival in a primitive environment. Hillis explains the existence of Mozart and Beethoven by evolving music first, with the songs competing for survival on the basis of musical quality. He explains the existence of Sophocles and Shakespeare by evolving speech second, with the songs competing for survival on the basis of meaning. A song with meaning attached to it evolves into speech, with the quality of the song evolving into quality of the speech. This version of our history explains the amazing fact that a bunch of apes only recently descended from the trees can produce great composers and poets. The evolution of great works of art depends on the quality of the music and the poetry, not on the digital language of the notes and syllables.

I am suggesting that the brain is mainly an analog device, with certain small regions specialized for digital processes. It is certainly not true, as is sometimes claimed by pundits talking on television, that the left hemisphere is digital and the right hemisphere is analog. It seems to be true that most of the digital processing is done on the left side, but the division of labor between the two hemispheres is still largely unexplored.

6. The Endicott House Meeting

Thirty-two years ago, in May 1981, we had a meeting at Endicott House, a clubhouse owned by MIT, talking about the future of computing. The meeting was organized by Tom Toffoli, and a lot of bright people were there, including Danny Hillis, Richard Feynman, Charles Bennett, and two people that I had not heard of before, Marian Pour-El and Ian Richards. Charles Bennett talked about his recent discovery, that digital computing could be done reversibly, without erasing information and without dissipating energy. Hillis talked about parallel computing and Feynman talked about quantum computing, two subjects that later turned out to be important. But to me the most exciting news at the meeting came from Pour-El and Richards, two mathematicians who were then at the University of Minnesota.

Marian Pour-El and Ian Richards proved a theorem that says, in a mathematically precise way, that analog computers are more powerful than digital computers [Pour-El and Richards, 1981]. They give examples of numbers that are proved to be non-computable with digital computers but are computable with a simple kind of analog computer. Alan Turing defined computable numbers to be those that can be computed with digital machines. The numbers discussed by Pour-El and Richards are analog-computable but not Turing-computable. An analog computer deals directly with continuous variables while a digital computer deals with discrete variables. Their analog computer is a classical field propagating though space and time and obeying a linear wave equation. The classical electromagnetic field obeying the Maxwell equations would do the job. Pour-El and Richards show that the field can be focussed on a point in such a way that the strength of the field at that point is not computable by any digital computer, but it can be measured by a simple analog device. The imaginary situations that they consider have nothing to do with biological information. The Pour-El–Richards theorem does not prove that analog brains are better than digital brains. It only makes this conjecture more plausible.

Charles Bennett recently sent me a picture of all the people at the Endicott House meeting except for himself. He is not in the picture because he took it himself. Pour-El and Richards are there in the picture, two young people who had done something extraordinary. I do not know what happened to them afterwards. I had the feeling that I was the only person at the meeting who took their work seriously. They had done for analog computing the same job that Alan Turing had done for digital computing forty-five years earlier. Turing published his classic paper on computable numbers in 1936 [Turing, 1936]. Pour-El and Richards published theirs in 1981. Both papers took a long time to attract attention. My son recently published a book with the title, "Turing's Cathedral", describing the magnificent edifice of knowledge and power that grew out of the 1936 paper. Pour-El and Richards's cathedral has not yet begun to grow.

Science-fiction writers gave us pictures of digital life and analog life, long before we could decide which picture comes closer to describing humans. To visualize digital-life, think of a human consciousness down-loaded from a brain into a digital computer, like the character HAL in the Stanley Kubrick movie "2001". To visualize analog-life, think of a human consciousness up-loaded into a Black Cloud as described in the science fiction novel by Fred Hoyle [Hoyle, 1957]. The Black Cloud is an analog device with its memory encoded in magnetic fields generated by dust grains moving though the cloud. If we are analog life, our downloading into a digital computer may involve a certain loss of our finer feelings and qualities. That would not be surprising. I certainly have no desire to try the experiment myself. Perhaps, when the time comes for us to adapt ourselves to a cold universe and abandon our extravagant flesh-and-blood habits, we should upload ourselves to black clouds rather than download ourselves to silicon chips. If I had to choose, I would go for the black cloud every time.

7. Mathematics as a Tool for Understanding Nature

I am supposed to be talking today about Quantitative Biology. The question is whether mathematics can provide a deep understanding of biology in the same way as it has provided a deep understanding of physics. Biology today is in a state of rapid and revolutionary development, as physics was a hundred years ago. In 1930, Sir James Jeans proclaimed to his audience in a radio broadcast, "The Great Architect of the Universe now begins to appear as a pure mathematician". Jeans was a physicist who wrote books about science for the general public. He also talked a lot on the radio. He was the Carl Sagan of the nineteen-thirties. He was famous enough to be the subject of a well-known Clerihew,

> Sir James Jeans
> Always says what he means.
> He is perfectly serious
> When he says the universe is mysterious.

Jeans, like Sagan, sometimes became intoxicated with the flamboyance of his phrases. But he was perfectly serious when he said God was a pure mathematician. He said this more than once. He said this because he was a physicist and because he had recently lived through the two most amazing revolutions that ever happened in physics, the discovery of general relativity by Einstein in 1915 and the discovery of quantum mechanics by Heisenberg and Schrödinger in 1925. General relativity and quantum mechanics carried physics into deep levels of mathematics that had never before found practical applications. General relativity made the universe of space and time a curved four-dimensional continuum, to be described in the language of differential geometry. Quantum mechanics made the states of an atom an infinite-dimensional complex vector space, to be described in the language of functional analysis. Whereas nineteenth-century physics described nature in terms of concrete models that anyone could visualize, twentieth-century physics described nature in terms of mathematical abstractions. The Riemann–Christoffel curvature tensor and the Schrödinger wave-function, the central concepts of general relativity and wave mechanics, existed in an abstract mathematical world, far removed from the familiar world of things that we can feel and touch.

The physicists of Jeans's generation, who had grown up in the nineteenth century and lived through the early years of the twentieth, were amazed to see that Nature was on the side of the revolutionaries. Observations showed that Nature behaved as Einstein, Heisenberg and Schrödinger predicted. Mathematical abstraction succeeded where mechanical models had failed. "Pure thought", said Einstein, "can grasp reality, as the ancients dreamed. The creative principle lies in mathematics". Experiments revealed Nature mirroring the abstract thoughts of Heisenberg and Schrödinger with amazing precision. We could say about Einstein what Ben Jonson said three hundred years earlier about Shakespeare,

> Nature herself was proud of his designs,
> And joyed to wear the dressing of his lines.

So, Jeans concluded, Nature dances to the tune of pure mathematics, not to the drumbeat of classical mechanics. If you want to understand Nature at a deep level, you must think like a pure mathematician. Since God by definition understands Nature better than we do, He too must be a pure mathematician.

The question is now whether mathematics could be as useful a tool for understanding the biological world as it has been for understanding the world of physics. Biology will be the dominant science of the twenty-first century, as physics was of the twentieth. Biology is forging ahead while physics is slowing down. It is important for those of us who are mathematicians to adapt ourselves to the brave new world of biology. Does the Great Architect of the Universe continue to be a pure mathematician when He turns his attention from physics to biology?

During the last twenty years, interesting mathematics has crept into biology through the study of the dynamics of populations. Populations of herbivores and carnivores, predators and prey, parasites and hosts, affect one another through strongly nonlinear interactions. Populations rise and fall in complicated ways described by nonlinear dynamical equations. Their behavior is often chaotic, and can be understood in terms of the mathematical theory of chaos. The theory of chaos does not go very deep. The theory is descriptive rather than analytical. It does not have the depth of differential geometry or functional analysis. Still, the theory is mathematically non-trivial. In population biology, mathematicians and biologists can work together with mutual respect.

Deep mathematics has not penetrated into the mainstream of biology, the biology of genes, enzymes, cells, tissues and immune systems. Mainstream biologists use mathematics only for the humble tasks of organizing data and simulating structures. They are interested in processing data but not in proving theorems. Is there any chance that deep mathematics could in future penetrate biology as it has penetrated physics? I believe there is a chance, but the chance is still remote. Deep mathematics might become important in biology in the understanding of the human mind and brain. So my advice to Rockefeller University is, if you want mathematics to remain useful for understanding Nature at a deep level, try to get some of your mathematicians interested in neurology.

8. Quantum Analog Computing

During the last thirty years, a big new enterprise has grown up with the brand-name Quantum Computing. This enterprise aims to imitate the architecture of ordinary digital computers, using quantum spins instead of classical on–off switches as carriers of information. A quantum spin means an atom or a collection of atoms with a spin obeying the rules of quantum mechanics. The machines built in this way are quantum digital computers. The holy grail of quantum digital computing is to implement the Shor algorithm. Eighteen years ago, Peter Shor astonished the world by inventing an algorithm, that a quantum computer can in principle use to factorize large integers in polynomial time [Shor, 1997]. "Polynomial time" means that the number of computer operations required is less than some power of the number of digits in the integer to be factorized. The finding of factors of a large integer is a notoriously hard problem for classical computers not using quantum processes. It is generally believed, but not proved, that a classical computer cannot factorize large integers in polynomial time. The security of various public-key encryption systems depends on the impossibility of rapid factorization. The Shor algorithm constitutes a threat to the security of such systems. But the threat to secure encryption becomes real only if the theoretically superior power of quantum digital computers can be embodied in real hardware. No existing machine comes close to the holy grail.

Quantum computing today is at roughly the same stage of practicality as classical computing in the days of Charles Babbage in the early nineteenth century. Babbage had the right concepts, but he lacked the physical tools to make his concepts useful. He tried to build an Analytical Engine out of mechanical clockwork, a project that absorbed his time, money and strength for many years, and inevitably failed. When I was a boy in London, I used to gaze in fascination at the relics of Babbage's Analytical Engine preserved in a glass case at the Science Museum. If we should try to build a quantum computer today, its fate would probably be similar. The quantum devices that we have available today are not much better than Babbage's clockwork.

I am asking whether our brains, or parts of our brains, might be quantum analog computers. I believe that this possibility was first suggested by Richard Feynman. The idea is that the brain might be an amplifier, sensitive to the quantum states of some special molecules constituting a molecular memory, and amplifying the molecular information until it becomes a signal strong enough to drive motor neurons to action. In this way, the brain would be able to use quantum jumps in the memory, which are strictly unpredictable according to the laws of quantum mechanics, to control executive decisions. The brain would be a device for amplifying quantum unpredictability so as to achieve freedom of will. As soon as I begin talking about freedom of will, I am entering a philosophical hornet's nest in which many famous thinkers and writers have been stung. But I consider philosophical discussions about the interpretation of quantum mechanics to be a waste of time. I am asking a practical question, to be answered by observation and experiment and not by philosophical speculation. I am asking whether there exists in the brain a structure embodying information in the quantum states of memory molecules, and hooking up these quantum states to an amplifier that translates them into classical neurological signals. If such a structure exists, then the brain is a quantum analog device, and the philosophers can continue to argue about whether it gives us free will.

How should we study experimentally the working of a human brain? The best experimental tool is probably a human baby. What goes on inside the head of a three-month-old baby? Anyone who has raised a baby knows the problem. How can we imagine what is going on inside that little head? How does it happen that that little head grows to almost adult size and capabilities in two years? In two years it sorts out the bewildering variety of neural inputs from eyes and ears, it recognizes faces and voices, it masters grammar and syntax, it knows the difference between nouns and verbs, it learns how to exploit the weaknesses of grown-ups.

In recent years, experts in child development have found reliable ways of communicating with younger and younger babies. A baby as young as three months old can communicate interest or lack of interest in something that it sees. If the baby is interested, it will move its eyes to follow the object. If it is not interested, it will disregard the object. Using this channel of communication, it is possible to prove that very young babies can already tell the difference between speech and noise, between known and unknown voices and faces, between familiar and unfamiliar syllables and shapes. As we move through the twenty-first century, it is likely that new tools will become available, scanning non-intrusively the brains of babies with high resolution in space and time. By the end of the century, we should be able to answer the question, how do babies begin to think? After that question is answered, we will have a better chance of answering the more difficult questions. How do babies turn into grown-ups? Where does grown-up imagination come from? How does a brain become creative? Where do music, poetry, art and science come from? How did a monkey recently descended from the trees become aware of itself and of the mysteries of the universe around it?

My dream is that some time in the next hundred years we shall answer these questions and understand how the brain works as a quantum analog device. Then we will be able to copy nature's architecture and build quantum analog machines ourselves. We will finally solve the problem of artificial intelligence as James Lighthill defined it. We will understand why digital artificial intelligence did not work. We will have the power in our hands to build machines that think like humans. Before that happens, we must think very hard, how to use that power for good rather than for evil.

References

F. J. Dyson [1979] "Time without end: physics and biology in an open universe", *Rev. Mod. Phys.* **51**, 447–460.

W. D. Hillis [1988] "Intelligence as an emergent behavior; or, the songs of eden", *Daedalus* **117**, 177.

F. Hoyle [1957] *The Black Cloud* (Harper and Brothers, New York).

M. B. Pour-El and I. Richards [1981] "The wave-equation with computable initial data such that its unique solution is not computable", *Adv. in Math.* **39**, 215–239.

P. W. Shor [1997] "Polynomial-time algorithms for prime factorization and discrete logarithms on a quantum computer", *SIAM J. Comput.* **26**, 1484–1509.

A. Turing [1936] "On computable numbers, with an application to the entscheidungsproblem", *Proc. London Math. Soc.* **42**, 230.

M. Weinstein *et al.* [2013] "Analyzing big data with dynamic quantum clustering", preprint.

Section 3

Memoirs

CHAPTER 3.1

HOMAGE TO GEORGE GREEN: HOW PHYSICS LOOKED IN THE 1940s*

1. Concepts and Tools

I am delighted and honoured to be chosen to speak here at the birthplace of Nottingham's great mathematician George Green, together with my old friend and benefactor Julian Schwinger. Julian Schwinger is my benefactor because he allowed me forty-five years ago to become famous by publishing his ideas. His generosity at that time was just as remarkable as the equal generosity of Richard Feynman whose death we are now mourning. Let me now say simply, thank you Julian, for without your help I would not be standing here today.

I am delighted to be talking about George Green because his life and work exemplify two of my favorite themes, two themes that are still today highly relevant both to the progress and to the public understanding of science. The first theme is the perennial and usually unsuccessful struggle to keep the doors of the temple of science open to amateurs and outsiders. George Green was a prime example of an amateur and outsider, without any official academic credentials, beating the insiders at their own game. He was lucky to have lived in the early nineteenth century rather than in the late twentieth century. He was, in spite of his social and educational deficiencies, allowed to enter the temple, and his achievements were recognized by the insiders. If George Green were living today, since science has become professionalized and the Ph.D. has become a necessary ticket for admission to the temple, he would have encountered much more formidable barriers to his ambitions. The insiders are now defending their turf against outsiders with bureaucratic weapons unknown in the 1830s.

---

*Talk given at the George Green Bicentenary Celebrations Nottingham University, England, 14 July 1993.

The second theme that George Green's work exemplifies is the historical fact that scientific revolutions are more often driven by new tools than by new concepts. Thomas Kuhn in his famous book, "The Structure of Scientific Revolutions", talked almost exclusively about concepts and hardly at all about tools. His idea of a scientific revolution is based on a single example, the revolution in theoretical physics that occurred in the 1920s with the advent of quantum mechanics. This was a prime example of a concept-driven revolution. Kuhn's book was so brilliantly written that it became an instant classic. It misled a whole generation of students and historians of science into believing that all scientific revolutions are concept-driven. The concept-driven revolutions are the ones that attract the most attention and have the greatest impact on public awareness of science, but in fact they are comparatively rare. In the last five hundred years we have had six major concept-driven revolutions, associated with the names of Copernicus, Newton, Darwin, Maxwell, Einstein and Freud, besides the quantum-mechanical revolution that Kuhn took as his model. During the same period there have been about twenty tool-driven revolutions, not so impressive to the general public but of equal importance to the progress of science. I will not attempt to make a complete list of tool-driven revolutions. Two prime examples are the Galilean revolution resulting from the use of the telescope in astronomy, and the Crick–Watson revolution resulting from the use of X-ray diffraction to determine the structure of big molecules in biology. The effect of a concept-driven revolution is to explain old things in new ways. The effect of a tool-driven revolution is to discover new things that have to be explained. In physics there has been a preponderance of tool-driven revolutions. We have been more successful in discovering new things than in explaining old ones. George Green's great discovery, the Green's function, is a mathematical tool rather than a physical concept. It did not give the world a new theory of electricity and magnetism or a new picture of physical reality. It gave the world a new bag of mathematical tricks, useful for exploring the consequences of theories and for predicting the existence of new phenomena that experimenters could search for. The Green's function was a tool of discovery, like the telescope and the microscope, but aimed at mathematical models and theories instead of being aimed at the sky and the microbe.

The invention of the Green's Function brought about a tool-driven revolution in mathematical physics, similar in character to the more famous tool-driven revolution caused by the invention of electronic computers a century and a half later. Both the Green's Function and the computer increased the power of physical theories, particularly in the fields of electromagnetism, acoustics and hydrodynamics. The Green's Function and the computer are prime examples of intellectual tools. They are tools for clear thinking. They helped us to think more clearly by enabling us to calculate more precisely.

But I did not come here today to preach to you about the evils of the Ph.D. system and the importance of tools in science, two subjects that I have frequently talked about on other occasions. I was invited here by Professor Challis, (and I quote), "to give a lecture on the events associated with the first introduction of Green's functions to a quantum-mechanical treatment of electrodynamics." So I am here as a historical monument, like the windmill on the hill nearby, as a witness of events long past in which both I and George Green played a role. I am supposed to tell you what happened in 1948 when the words "Green's function", which had been part of the accepted language of classical electrodynamics and fluid mechanics for a hundred years, suddenly began to be spoken by quantum theorists and became part of the fashionable jargon in the new field of quantum electrodynamics. I decided not to do precisely what Professor Challis asked. If I should try to describe the events of 1948 in detail, I would have to fill several blackboards with equations, and my story would be intelligible only to those members of the audience who already know it. I decided to interpret Professor Challis's request broadly. Instead of giving you a lecture on quantum electrodynamics, I will talk in a more general way about the physics of the 1940s. I will try to give you an impression of the scientific communities in Europe and the United States as they then existed. I hope to explain how it happened that my own modest activities as a messenger transmitting knowledge from Europe to America turned out to be useful.

2. Classical and Quantum Green's Functions

To understand what happened in the 1940s we must begin with some historical background. There are two kinds of physics, classical physics beginning with Galileo and Newton in the seventeenth century, and quantum physics beginning with Planck and Bohr in the twentieth century. Classical physics describes big things such as rocks and planets. Quantum physics describes small things such as atoms and electrons. Next, cutting across the division of physics into classical and quantum, there is a division of physical objects into discrete and continuous. A rock is a discrete object. A flowing liquid or a magnetic field is a continuous object. Discrete objects are described by a finite set of numbers specifying their positions and velocities. The physics of discrete objects is called mechanics. Continuous objects are described by fields specifying their distribution and movement in space and time. The physics of continuous objects is called field theory. We have then four varieties of physical theories, classical mechanics, classical field theory, quantum mechanics and quantum field theory.

A highly compressed account of the history of theoretical physics goes like this. Physics is a drama in six acts. Act one, classical physics of discrete objects was worked out by Galileo and Newton. Act two, a hundred years later, classical physics of continuous objects was worked out by Euler and Coulomb and Oersted. Euler did hydrodynamics, Coulomb did electrostatics and Oersted did magnetism. So it happened that, at the beginning of the nineteenth century, the classical field-theories of hydrodynamics and electrostatics and magnetism were well established. Act three, George Green in 1828 revolutionized classical field theory by introducing his new tool, the Green's Function, which described directly the causal relationship between the behavior of a field at any two points in space and time. The Green's Function measures the local response of the field at a given point at a later time to a local disturbance of the field at another given point at an earlier time. Green used the Green's Function to clarify in a fundamental way the causal relationships between electric and magnetic fields. Helmholtz subsequently used Green's Functions to clarify in an equally fundamental way the causal relationships between pressure and velocity in acoustics. Act four, Heisenberg and Schrödinger worked out the quantum physics of discrete objects, describing the behavior of atoms and electrons with the theory that became known as quantum mechanics. Act five, Fermi, Heisenberg and Dirac invented quantum field theory to describe the quantum physics of continuous objects. The quantum field theory that described electricity and magnetism was called quantum electrodynamics. But quantum field theory did not work well as a practical tool. It was unreliable and tended to give absurd answers to simple questions. You asked a quantum field theory the question, "What is the mass of an electron?" and the answer came back, "Infinity". That was not very helpful. As a result of these well-publicized absurdities, the majority of practical physicists, especially in America, wrote off quantum field theory as useless and probably wrong. So at the end of act five in the 1930s we had physics divided into two disconnected parts, the classical field theories which worked beautifully in the classical domain, and the quantum mechanics of particles which worked beautifully in the quantum domain. There was not much connection between the two domains. Green's Functions were a convenient working tool in the classical domain, but there were no Green's Functions in the quantum domain. The quantum field theories, which should have been the link between the two domains, were discredited and generally believed to be useless. That was the situation at the beginning of the 1940s. Act six, the resurrection of quantum field theories and the introduction of quantum Green's Functions at the end of the 1940s. Act six is the main subject of this lecture.

In order to set the stage for act six, I looked at the four books out of which I learned physics as a student, to see how often the name of George Green appears in them. The four books, which I still have as treasured possessions on my shelves, are "Theoretical Physics" by Georg Joos, written in 1932, "The Principles of Quantum Mechanics" by Paul Dirac, written in 1930, "The Quantum Theory of Radiation" by Walter Heitler, written in 1935, and "Quantentheorie der Wellenfelder" by

Gregor Wentzel, written in 1942. I will have more to say in a moment about the Wentzel book. All four books are classics, full of beautiful writing and clear thinking. I still refer to them frequently as sources of useful information, not only as historical relics. It turns out that George Green is mentioned only twice, once by Joos and once by Dirac, in both cases in connection with Green's Theorem. Green's Theorem is one of Green's major contributions to science, establishing an exact relation between the sources and the fluxes of two fields. It relates the sources of two fields inside any region of space to the fluxes of the same fields through the surface bounding the region. The theorem is applied by Joos to a problem in classical electrostatics, by Dirac to a problem in quantum scattering of a particle. But Green's more important discovery, the Green's Function, is not mentioned by name in any of the books. If you look closely at the Heitler and the Wentzel books, you will see that Green's functions are lurking on many of their pages, but they are not labeled as such. The Green's functions appear mainly in equations and are called commutators or potentials. I only found out, after I became a professor and had to learn something of the history of physics in order to teach it, that these elegant and useful tools had been borrowed from George Green.

3. Nicholas Kemmer

I now begin my narrative of act six as I experienced it, first in England and then in America. In 1946, I came to Cambridge (the real Cambridge, not the American imitation) with the intention of learning modern physics. When I arrived there, I found that experimental physics was at a low ebb. The experimenters had been away during the war. In 1946, they were still struggling to get started on new enterprises which were to achieve huge success within a few years, the beginnings of the new sciences of radio-astronomy and molecular biology. I understood that Martin Ryle with his radio receivers and Max Perutz with his haemoglobin crystals were doing exciting stuff, but the stuff they were doing was clearly not physics. If I had wanted to be in a place where world-class experimental physics was being done, I should have gone to Bristol. In Bristol, Cecil Powell and his team of scanners with their microscopes and photographic emulsions were developing the techniques which led them within two years to the discovery of the pi-meson. But I was by training a mathematician, a student of Besicovitch, who had taught me the fine art of combining geometrical with analytical reasoning. I enjoyed talking with experimenters, but my more urgent need was to talk to a competent mathematical physicist. I needed to find somebody in Cambridge who could tell me what the important unsolved problems in theoretical physics were, and how I might use my mathematical skills to solve them. Kemmer belonged to the generation of scientists whose careers were maximally disrupted by the war. As a young man in 1938 he had published a theory of nuclear forces mediated by a symmetric triplet of meson fields, one positive, one negative and one neutral. The purpose of the symmetric triplet was to achieve equality of the neutron–proton and proton–proton forces. In 1938 not one of the three hypothetical mesons had yet been discovered. The symmetric meson theory was considered a wild speculation. Ten years later, all three mesons were found, and the theory was proved to be a correct description of a new symmetry of nature. It is one of the most brilliant predictions in the history of physics, comparable in brilliance with Yukawa's original prediction of the existence of the meson. But Kemmer received little public acclaim when his theory was confirmed. To blow his own trumpet was not in his nature. In the meantime, he had spent most of the wartime years working on the Canadian atomic energy project at Chalk River.

During my year at Cambridge I decided to go to America and make a fresh start there. In spite of my friendship with Kemmer, I found Cambridge depressing. I wanted to be in a place where I would be involved in an active group of young people doing research. By chance I met Sir Geoffrey Taylor, who had a little hand-made wind-tunnel in a cellar under the Cavendish Laboratory and did classic experiments on turbulence in the classic Cavendish string-and-sealing-wax style. He was also the world's greatest expert on blast-waves, and had been at Los Alamos during the war to make sure that the bombs were exploded at the correct height to achieve the maximum blast damage. I told him I was planning to go to America and asked where I should go. He said at once, "Oh, you should go to Cornell and work with Bethe. That is where all the brightest people from Los Alamos went when the war was over." The conversation was over in one minute. At that time I hardly knew that Cornell existed, but I took Sir Geoffrey's advice, and a year later I was a student of Hans Bethe and a friend of Richard Feynman. That was my second stroke of luck.

4. Physics in 1948

I said enough about my personal history. Let me now turn to a description of the state of physics as I found it during my first year in America, the year 1948. My personal memories of the events of that year are hopelessly unreliable after forty-five years have gone by. A much more informative view of the state of American physics is obtained by looking at Volume 73 of the Physical Review, the leading American physics journal which was started at Cornell University in 1893. In the 1940s the journal came out twice every month. Volume 73 contains the issues from January to June 1948. It covers all areas of physics, experimental and theoretical, atomic and nuclear and astronomical, quantum and classical.

Looking through Volume 73 today, I see a great number of familiar faces belonging to old friends. Almost every paper in it is interesting, and many of them are memorable. In 1948, the issues of the journal were thin enough so that we could read them from cover to cover. Many of us did. Nowadays the journal is fragmented into six parts, each of which is so fat that nobody even attempts to read it. In 1948, it was possible to read the whole journal and obtain an overview of everything that American physicists were doing.

The paper that impressed me most strongly in 1948 and still impresses me today is entitled "Relaxation effects in nuclear magnetic resonance absorption", a monumental piece of work, thirty-four pages long, by Bloembergen, Purcell and Pound, three Harvard physicists. All three of them are still going strong and two of them are still at Harvard. I worked through all the details of this massive work with intense pleasure. Nuclear magnetic resonance means the tickling of nuclear magnets inside a piece of solid or liquid material by alternating magnetic fields applied from the outside. Nuclear magnetic resonance was in 1948 a recently discovered phenomenon. It is now, forty-five years later, the basis of the medical technique known as MRI, magnetic resonance imaging, which enables doctors to obtain clear pictures of brain tumours and other soft tissue abnormalities in their patients. The paper of Bloembergen, Purcell and Pound addresses the question, what are the effects of the environment in which the nuclei are embedded on the detailed behavior of their magnetic resonance. The paper reports a comprehensive series of experiments together with an equally comprehensive theoretical analysis. It is one of the finest examples of the American style of physics, with experiment and theory working together as inexorably as a steam-roller and squashing a problem flat. The paper provided a fundamental understanding of the various ways in which the physical and chemical properties of the environment are linked with the shape of the nuclear resonance. It demonstrated that the nuclear resonance could be made into a powerful new tool for exploring the properties of matter. It provided the essential foundation of knowledge on which the development of MRI as a tool of medical diagnosis could be built thirty years later.

In the same volume of the Physical Review are many other wonderful papers on the most diverse subjects. Alpher, Bethe and Gamow on the origin of chemical elements. Gleb Wataghin on the formation of chemical elements inside stars. Edward Teller on the change of physical constants. Lewis, Oppenheimer and Wouthuysen on the multiple production of mesons. Foley and Kusch on the experimental discovery of the anomalous magnetic moment of the electron. Julian Schwinger on the theoretical explanation of the anomalous moment. Paul Dirac on the quantum theory of localizable dynamical systems. This was a vintage year for historic papers. I mention these seven just to give you the flavour of what physicists were doing in the early post-war years. The Alpher–Bethe–Gamow paper proposed that the chemical elements were formed by successive capture of neutrons on protons during the initial expansion of the universe from a hot dense beginning. Bethe had nothing to do with the writing of the paper. He allowed his name to be put on it to fill the gap between Alpher and Gamow. This joke, which was Gamow's idea, made the paper famous. Meanwhile, the paper of Wataghin, which proposed that the elements were formed in neutron stars, or more precisely in the process of rapid expansion of neutron stars into interstellar space, received much less attention. Wataghin was then living in Brazil and was not widely known. Unfortunately, it took us many years to collect the evidence which proved that, at least for the great majority of the elements, Alpher–Bethe–Gamow were wrong and Wataghin was right. I do not have time today to discuss the other papers on my list. Volume 73 contained several hundred papers, almost all of them worth reading.

One fact which I found remarkable in 1948 and still find remarkable today is that, among the hundreds of papers in Volume 73, the paper of Dirac is the only one concerned with quantum field theory. Dirac was a voice from another world. The vast majority of the papers are like the paper of Bloembergen, Purcell and Pound, sticking close to experiments and using a minimum of theory. The American scientific tradition was strongly empirical. Theory was regarded as a necessary evil, needed for the correct understanding of experiments but not valued for its own sake. Quantum field theory had been invented and elaborated in Europe. It was a sophisticated mathematical construction, motivated more by considerations of mathematical beauty than by success in explaining experiments. The majority of American physicists had not taken the trouble to learn it. They considered it, as Samuel Johnson considered Italian opera, an exotic and irrational entertainment. The paper of Dirac, although it was published in America, found few readers. It was read by the small community of experts in general relativity who were themselves isolated from the mainstream of American physics.

Thus it happened that I arrived at Cornell as a student, and found myself, thanks to Nicholas Kemmer, the only person in the whole university who knew about quantum field theory. The great Hans Bethe and the brilliant Richard Feynman taught me a tremendous lot about many areas of physics, but when we were dealing with quantum field theory I was the teacher and they were the students. Bethe and Feynman had been doing physics successfully for many years without the help of quantum field theory, and so they were not eager to learn it. It was my luck that I arrived with this gift from Europe just at the moment when the new precise experiments of Lamb and others on the fine details of atomic energy levels required quantum field theory for their correct interpretation. When I used quantum field theory to calculate an experimental number, the Lamb shift separating the energy levels of two of the states in a hydrogen atom with a spinless electron, Bethe was impressed. He said it was the first time he had seen quantum field theory do anything useful. For Bethe, formal mathematical machinery was pointless unless it could be used for calculating numbers. For him, and for almost all American theorists at that time, calculating numbers was the object of the game. Since the little gift that I brought from Europe to America could be used for calculating numbers, and the numbers could be checked by experiment, the gift received a friendly reception.

5. Quantum Field Theory and Green's Functions

Julian Schwinger had known all about quantum field theory long before. But he shared the American view that it was a mathematical extravagance, better avoided unless it should turn out to be useful. In 1948, he understood that it could be useful. He used it for his calculations of the fine details of atomic physics revealed by the experiments of Lamb and Retherford, Foley and Kusch at Columbia. But he used it grudgingly. In his publications he preferred not to speak explicitly about quantum field theory. Instead, he spoke about Green's Functions. It turned out that the Green's Functions which Schwinger talked about and the quantum field theory that Kemmer talked about were fundamentally the same thing. In Schwinger's papers I could recognize some of my old friends, functions that I had seen before in Wentzel's book. This was one of the ways that Green's Functions came to occupy a central place in the particle physics of the 1950s.

The second way that Green's Functions emerged in particle physics was through the work of Richard Feynman at Cornell. Feynman had never been interested in quantum field theory. He had his own private way of doing calculations in particle physics. His way was based on things that he called "Propagators", which were probability amplitudes for particles to propagate themselves from one space-time point to another. He calculated the probabilities of physical processes by adding up the propagators. He had rules for calculating the propagators. Each propagator was represented graphically by a collection of diagrams. Each diagram gave a pictorial view of particles moving along straight lines and colliding with one another at points where the straight lines met. When I learned this technique of drawing diagrams and calculating propagators from Feynman, I found it completely baffling, because it always gave the right answers but did not seem to be based on any solid mathematical foundation. Feynman called his way of calculating physical processes "the space-time approach", because his diagrams represented events as occurring at particular places and at particular times. The propagators described sequences of events in space-time. It later turned out that Feynman's propagators were identical with Green's Functions. Feynman had been talking the language of Green's Functions all his life without knowing it.

The third way that Green's Functions appeared in the particle physics of the 1940s was in the work of Sin-Itiro Tomonaga, who had developed a new and elegant version of relativistic quantum field theory. His work was done in the complete isolation of war-time Japan, and was published in Japanese in 1943. The rest of the world became aware of it only in the spring of 1948, when an English translation of it arrived in Princeton, sent by Hideki Yukawa to Robert Oppenheimer. Tomonaga was a physicist in the European tradition, having worked as a student with Heisenberg at Leipzig before the war. For him, in contrast to Schwinger and Feynman, quantum field theory was a familiar and natural language in which to think. Tomonaga and Dirac were on the same wave-length. In the paper of Dirac which I mentioned earlier, published in the 1948 Physical Review, Dirac mentions Tomonaga in the text and in a footnote, but does not refer either to Schwinger or to Feynman.

After the war, Tomonaga's students in Japan had been applying his ideas to calculate the properties of atoms and electrons with high accuracy, and were reaching the same results as Schwinger and Feynman. When Tomonaga's papers began to arrive in America, I was delighted to see that he was speaking the language of quantum field theory that I had learned from Kemmer. It did not take us long to put all the various ingredients of the pudding together. When the pudding was cooked, all three versions of the new theory of atoms and electrons turned out to be different ways of expressing the same basic ideas. The basic idea of all three ways was to calculate Green's Functions for all atomic processes that could be directly observed. Green's Functions appeared as the essential link between the methods of Schwinger and Feynman, and Tomonaga's relativistic quantum field theory provided the firm mathematical foundation for all three versions of quantum electrodynamics.

6. The Later History

The history of physics did not end in 1950. In this talk I will only sketch briefly some highlights of the later history. One of the major early advances beyond Tomonaga, Schwinger and Feynman was made by Rudolf Peierls in Birmingham in 1951 and published in a paper, "The commutation laws of relativistic field theory", in the Proceedings of the Royal Society. Peierls is like Kemmer, an exceptionally unselfish person as well as a brilliant physicist. When I returned from America to England in 1949, he welcomed me as a member of his department in Birmingham and gave me all the privileges that he never asked for himself. He had a heavy teaching load; mine was minimal. He and his wife Genia were responsible for a big household and four children; I was a guest in their home. The idea was that he would take care of all the mundane chores while I would have freedom and leisure to make important discoveries in physics. Of course, things did not work out the way they were planned. During my two years in Birmingham, the most brilliant discovery made in the Peierls department was the general commutation law of relativistic field theories. The discovery was his, not mine. I remember the surprise and delight when he told me about his discovery in the garden of his house on Carpenter Road. It gave us for the first time a deep and general understanding of the connection between Green's Functions and commutation relations between fields. It also clarified the meaning of the correspondence principle which connects classical with quantum field theories. It put Green's Functions where they belong, in the logical foundation of classical and quantum physics.

During the 1950s, the new methods of calculation, using Green's Functions to describe the behavior of quantum fields, were successfully applied to a variety of problems in electrodynamics. Experiments and theory were pushed to higher and higher levels of accuracy. When we began these calculations, we had hoped to find a clear discrepancy between theory and experiment. A discrepancy would reveal some fundamental information about the many known and unknown particles and interactions that the theory of quantum electrodynamics does not take into account. Since quantum electrodynamics does not pretend to be a complete theory of everything, it must at some level of accuracy disagree with experiment. But all attempts to find a discrepancy ended in disappointment. As each quantity was measured and calculated to more and more places of decimals, the measured and calculated numbers remained obstinately equal. Quantum electrodynamics turned out to be a more accurate description of nature than anybody in the 1940s had imagined possible. It now agrees with experiment to ten or eleven places of decimals. We are left with an unsolved mystery to explain. How could all the important fields and particles that lie outside the scope of quantum electrodynamics have conspired to hide their influence on the processes that lie inside? To solve the mystery, even more accurate experiments and calculations will be required.

After the Green's Function method had been successfully applied to quantum electrodynamics, the next big step was to apply the same method to the description of many-electron systems in the physics of condensed matter. In 1955, working with Charles Kittel in Berkeley, I began the application to condensed matter physics with a study of spin-waves in ferromagnets. I found that all the Green's Function tricks that had worked so well in quantum electrodynamics worked even better in the theory of spin-waves. The spin-wave is the simplest propagating mode of disturbance in the condensed assemblage of electrons inside a ferromagnet, just as the photon is the simplest propagating mode in the electromagnetic field in free space. I was able to calculate the scattering of one spin-wave by another, using the same tricks that Feynman had used in quantum electrodynamics to calculate the scattering of light by light. Meanwhile, the Green's Function method was applied systematically by Bogolyubov and other people to a whole range of problems in condensed matter physics. The main novelty that arose when we moved into condensed matter physics was the appearance of temperature as an additional variable. In quantum electrodynamics we had considered atoms and electrons in free space, in an environment with zero temperature. In condensed matter the temperature can never be zero, and many of the most interesting questions concern the effect of temperature on the properties of the system. The appropriate tools for analyzing condensed matter properties are therefore thermal Green's Functions, Green's Functions describing matter in thermal equilibrium at a given temperature. A beautiful thing happens when you make the transition from ordinary Green's Functions to thermal Green's Functions. To make the transition, all you have to do is to replace the real frequency of any oscillation by a complex number whose real part is the frequency and whose imaginary part is the temperature. Thus thermal Green's Functions are just as easy to calculate as ordinary Green's Functions. To put in the temperature, you simply give the frequency an imaginary component. This is mathematical magic which I will not attempt to explain. Green's Functions make such magic possible. That is one of the sources of their power and their beauty.

Soon after thermal Green's Functions were invented, they were applied to solve the outstanding unsolved problem of condensed matter physics, the problem of superconductivity. They allowed Bardeen, Cooper and Schrieffer to understand superconductivity as an effect of a particular thermal Green's Function expressing long-range phase-coherence between pairs of electrons. The Bardeen–Cooper–Schrieffer theory of 1957 explained satisfactorily all the observed features of superconductors as they were then known. The only thing the theory did not do was to give any hint of the existence of the high-temperature superconductors that were discovered unexpectedly thirty years later.

In the 1960s, after Green's Functions had become established as the standard working tools of theoretical analysis in condensed matter physics, the wheel of fashion in particle physics continued to turn. For a decade, quantum field theory and Green's Functions in particle physics were unfashionable. The prevailing view was that quantum field theory had failed in the domain of strong interactions, and that only phenomenological models of strong interaction processes could be trusted. Then, in the 1970s, the wheel of fashion turned once more. Quantum field theory was back in the lime-light with two enormous successes, the Weinberg–Salam unified theory of electromagnetic and weak interactions, and the gauge theory of strong interactions now known as quantum chromodynamics. Green's Functions were once again the working tools of calculation, both in particle physics and in condensed matter physics. And so they have remained up to the present day.

In the 1980s, quantum field theory moved off into new directions, to lattice gauge theories in one direction and to superstring theories in another. The Wilson Loop is the reincarnation of a Green's Function in lattice gauge theory. I have no doubt that there is a corresponding reincarnation of Green's Functions in superstring theory, but I stop my narrative at this point in order not to expose my ignorance further. As we move into the 1990s, Green's Functions are alive and well, still going strong, ready to help us again as soon as the wheel of fashion turns once more and the next new theory of everything emerges. That is not the end of the story, but it is the end of my talk.

CHAPTER 3.2

JAMES BRADLEY, THE INVENTOR OF MODERN SCIENCE[*]

If we say that science is modern when it is based on measurements of high precision, then modern science began in 1729. James Bradley did the first high-precision measurements. He was the first to understand that accurate measurement requires meticulous monitoring and control of possible sources of error. He was the first to record temperature and barometric pressure whenever he made an observation.

In 1729, Newton had been dead for two years. Only a hundred and twenty years had passed since Galileo's little telescope had first been pointed at Jupiter and its satellites. James Bradley, having abandoned a promising career as a Church of England priest and chaplain to the bishop of Hereford, was aged thirty-six and a rising star in the world of astronomy. Bradley had the good luck to have a widowed aunt who owned the house where she lived in Wanstead. The house had a high roof, and the aunt allowed him to cut a hole in her roof and use the upper part of the house as an observatory. Her deceased husband had also been a devotee of astronomy, so she was used to this sort of madness. Under the hole, Bradley suspended his long and slender three-inch refracting telescope, looking up into the sky with a focal length of twelve and a half feet. The telescope was nothing unusual. The revolutionary part of his equipment was a brass circular arc along which the eyepiece of the telescope could be moved with a micrometer screw. The eyepiece was focussed simultaneously on a star overhead and on a fine crosswire. Bradley could move the star exactly onto the crosswire with the micrometer screw, and then read off the angle on the brass arc. At one place on the arc there was a notch leaving space for a vertical plumb-line made of wire with one hundredth of an inch diameter. We must assume that he had his aunt well trained so that she did not slam doors or do the

*Extract from talk to American Philosophical Society, November 1993.

laundry while observations were in progress. In those days he did not need to worry about vibrations of the house caused by passing trucks or trains.

The telescope and the circular arc were not built by Bradley. They were built by George Graham, a famous instrument-maker who was also a Fellow of the Royal Society. Here are some sentences from a recent article about Bradley [Chapman, 1993].

"Bradley and Graham were not only concerned with making and using instruments, but also with monitoring their behavior over time, to detect and compensate for errors. Never before had an astronomer examined instruments so rigorously in a thorough-going quest for errors. Bradley suspected that the quadrantal arc of his instrument was gradually diminishing over time, probably because of the effects of variations of temperature and humidity on the iron-brass structure. Bradley was the first to insist that an instrument had to be constantly monitored for errors if it was to be capable of detecting delicate celestial motions. He made extensive use of cross-checks to calibrate instruments. He checked the scale errors of Graham's quadrant with a separate micrometer, and later, after he moved to the Greenwich Observatory, he checked all the Greenwich instruments against one another. He recorded barometric pressure and air temperature whenever he made an observation. All these innovations made Bradley's observations the earliest that can be relied on by modern astronomers".

Using his plumb-line and his micrometer screw, Bradley was able to measure the angle between a star and the local vertical with a precision consistently better than one second of arc. Measuring the angle to an accuracy of one second of arc meant measuring the movement of his eyepiece along the twelve-foot arc to an accuracy of one thousandth of an inch. Using modern language, we could say that he was measuring with six-figure accuracy, with an error of one or two parts per million. Bradley was the first astronomer who measured the sky with six-figure accuracy. More than that, he was the first scientist who successfully measured anything at all with six-figure accuracy. He measured angles about a hundred times more accurately than the astronomers of Newton's time. Newton himself had recognized Bradley's superiority. Newton remarked, when Bradley was a young man and Halley was still active, that Bradley was the best astronomer in Europe. Newton was not renowned for generosity toward aspiring young scientists. Probably his generous praise of Bradley was intended to annoy Halley rather than to please Bradley. Whether Newton intended it or not, his remark had the result that Bradley was appointed Savilian Professor at Oxford at the age of twenty-eight. Oxford required him to lecture but did not provide him with an observatory. When he was lecturing he was earning his living as a professional astronomer. When he was observing at his aunt's house, he was an amateur working at his own expense. He continued observing for several years, measuring star positions to six-figure accuracy, before he discovered aberration.

Aberration means the displacement of the image of a star in the sky due to the speed of the earth in its orbit around the sun. Since all stars in any limited region of sky are displaced together, you cannot measure aberration by measuring displacements of stars relative to one another. You have to measure absolute displacements. The most practical way to measure absolute displacements is to measure the angle between the image of a star and the zenith as the star crosses the northern or the southern meridian. That is what Bradley did, measuring the angles routinely to the nearest quarter of a second of arc. He began with the star Gamma Draconis, which he chose for a good reason, since in England it passes close to the zenith every night. Bradley knew that his star images might also be displaced by atmospheric refraction, and he knew that the effects of refraction would be worse as he moved further from the zenith. Eighteen nights after he started observing, he found that the image of Gamma Draconis had moved three and three-quarters of a second to the south, roughly fifteen times the uncertainty of his measurements. He continued observing Gamma Draconis and 200 other stars systematically. He found that all the stars moved in elliptical paths on the sky, coming back to their original positions after a year. The orientations and shapes of the ellipses depended in a simple way on the positions of the stars. The size of the ellipse as measured by its major axis was the same for all stars, namely forty-one seconds of arc. For two years he continued to collect data without understanding what it meant.

The obvious explanation for an annual movement of a star on the sky was parallax, the change in the viewing angle of a star at a finite distance due to the shift in the position of the earth. The observed parallax of any star should be inversely proportional to its distance. Bradley had begun his accurate measurements of star positions in the hope of measuring parallaxes and thereby measuring the distances of stars. But he quickly saw that the motions he observed could not be due to parallax. The observed motions were ninety degrees out of phase with the motions expected to be produced by parallax. The observed displacements were roughly perpendicular to the earth–sun direction instead of being parallel to it. According to the legend that we find in Agnes Clerke's biography of Bradley [Clerke, 1886],

"A year's assiduous use of this instrument gave him a set of empirical rules for the annual apparent motions of stars in various parts of the sky; but he had almost despaired of being able to account for them, when an unexpected illumination fell upon him. Accompanying a pleasure party in a sail on the Thames one day about September 1728, he noticed that the wind seemed to shift each time that the boat put about, and a question put to the boatman brought the significant reply that the changes in direction of the vane at the top of the mast were merely due to changes in the boat's course, the wind remaining steady throughout. This was the clue he needed. He divined at once that the progressive transmission of light, combined with the advance of the earth in its orbit, must cause an annual shifting of the direction in which the heavenly bodies are seen, by an amount depending upon the ratio of the two velocities. Working out the problem in detail, he found that the consequences agreed perfectly with the rules already deduced from observation".

Let me explain Agnes Clerke's account in a little more detail. Suppose that the boat with Bradley on board was trying to sail up-river to the west against a steady westerly wind. The boat would then be tacking, setting its course alternately to north-west and south-west. Each time the boat steered north-west, the wind would seem to be blowing from a direction a little north of west. In the same way, when Bradley first observed Gamma Draconis in December 1727, the earth was moving around the sun in an easterly direction, and so the light from the star appeared to come from a direction a little to the east of the star's true direction. Three weeks later, the earth was moving in a direction south of east, and so the light from the star appeared to come from a direction displaced toward the south. The magnitude of the observed displacement, when combined with the known orbital speed of the earth, gave at once a new determination of the velocity of light. The maximum displacement is equal to the ratio of the orbital speed to the velocity of light. Bradley's value for the velocity of light is within one percent of the modern value.

Bradley's discovery was immediately acclaimed, not only by astronomers but by the educated public all over Europe, as the first direct demonstration that the Copernican view of the universe was right. The elliptical motions of the stars in the sky are visual proof that the earth is moving around the sun. If the earth were stationary with the sun moving around it, aberration could not occur. But Bradley was not content to bask in the glory of his first discovery. He had already noticed that there were small discrepancies between the observed stellar displacements and the theory of aberration. He continued for many years to pursue these discrepancies with meticulous care. He suspected that they were due to a periodic perturbation of the earth's axis of rotation by the gravitational effect of the moon. But he was unwilling to claim an explanation for the discrepancies until he had observed them for nineteen years and verified his prediction that after nineteen years they would come back to zero. It happens that nineteen years is the period of precession of the nodes of the moon's orbit around the earth. After the nineteen years had gone by and the stars were back in their proper places, Bradley announced his second great discovery, the nodding motion of the earth's axis that he called nutation. By that time he had succeeded Halley as astronomer royal and had moved his instruments to the Greenwich Observatory. He continued for the rest of his life to push the limits of accuracy in astronomical work of all kinds. He set the standards of positional astronomy for the next two hundred years. His published works amount altogether to seventy pages. He was a man of few words. According to Agnes Clerke: "Scarcely an astronomer in Europe but sought a correspondence with him, which he usually declined, being averse to writing, and leaving many letters unanswered."

Bradley did not have graduate students in the modern style, but he had assistants, young men that he trained in the art of high-precision measurement and calculation. One of his assistants was Charles Mason, who afterwards achieved immortality as surveyor of the Mason–Dixon line separating Maryland from Pennsylvania. The line ran for 233 miles along the parallel of latitude 39 degrees 43 minutes, across land already claimed by greedy land-owners on both sides. It was important to stake out the line on the ground with astronomical precision. It was important that Charles Mason knew about the disturbing effects of nutation on measurements of latitude. If nutation had not been taken into account, the Mason–Dixon line might have deviated from its proper position by as much as 900 feet north or south, and the American Civil War might have started a hundred years earlier than it did.

The influence of Bradley extended far beyond the people that he trained. The revolution that he started by insisting on six-figure accuracy spread to other countries besides England and to other sciences besides astronomy. A succession of great mathematicians from Laplace to Gauss and Poincaré created the new science of analytical dynamics to impose a coherent intellectual order on the precisely observed celestial motions. A hundred and fifty years after Bradley, Michelson and Morley attempted to measure the second-order aberration of light caused by the earth's motion through the aether, and by their negative result led the way to the theory of relativity. Finally, after two hundred years, the tradition of six-figure accuracy begun by Bradley led to the modern flowering of experimental physics, with Rabi's molecular beam apparatus probing the resonances of atoms as precisely as Bradley's micrometer probed the displacements of Gamma Draconis.

CHAPTER 3.3

A CONSERVATIVE REVOLUTIONARY*

I am delighted to have this opportunity to sing the praises of my old friend and colleague Frank Yang. The title of my talk is "A Conservative Revolutionary". The meaning of the title will become clear at the end of the talk.

One of my favorite books is Frank's "Selected papers 1945–1980 with commentary", published in 1983 to celebrate his sixtieth birthday. This is an anthology of Frank's writings, with a commentary written by him to explain the circumstances in which they were written. There was room in the book for only one third of his writings. He chose which papers to include, and his choices give a far truer picture of his mind and character than one would derive from a collection chosen by a committee of experts. Some of the chosen papers are important and others are unimportant. Some are technical and others are popular. Every one of them is a gem. Frank was not trying to cram as much hard science as possible into five hundred pages. He was trying to show us in five hundred pages the spirit of a great scientist, and he magnificently succeeded. The papers that he chose show us his personal struggles as well as his scientific achievements. They show us the deep sources of his achievements, his pride in the Chinese culture that raised him, his reverence for his teachers in China and in America, his love of formal mathematical beauty, his ability to bridge the gap between the mundane world of experimental physics and the abstract world of groups and fiber bundles. He wisely placed the eighty pages of commentaries together at the beginning of the book instead of attaching them to the individual papers. As a result, the commentaries can be read consecutively. They give us the story of Frank's life in the form of an intellectual autobiography. The autobiography is a classic. It describes the facts of his life in clear and and simple words. It quietly reveals the intense feelings and loyalties that inspired his work and made him what he is.

---

*Remarks at the banquet in honor of the retirement of C. N. Yang at Stony Brook, May 21, 1999.

One of the smallest and brightest of the gems in Frank's book is a two-page description of Fermi, written as an introduction to a paper by Fermi and Yang that was included in a volume of Fermi's collected papers. Frank studied with Fermi in Chicago from 1946 to 1949. He learned more physics from Fermi than from anybody else, and Fermi's way of thinking left an indelible impression in his mind. Frank writes, "We learned that physics should not be a specialist's subject. Physics is to be built from the ground up, brick by brick, layer by layer. We learned that abstractions come after detailed foundation work, not before". Fermi's practical spirit can be seen in the title of the great Yang–Mills paper published in 1954. Anyone speaking about the paper today would call it the paper that introduced non-Abelian gauge fields. But the title does not mention non-Abelian gauge fields. The title is "Conservation of isotopic spin and isotopic gauge invariance". The physical question, how to understand the conservation of isotopic spin, came first, and the mathematical abstraction, non-Abelian gauge fields, came second. That was the way Fermi would have approached the problem, and that was the way Frank approached it too. Fermi was great because he knew how to do calculations and also knew how to listen to what nature had to say. All through his life, Frank has balanced his own gift for mathematical abstraction with Fermi's down-to-earth attention to physical details.

Frank has not been idle during the fifteen years since his selected papers were published. Another book was published in 1995, this time not written by Frank but by his friends, a festschrift to celebrate his seventieth birthday, with the title "Chen Ning Yang, a great physicist of the twentieth century". This book contains, hidden among the technical contributions, a number of personal tributes and recollections. It describes Frank's active involvement, continuing up to the present day, helping science to grow and flourish in three Chinese communities, in the People's Republic of China, in Taiwan and in Hong Kong. Frank is happy to be able to pay back the debt that he owes to his native land and culture.

Not included in either of the two books is a paper written by Frank two years ago with the title "My father and I". This is a tribute to his father, who was a professor of mathematics and died in 1973. It is a wonderfully sensitive account of his relationship to his father and of the pain that each of them suffered as a result of their separation. His father stayed in China through the hard years while Frank grew to greatness in America. Both of them knew that it was better so. Without America, Frank could not have become a world-class scientist. Exiled from China, his father would have been a tree without roots. And yet, the separation hurt both of them deeply. For Frank, his personal separation from his father and the political separation of America from China were two parts of a single tragedy. Luckily, President Nixon decided to recognize the People's Republic just in time, so that Frank was able to visit China twice before his father died and to sit by his bedside during his last illness. In the commentary to his selected papers, Frank describes the difficult decision that he made in 1964 to become a citizen of the United States. This was a formal recognition of his separation from China and from his father. He writes, "My father . . . had earned a Ph.D. degree from the University of Chicago in 1928. He was well traveled. Yet I know, in one corner of his heart, he did not forgive me to his dying day for having renounced my country of birth."

The memoir "My father and I" ends on a happier note. It ends with a glorious moment of reunion. Frank describes how he stood at midnight on July 1, 1997, at the Hong Kong Convention and Exhibition Center, to watch the Union Jack being lowered and the flag of the People's Republic being slowly raised, while the band played "Arise, you who would not be enslaved". Frank writes, "Had father observed this historical ceremony marking the renaissance of the Chinese people, he would have been even more moved than I. . . . The intellectuals of his generation had to personally experience the humiliating exploitations in the Foreign Concessions . . . and countless other rampant foreign oppressions. . . . How they had looked forward to the day when a prosperous China could stand up, when the British Empire had to lower the Union Jack and withdraw troops, when they can see for themselves the Chinese flag proudly announce to the world: This is Chinese Territory! That day, July the first, 1997, is the day their generation had dreamed of throughout their lives".

We can all rejoice that Frank was standing there to give his blessing and his father's blessing to the reunion. For me, that pride and that feeling of fulfilment that Frank expresses have a special resonance. I too belong to a great and ancient civilization. My home-town in England was also the home-town of Alfred the scholar king, who made our town into a great center of learning eleven hundred years ago, while the Tang dynasty was establishing the system of government by scholars that endured for a thousand years in China. Our king Alfred was translating scholarly texts from Latin into English, soon after the Tang poet Tu Fu wrote the poem that Frank quotes at the beginning of his selected papers: "A piece of literature is meant for the millennium. But its ups and downs are known already in the author's heart".

Like Frank, I too left my homeland and became an American citizen. I still remember the humiliation of that day in Trenton when I took the oath of allegiance to the United States, and the ignoramus who performed the ceremony congratulated me for having escaped from the land of slavery to the land of freedom. With great difficulty I restrained myself from shouting out loud that my ancestors freed our slaves long before his ancestors freed theirs. I share Frank's ambivalent feelings toward the United States, this country that has treated us both with so much generosity and has treated our ancient civilizations with so little understanding. And I share Frank's pride in the peaceful lowering of the Union Jack and raising of the Chinese flag that he witnessed in Hong Kong, the place where our two ancient civilizations briefly came together and gave birth to something new.

Five years ago, I had the honor of speaking at the ceremony in Philadelphia, when the Franklin Medal was awarded to Frank Yang by the American Philosophical Society. We were assembled in the historic meeting-room of the society, with the portraits of Benjamin Franklin, the founder of the society, and Thomas Jefferson, one of its most active members, looking down at us. It was self-evident that Franklin and Jefferson approved of the award. We know that Frank Yang feels a special admiration for Franklin, since he gave the name of Franklin to his elder son. I would like to end this little talk with the same words that I used to praise Frank on that happy occasion.

Professor Yang is, after Einstein and Dirac, the preeminent stylist of twentieth-century physics. From his early days as a student in China to his later years as the sage of Stony Brook, he has always been guided in his thinking by a love of exact analysis and formal mathematical beauty. This love led him to his most profound and original contribution to physics, the discovery with Robert Mills of non-Abelian gauge fields. With the passage of time, his discovery of non-Abelian gauge fields is gradually emerging as a greater and more important event than the spectacular discovery of parity non-conservation which earned him the Nobel Prize. The discovery of parity non-conservation, the discovery that left-handed and right-handed gloves do not behave in all respects symmetrically, was a brilliant act of demolition, a breaking-down of intellectual barriers that had stood in the way of progress. In contrast, the discovery of non-Abelian gauge fields was a laying of foundations for new intellectual structures that have taken thirty years to build. The nature of matter as described in modern theories and confirmed by modern experiments is a soup of non-Abelian gauge fields, held together by the mathematical symmetries that Yang first conjectured forty-five years ago.

In science, as in urban renewal and international politics, it is easier to demolish old structures than to build enduring new ones. Revolutionary leaders may be divided into two kinds, those like Robespierre and Lenin who demolished more than they built, and those like Benjamin Franklin and George Washington who built more than they demolished. There is no doubt that Yang belongs to the second kind of revolutionary. He is a conservative revolutionary. Like his fellow-revolutionaries Franklin and Washington, he cherishes the past and demolishes as little as possible. He cherishes with equal reverence the great intellectual traditions of Western science and the great cultural traditions of his ancestors in China.

Yang likes to quote the words of Einstein, "The creative principle lies in mathematics. In a certain sense, therefore, I hold it true that pure thought can grasp reality, as the ancients dreamed". On another occasion Yang said, "That taste and style have so much to do with one's contribution in physics may sound strange at first, since physics is supposed to deal objectively with the physical universe. But the physical universe has structure, and one's perceptions of this structure, one's partiality to some of its characteristics and aversion to others, are precisely the elements that make up one's taste. Thus it is not surprising that taste and style are so important in scientific research, as they are in literature, art and music". Yang's taste for mathematical beauty shines through all his work. It turns his least important calculations into miniature works of art, and turns his deeper speculations into masterpieces. It enables him, as it enabled Einstein and Dirac, to see a little further than other people into the mysterious workings of nature.

CHAPTER 3.4

A MEETING WITH ENRICO FERMI*

One of the big turning points in my life was a meeting with Enrico Fermi in the spring of 1953. In a few minutes, Fermi politely and ruthlessly demolished a program of research that I and my students had been pursuing for several years. He probably saved us from several more years of fruitless wandering along a road that was leading nowhere. I am eternally grateful to him for destroying our illusions and telling us the bitter truth.

Fermi was one of the great physicists of our time, outstanding both as a theorist and as an experimenter. In 1942, he led the team that built the first nuclear reactor in Chicago. In 1953, he was leading the team that built the Chicago cyclotron and used it to explore the strong forces that hold nuclei together. He made the first accurate measurements of the scattering of mesons by protons. His meson-proton scattering experiment gave the most direct evidence then available of the nature of the strong forces.

In 1953, I was a young professor of theoretical physics at Cornell, responsible for directing the research of a small army of graduate students and post-docs. I had put them to work calculating meson-proton scattering, so that their theoretical calculations could be compared with Fermi's measurements. In 1948 and 1949, we had made similar calculations of atomic processes, using the theory of quantum electrodynamics, and found spectacular agreement between experiment and theory. Quantum electrodynamics is the theory of electrons and photons interacting through electromagnetic forces. Because the electromagnetic forces are weak, we could calculate the atomic processes precisely. By 1951, we had triumphantly finished the atomic calculations and were looking for fresh fields to conquer. We decided to

---

*Contribution to "Turning Points" series in Nature magazine.

use the same techniques of calculation to explore the strong nuclear forces. We began by calculating meson-proton scattering, using a theory of the strong forces known as pseudoscalar meson theory. By the spring of 1953, after heroic efforts, we had plotted theoretical graphs of meson-proton scattering. We joyfully observed that our calculated numbers agreed pretty well with Fermi's measured numbers. So I made an appointment to meet with Fermi and show him our results. Proudly, I rode the Greyhound bus from Ithaca to Chicago with a package of our theoretical graphs to show to Fermi.

When I arrived in Fermi's office, I handed the graphs to Fermi, but he hardly glanced at them. He invited me to sit down, and asked me in a friendly way about the health of my wife and our new-born baby son, now fifty years old. Then he delivered his verdict in a quiet, even voice. "There are two ways of doing calculations in theoretical physics", he said. "One way, and this is the way I prefer, is to have a clear physical picture of the process that you are calculating. The other way is to have a precise and self-consistent mathematical formalism. You have neither". I was slightly stunned, but ventured to ask him why he did not consider the pseudoscalar meson theory to be a self-consistent mathematical formalism. He replied, "Quantum electrodynamics is a good theory because the forces are weak, and when the formalism is ambiguous we have a clear physical picture to guide us. With the pseudoscalar meson theory there is no physical picture, and the forces are so strong that nothing converges. To reach your calculated results, you had to introduce arbitrary cut-off procedures that are not based either on solid physics or on solid mathematics".

In desperation I asked Fermi whether he was not impressed by the agreement between our calculated numbers and his measured numbers. He replied, "How many arbitrary parameters did you use for your calculations?" I thought for a moment about our cut-off procedures and said, "Four". He said, "I remember my friend Johnny von Neumann used to say, with four parameters I can fit an elephant, and with five I can make him wiggle his trunk". With that, the conversation was over. I thanked Fermi for his time and trouble, and sadly took the next bus back to Ithaca to tell the bad news to the students. Because it was important for the students to have their names on a published paper, we did not abandon our calculations immediately. We finished the calculations and wrote a long paper that was duly published in the Physical Review with all our names on it. Then we dispersed to find other lines of work. I escaped to Berkeley, California, to start a new career in condensed matter physics.

Looking back after fifty years, we can see clearly that Fermi was right. The crucial discovery that made sense of the strong forces was the quark. Mesons and protons are little bags of quarks. Before Murray Gell-Mann discovered quarks, no theory of the strong forces could possibly have been adequate. Fermi knew nothing about quarks, and died before they were discovered. But somehow he knew that something essential was missing in the meson theories of the 1950s. His physical intuition told him that the pseudoscalar meson theory could not be right. And so it was Fermi's intuition, and not any discrepancy between theory and experiment, that saved me and my students from getting stuck in a blind alley.

CHAPTER 3.5

EDWARD TELLER, 1908–2003*

At the end of his long life, Edward Teller with the help of his editor Judith Shoolery published his memoirs [Teller and Shoolery, 2001], a lively and poignant account of his adventures in science and politics. Hostile reviewers of the memoirs pointed out that some details of his stories are inaccurate. But Teller writes in his introduction, "Our memories are selective; they delete some events and magnify others. Just the simple act of recalling the past affects the recollection of what happened. That some of my remembrances are not the commonly accepted version of events should not be surprising". Memoirs are not history. Memoirs are the raw material for history. Memoirs written by generals and politicians are notoriously inaccurate. A writer of memoirs should make an honest attempt to set down the course of events as they are recorded in memory. This Teller did. If some of the details are wrong, this detracts little from the value of his book as a panorama of a historical epoch in which he played a leading role. I have used the memoirs as the basis for this brief summary of his career.

Teller was born in 1908 into a prosperous middle-class Jewish family in Budapest. He lived through the turbulent years of World War I, the dismemberment of the Austro-Hungarian empire, the short-lived Communist regime of Bela Kun, and the devastating currency inflation that followed, protected by loving and resourceful parents. All through his life, from childhood to old age, he had a gift for friendship. His memoirs are full of stories about his friends and the tragic fates that many of them encountered. He cared deeply for them as individuals, and described them with sympathetic understanding. He escaped from sharing their fate when he emigrated from Hungary to Germany in 1926, from Germany to Denmark in 1933, to England in 1934 and to America in 1935. He always retained an acute sense of the precariousness of human life and the fragility of political institutions. He was

*Memoir for US National Academy of Sciences.

127

one of the lucky survivors of a great tragedy, when barbarians overran Europe and destroyed the world of his childhood. He saw America as the last refuge, for himself and for the civilization that he cherished. That was why he saw it as his inescapable duty to keep America armed with the most effective weapons, with bombs for deterring attack and with missiles for active defense.

The springtime years of Teller's professional life, the years when he was happiest with his work and his friends, were the seven years 1926–1933 that he spent as a student in Germany. His stay in Germany started badly. Riding a trolley-car in Munich to meet some friends for a hiking trip, he overshot the meeting-place, jumped off the moving trolley-car and fell under the wheels. His right foot was chopped in half. As he lay on the road assessing the damage, he thought how lucky he was not to be one of the millions of young men who had lain wounded on the muddy battlefields of World War I a few years earlier. At least he was alive, with a clean wound and the certainty of being rescued. The surgeon in Munich reconstructed what was left of his foot so that he could still walk on the heel. With the help of a prosthesis, he became agile enough to go hiking in the mountains and to play a respectable game of ping-pong. He observed that his mother suffered more than he did from the accident. As she sat grieving by his bedside at the hospital, he tried unsuccessfully to cheer her up. For her it was a deeply tragic event, while for him it was merely a nuisance that did not touch the important things in his life. The accident gave him confidence. As he liked to say when I got to know him twenty years later, if it doesn't kill you it makes you stronger. The accident gave him a good excuse to leave Munich and go to Leipzig to work with Heisenberg.

The years 1926–1933 were the time when German science was blazing with creative activity while the Weimar Republic was crumbling. When Teller joined the group of young people working at Leipzig with Heisenberg as leader, Heisenberg was 28. He had invented quantum mechanics in 1925, and then invited all and sundry to join him in using quantum mechanics to understand the workings of nature. Quantum mechanics described the behavior of atoms, and so it should be able to explain everything that atoms do. It should be possible with quantum mechanics to explain all of atomic physics, most of solid-state physics, most of astrophysics, and all of chemistry. There were enough good problems, so that every student could find something important to do. Teller, having been trained as a chemist, chose chemistry as the subject to be explained with quantum mechanics. He started well by beating Heisenberg at ping-pong. He and Heisenberg remained friends for life. After World War II, when many American physicists condemned Heisenberg for staying in Germany through the Hitler years, Teller went out of his way to befriend him. He knew that Heisenberg had never been a Nazi, and he respected Heisenberg's decision to stay loyal to his country and share its fate.

In Leipzig, Teller wrote a Ph.D. thesis [Teller, 1930] on the hydrogen molecule ion, the simplest of all molecules. He was able to calculate not only the ground state but also the excited quantum states of the molecule, using an old-fashioned mechanical calculator. But he did not enjoy working alone. He much preferred the give-and-take of working together with friends. Almost all his work after the thesis was done jointly with others. During his years in Germany he collaborated fruitfully with Lev Landau, George Placzek and James Franck, solving various problems on the borderline between physics and chemistry. As he himself said, he was a problem-solver rather than a deep thinker. He enjoyed solving problems, whether or not they were important. The years 1926–1933 were harvest-time for problem-solvers. In those years, the problem-solvers laid the foundations for most of modern physics and chemistry. Teller's main contributions during this time were to explain diamagnetism in solids [Teller, 1931], and to explain spectra of polyatomic molecules [Herzberg and Teller, 1933]. Both these problems required the application of quantum mechanics to systems involving many electrons.

When Hitler took power in 1933, Teller moved to Copenhagen. There he met George Gamow, a young Russian who had been the first to apply quantum mechanics to nuclear physics. In 1934, Gamow moved to George Washington University in America, and Teller moved to London with his newly-wed wife Mici. She was a childhood friend from Budapest, who loved and sustained him through all his joys and sorrows, and remained by his side for 66 years until her death in 2000. In 1935, Gamow invited Teller to join him at George Washington University, and Mici, who had spent two years in America as a student, encouraged him to accept. During the years 1935–1939 that Gamow and Teller were together in Washington, they almost recreated the golden age of German physics in America. They found many of their European friends already in America, and quickly made new friends among the natives. Gamow was four years younger than Heisenberg and almost as brilliant. But Gamow had no skill as an organizer and no desire to be a leader like Heisenberg. He produced brilliant new ideas at a rapid rate, and left it to Teller to work out the details. He also left to Teller the chores of administration, organizing meetings and taking care of students. Teller worked happily with Gamow and also with other collaborators. The most important results of Teller's research during this time were the Gamow–Teller theory of weak interactions [Gamow and Teller, 1936], and the Jahn–Teller theory of polyatomic molecules with electrons in degenerate states [Jahn and Teller, 1937]. The Gamow–Teller theory was in competition with an alternative theory due to Fermi. This was one of the very few occasions on which Fermi guessed wrong. The Gamow–Teller theory was Teller's first venture into nuclear physics. Twenty years later, it became the basis for a unified theory of weak interactions.

One of Teller's friends in Washington was Merle Tuve, a native American and a first-rate physicist, who built particle accelerators and used them to do nuclear experiments at the Department of Terrestrial Magnetism of the Carnegie Institution. Tuve was one of the pioneers of accelerator physics, and made the first accurate measurements of the nuclear interaction between two protons. After Teller had spent a summer teaching in Chicago, the University of Chicago was thinking of offering him a permanent job. The Chicago physicists wrote to Tuve asking for his opinion of Teller. Tuve wrote back: "If you want a genius for your staff, don't take Teller, get Gamow. But geniuses are a dime a dozen. Teller is something much better. He helps everybody. He works on everybody's problem. He never gets into controversies or has trouble with anyone. He is by far your best choice." Teller quotes this letter in his memoirs and remarks, "I do believe it described me as I was during those happy years in Washington". He looked back on those years with nostalgia as a time when he could do science with everyone and be friends with everyone, before the bitter struggles over nuclear politics took him away from science and tore apart his friendships.

The record of Teller's publications confirms Tuve's statement. Teller in the first half of his life had an unusual gift for fruitful collaborations. I have taken 1952 as the point of division between the two halves of his life. In 1952, he moved from the University of Chicago to the new weapons laboratory that he founded at Livermore in California. That was the year when he stopped being an academic scientist and became a full-time nuclear entrepreneur. In the bibliography of his technical publications there are a hundred and forty six papers. Before 1952, he wrote seven papers alone and seventy seven with collaborators. In that period, most of his papers describe research done with one collaborator. Many of the leading physicists of that time appear as collaborators. After 1952, he wrote forty two papers alone and twenty with collaborators. In that period, most of the papers are reviews or lectures, describing plans for the future or surveys of the past. The transition from a gregarious to a solitary pattern of intellectual life is painfully clear.

In January 1939, Gamow and Teller were hosts at the meeting of theoretical physicists which was held annually at George Washington University. That year's meeting was supposed to be devoted to low-temperature physics. On the first morning of the meeting, Gamow introduced Niels Bohr, who had just arrived on a boat from Denmark, and Bohr told the assembled physicists the news of the discovery of fission of uranium in Germany a month before. In the evening of the same day, Merle Tuve invited everyone to his laboratory to see a demonstration of the intense bursts of ionization produced by uranium fission in a Geiger counter. The age of nuclear energy had arrived, and Teller was involved in it from the first day. In February 1939, Teller's friend Leo Szilard called him from New York to announce that he had found abundant secondary neutrons emitted in uranium fission. This meant that an explosive nuclear chain reaction was certainly possible. In March 1939 an informal strategy meeting was held in Princeton. Present were Bohr, Wheeler,

Wigner, Weisskopf, Szilard and Teller. One American, one Dane, one Austrian and three Hungarians. Two decisions were made, first to keep further discoveries about fission secret as far as possible, second to try to bring the situation to the attention of responsible people in the American government.

In June 1939, Teller moved from Washington to Columbia University to help Fermi and Szilard with their project to build the first nuclear reactor. In New York, a few weeks before the outbreak of World War II, Heisenberg came to visit Teller. He was on his way back to Germany from a lecture tour in America. He had many offers of jobs in America and could easily have stayed. Teller asked him why he was going back to a country that was clearly headed for disaster. Heisenberg replied, "Even if my brother steals a silver spoon, he is still my brother". Teller understood that nothing he could say would cause Heisenberg to change his mind. A few days later Szilard, who could not drive a car, came to see Teller and asked him for a ride. Szilard had written a letter to President Roosevelt informing him of the discovery of fission and the possibility of nuclear bombs. The letter asked the President to set up a channel of communication between the government and the physicists working on nuclear chain reactions in America. Szilard's plan was to persuade Einstein to sign the letter. Teller was needed as a chauffeur to bring Szilard and the letter to Einstein's summer home on Long Island. Einstein signed the letter, and Szilard successfully delivered it to Roosevelt. As a result, an official Advisory Committee on Uranium was established, and the bureaucratic machinery that later grew into the Manhattan Project slowly began to grind.

Teller worked on nuclear energy from 1939 to 1945, two years at Columbia University helping Fermi design the first nuclear reactor, two years at the Metallurgical Laboratory in Chicago helping to design the Hanford plutonium production reactors, and two years at Los Alamos working on bombs. In all three places, he worked on a variety of projects. His wide knowledge of physics and chemistry made him useful as a liaison between different parts of the enterprise. The one thing that he could not and would not do was to sit down and do precise theoretical calculations. His thesis work, calculating the states of the hydrogen molecule ion, had given him a lifelong distaste for lengthy calculations. At Los Alamos this brought him into collision with Hans Bethe, the head of the Theoretical Division, who was Teller's boss. Bethe asked him to do a massive calculation of the physics and hydrodynamics of an imploding bomb. Teller refused, saying that if he tried to do such a calculation he would not make any useful contribution to the war effort. Teller's friendship with Bethe never recovered from this disagreement. Oppenheimer moved Teller out of Bethe's division and made him leader of an independent group. After that, Teller reported directly to Oppenheimer, and Oppenheimer kept him busy with a variety of assignments more suited to his temperament. Teller enjoyed working for Oppenheimer and considered him an excellent director.

During the wartime years, Teller worked only intermittently on hydrogen bombs. This work started in the summer of 1942 when Oppenheimer held a meeting in Berkeley to explore the possibilities. The meeting concluded that, if a fission bomb could be made to work, it could probably be used to ignite a hydrogen bomb. After the meeting, Teller found reasons why the ignition would not work. He became seriously interested in the problem and continued to think about it. During his two years in Los Alamos, he spent about one third of his time working on hydrogen bombs. The result of his efforts was a very sketchy design called the "Classical Super". The question whether the Classical Super would work could only be decided by massive calculations, using electronic computers which did not yet exist. There matters stood from 1945 to 1950.

From 1946 to 1952, Teller was a professor at the University of Chicago. He enjoyed the return to academic life and especially enjoyed interacting with a brilliant bunch of students including Chen Ning Yang, Tsung Dao Lee, Marshall Rosenbluth and Marvin Goldberger. Two of his closest friends, Enrico Fermi and Maria Mayer, were colleagues. During these years he worked with Fermi on the capture of negative mesons in matter [Fermi and Teller, 1947], with Mayer on the origin of the chemical elements [Mayer and Teller, 1949], and with Robert Richtmyer on the origin of cosmic rays [Richtmyer and Teller, 1949]. I met Teller for the first time in March 1949 when I gave a colloquium in Chicago with Fermi and Teller sitting side by side in the front row. I spoke about the new theories of quantum electrodynamics. I made some very polite remarks about Schwinger's theory and then explained why Feynman's theory was better. As soon as I finished my talk, Teller asked a question and answered it himself. "What would you think of a man who cried, 'There is no God but Allah, and Mohammed is his prophet', and then at once drank down a great tankard of wine? I would consider him a very sensible fellow". Afterwards I was able to meet with Teller alone and he talked happily about all the things he was doing.

I quote now from a letter which I wrote to my parents in England, dated March 11, 1949. "Teller to me has always been an enigma. He has done all kinds of interesting things in physics, but never the same thing for long, and he seems to do physics for fun rather than for glory. However, during the last few years there have been reports that he has been engaged in perfecting the most fiendish engines of destruction; and I have always wondered how such a man could do such things. In Chicago I found without difficulty the answer. I started a long argument with him about political questions, and it appears that he is an ardent supporter of the 'World Government' movement, an organization which preaches salvation in the form of a world government, to be set up in the near future with or without Russia, and to have sovereign powers over the economic and social policies of its member nations. Teller evidently finds this faith soothing to his conscience; he preaches it with great charm and intelligence; all the same, I feel that he is a good example of the saying that no man is so dangerous as an idealist."

In the same letter there is a passage describing the community of physicists in Chicago: "The most striking thing about all these people, and also their wives whom I met as I went from house to house and from family to family, is how happy they seem to be. All of them say they have never found any place on earth so pleasant to be in as Chicago. There seems to be an exceptionally free and easy atmosphere, rather like Cornell, and with the added advantages of a metropolitan city." These were the golden years of physics in Chicago, when Fermi was king and Teller was his court jester. Teller enjoyed those years to the full. But during those same years he could not stop thinking about the question that he had left unanswered when he left Los Alamos in 1946. Could a hydrogen bomb be made to work? In June 1949, he returned to Los Alamos to continue his lonely effort to understand what Nature allows us to do. In August the first Russian nuclear bomb was tested, and in January 1950, President Truman announced that work on the "so-called hydrogen or super bomb" would continue. After the President's announcement, Teller wrote to Maria Mayer from Los Alamos: "Whatever help and whatever advice I can get from you . . . I need it. Not because I feel subjectively that I must have help, but because I know objectively that we are in a situation in which any sane person must and does throw up his hands and only the crazy ones keep going".

In 1950, electronic computers were able to simulate in a rough fashion the Classical Super design for a hydrogen bomb and showed that it did not work. George Gamow drew a famous cartoon of Teller trying to set fire to a wet piece of rock with a match. But to Teller the downfall of the Classical Super came as a liberation. For eight years his thoughts had been fixed on the Classical Super, which required deuterium to burn at low density, so that radiation could escape from the burning region and not come to thermal equilibrium with the matter. The idea was to achieve a "runaway burn", with the temperature of the matter remaining much higher than the temperature of the radiation. The computers showed that runaway burn did not work. So Teller started to look seriously at the opposite situation, with deuterium at high density and the radiation trapped in thermal equilibrium with the matter. He quickly found that at high density deuterium could burn well in thermal equilibrium. From that point it was a short step to design an arrangement by which a fission bomb could compress deuterium to high density and then ignite it. Teller's colleague Stanislas Ulam at Los Alamos thought of a similar arrangement at the same time, and so the idea became known as the Teller–Ulam design. It was successfuly tested in 1952 and has been the basis for American hydrogen bombs ever since. Andrei Sakharov had the same idea in 1954, and it quickly became the basis for Russian hydrogen bombs too. Many years later Teller and Sakharov met. They did not agree about political questions, but expressed a deep respect for each other. Sakharov remarked in his memoirs that Teller's treatment at the hands of his American colleagues was "unfair and even ignoble".

In 1951, Teller returned briefly to his academic life in Chicago, but in 1952, he moved permanently to the Livermore laboratory, a brand-new weapons laboratory which his friend Ernest Lawrence had organized in California to give some competition to Los Alamos. He stayed at Livermore for 23 years, attracted a brilliant group of young collaborators, and saw the laboratory quickly rise to become an equal partner with Los Alamos in weapons development and in many other enterprises. Livermore was more adventurous than Los Alamos and more willing to try out crazy ideas. A much larger fraction of Livermore bomb-tests failed, but Teller considered failed tests a badge of honor rather than a disgrace. In the end, the Livermore-designed weapons proved to be as rugged and reliable as those designed by Los Alamos.

Soon after Teller moved to Livermore, he was invited to testify at the Oppenheimer security hearings in Washington. At the hearings he was asked whether he considered Oppenheimer to be a security risk, and answered yes. For this, the majority of physicists, including many of his friends, never forgave him. The estrangement caused Teller tremendous grief. The community of physicists was split in two, and Teller became a symbol of the division. At the time when this happened, I was puzzled and shocked by the violence of the reaction against Teller. To me it seemed that the main question was whether the security rules should be applied impartially to famous people and unknown people alike. It was a question of fairness. If any unknown person had behaved as Oppenheimer behaved, telling a lie to a security officer about an incident that involved possible spying, he would certainly have been denied clearance. The question was whether Oppenheimer, because he was famous, should be treated differently. Should there be different rules for peasants and princes? This was a question concerning which reasonable people could disagree. I tended to agree with Teller that the rules ought to be impartial. And I saw no reason why people who disagreed with him should condemn him for speaking his mind. Teller's estrangement from the community of physicists became worse when three of his closest friends, Enrico Fermi, John von Neumann and Ernest Lawrence, happened to die prematurely within a few years after the Oppenheimer hearings. Each of them died in his fifties and should have remained vigorously active for at least another twenty years. The loss of all three made Teller even more isolated as he started his new life at Livermore.

In the summer of 1956, I had one of my happiest experiences, working with Teller on the design of a safe nuclear reactor. Teller's friend Frederick de Hoffmann, a young physicist from Los Alamos, had started a company called General Atomic in San Diego to manufacture reactors for civilian use. Teller and I came for the summer with a group of physicists, chemists and engineers to help the company get started. Teller had been saying for many years that the essential problem for public acceptance of nuclear power was safety. He proposed that General Atomic should start by building a spectacularly safe reactor. His definition of safe was that you could give the reactor to a bunch of children to play with and be sure that they would not get hurt. Safety must be guaranteed by the laws of nature and not by engineered safeguards. For three months, Teller and I argued furiously about the design. Every day, Teller would think of some brilliant new idea and the rest of us would do calculations to show why it would not work. Finally we found a scheme that worked, and used it to design a small reactor called TRIGA, short for Training, Research and Isotope-production, General Atomic. The TRIGA was designed, built, licensed and sold within two years. The company sold seventy five of them, mostly to hospitals for making short-lived isotopes, and they have never run into any safety problems. Teller and I had hoped that big power reactors using the TRIGA design could give rise to a nuclear power industry without safety problems. Unfortunately, the nuclear power industry was stuck with designs borrowed from the submarine-propulsion reactor program of Admiral Rickover, and never considered the TRIGA design as a serious competitor. Many years later, Teller and his colleagues at Livermore developed designs for safe nuclear power reactors that could be buried deep underground, operated with a single loading of fuel for fifty years, never refueled and never unloaded. Teller remained always hopeful that nuclear power would one day be so safe that the public would finally accept it.

Teller pushed hard to develop at Livermore other programs besides weapon-development. He started a very successful educational program informally known as Teller Tech, which brought graduate students to the University of California campus at Davis. The students were enrolled in the College of Engineering at Davis and received Ph.D. degrees in Applied Science from Davis, but spent half their time at Livermore. Courses were taught by leading scientists at Davis and at Livermore. Teller enjoyed doing his share of the teaching. Many of the graduates remained at Livermore as members of the staff, while others went on to distinguished careers in universities and in industry.

Roughly one half of the Livermore budget went into weapons. In addition, there was a large program to build controlled fusion reactors, both magnetic and inertial. There was a program to develop a supersonic nuclear ram-jet that could fly non-stop around the world at low altitude. And there were two projects that were particularly dear to Teller's heart, the PLOWSHARE program to use nuclear explosions for peaceful purposes, and the strategic defense program to shoot down enemy missiles using X-ray lasers and brilliant pebbles. The PLOWSHARE program aimed to use nuclear explosions to excavate large masses of dirt or rock cheaply, the main purpose being to create artificial harbors and canals. To minimize the contamination of the landscape by radioactive fall-out, the PLOWSHARE experts designed bombs whose explosive yield came mostly from fusion and as little as possible from fission. The X-ray laser was a device that could convert a substantial fraction of the energy of a fission bomb into a collimated beam of X-rays. It was supposed to kill missiles a long way away by firing X-rays at them with extreme accuracy. The brilliant pebble was a small interceptor rocket that was supposed to kill a missile by direct impact.

In the end, neither the PLOWSHARE program nor the strategic defense program fulfilled Teller's hopes. None of the places that were candidates for PLOWSHARE excavations welcomed the idea with enthusiasm. Nobody had any urgent need for new harbors and canals, and as environmental regulations became more stringent, the chance that any PLOWSHARE project would ever be approved became increasingly remote. Livermore's proposals for strategic defense also ran into difficulties. The X-ray laser was designed to destroy missiles in the boost phase while they were still accelerating with rocket power, but the X-rays could not penetrate any considerable depth of atmosphere. As a result, the missiles could defeat the defense by accelerating more rapidly and shortening the boost phase. The brilliant pebbles were supposed to weigh a couple of pounds and turned out to weigh a couple of hundred pounds. Extravagantly large numbers of them would be required to be sure of having one at the right place and time to intercept a missile. However, the Strategic Defense Initiative that President Reagan started in 1983 embodied some of the Livermore proposals, and Teller gave it strong support.

After the Strategic Defense Initiative had spent a lot of money and accomplished very little, Teller and I went together to the Pentagon to talk with General Abrahamson who was then running the program. Teller and I agreed that strategic defence was in principle a good idea, and that secrecy was in principle a bad idea. The SDI was a technically flawed program whose failures were concealed by excessive secrecy. Teller and I went to the General to tell him that the only way to make SDI technically effective was to abolish the secrecy and bring it out into the open. If the program were open, it might receive the expert criticism and the influx of new ideas from the outside that it desperately needed. Teller delivered the message with his usual eloquence, and the General responded by saying that of course he agreed with us, and he would be removing the secrecy within a few weeks. Needless to say, nothing of the kind ever happened. Teller remained publicly supportive of SDI but privately furious at the General for deceiving us.

In 1975, Teller retired from Livermore and became a Senior Fellow at the Hoover Institution on the campus of Stanford University. Here he spent the sunset years of his life, in close touch with the work of the laboratory at Livermore, writing books and giving lectures, politically active to the end, still fighting for strategic defense and nuclear energy. At the end of his memoirs is a chapter entitled "Homecoming", describing his seven visits to Hungary between 1990 and 1996. In Hungary he felt immediately at home after an absence of fifty four years. He had never stopped speaking Hungarian with his wife, so that he remained fluent in the language. He was welcomed not only as a national hero but as a long-lost brother. He was as proud of Hungary as Hungary was proud of him. His homecoming gave his life the happy ending that was denied to him in America.

References

E. Teller [1930] "Über das wasserstoffmolekülion", *Zeits. f. Phys.* **61**, 458–480.

E. Teller [1931] "Der Diamagnetismus von freien elektronen", *Zeits. f. Phys.* **67**, 311–319.

G. Herzberg und E. Teller [1933] "Schwingungsstruktur der elektronenübergänge bei mehratomigen molekülen", *Zeits. f. Phys. Chem.* **21**, 410–446.

G. Gamow and E. Teller [1936] "Selection rules for the beta-disintegration", *Phys. Rev.* **49**, 895–899.

H. A. Jahn and E. Teller [1937] "Stability of polyatomic molecules in degenerate electronic states. I. Orbital degeneracy", *Proc. Roy. Soc. A* **161**, 220–235.

E. Fermi and E. Teller [1947] "The capture of negative mesotrons in matter", *Phys. Rev.* **72**, 399–408.

M. G. Mayer and E. Teller [1949] "On the origin of elements", *Phys. Rev.* **75**, 1226–1231.

R. D. Richtmyer and E. Teller [1949] "On the origin of cosmic rays", *Phys. Rev.* **75**, 1729–1731.

E. Teller and J. L. Shoolery [2001] Memoirs: *A Twentieth-Century Journey in Science and Politics* (Perseus Publishing Co., Cambridge).

CHAPTER 3.6

JOHN ARCHIBALD WHEELER*

John Archibald Wheeler (1911–2008) was a great physicist and also a great teacher. He famously said, "We all know that the real reason universities have students is in order to educate the professors". During his tenure (1938–1976) in the Princeton University physics department, he broke all the records by supervising forty six graduate students and forty six undergraduate seniors. Besides that, he supervised several more at the University of North Carolina (1935–1938) and the University of Texas (1976–1986). He was a central figure in the international community of physicists for seventy years. He was a fiercely patriotic American who took pride in his contributions to nuclear weaponry, and he was a warm personal friend of physicists all over the world.

Complementarity is the philosophical principle which Niels Bohr invented to make sense of quantum mechanics. Complementarity says that nature is too rich to be described by a single picture. To describe a quantum process, we need two complementary pictures which cannot be seen simultaneously. The classic example of complementarity is the behavior of light, which can be seen in one experiment as a wave and in another experiment as a collection of particles. The wave and the particles cannot be seen together in one experiment, but they are both essential for defining the nature of light. Complementarity is firmly established by a wealth of experiments in the domain of atomic physics. But Bohr extended complementarity to a wider domain where its validity is still controversial. He wrote of complementarity between molecular and organic descriptions in biology, between practical usage and strict definition in linguistics, between justice and mercy in ethics, and between quantum and classical physics in descriptions of the world at large.

---

*Memoir for Proceedings of the American Philosophical Society.

Complementarity played two separate roles in Wheeler's life. In his technical activity as a theoretical physicist, complementarity was a working tool on which he could rely to get the right answers to practical questions. In his broader activities as a philosophical thinker, complementarity enabled him to combine opposites, to be a conservative revolutionary, a prosaic poet, a calculating dreamer. Wheeler's technical mastery of physics is best seen in the classic paper of Bohr and Wheeler, "The mechanism of nuclear fission", [1939], published on the day when Hitler's armies marched into Poland and World War II began. Bohr and Wheeler wrote the paper in Princeton, where Bohr was visiting in the spring of 1939, a few months after the discovery of fission. The paper is a masterpiece of clear thinking and lucid writing. It reveals, at the center of the mystery of fission, a tiny world where everything can be calculated and everything understood. The tiny world is a nucleus of uranium 236, formed when a neutron is freshly captured by a nucleus of uranium 235.

The uranium 236 nucleus sits precisely on the border between classical and quantum physics. Seen from the classical point of view, it is a liquid drop composed of a positively charged fluid. The electrostatic force that is trying to split it apart is balanced by the nuclear surface tension that is holding it together. The energy supplied by the captured neutron causes the drop to oscillate in various normal modes that can be calculated classically. Seen from the quantum point of view, the nucleus is a superposition of a variety of quantum states leading to different final outcomes. The final outcome may be a uranium 235 nucleus with a re-emitted neutron, or a uranium 236 nucleus with an emitted gamma-ray, or a pair of fission-fragment nuclei with one or more free neutrons. Bohr and Wheeler calculated the cross-section for fission of uranium 235 by a slow neutron and obtained the right answer within a factor of two. Their calculation is a marvelous demonstration of the power of classical mechanics and quantum mechanics working together. By studying this process in detail, they showed how the complementary views provided by classical and quantum pictures are both essential to the understanding of nature. Without the combined power of classical and quantum concepts, the intricacies of the fission process could never have been understood. Bohr's notion of complementarity was triumphantly vindicated.

Besides the theory of nuclear fission, Wheeler contributed many other important ideas to physics. Here is a very incomplete list:

1. Invention of a mathematical scheme called the scattering matrix or S-matrix, for describing reactions between particles and nuclei [1937]. This was so far ahead of its time that it had to be rediscovered five years later by Heisenberg, and Heisenberg's name is usually attached to it. It became a standard tool for analyzing experiments in particle physics.

2. With Richard Feynman, who was Wheeler's student at Princeton, the discovery of a new approach to electrodynamics [1940–1942], which has been immensely fruitful in the later development of particle theories.
3. Discovery of many new solutions of Einstein's gravitational equations [1955]. These solutions, which Wheeler called "geons", started the new science which became relativistic astrophysics.
4. The first suggestion that a spinning neutron star was the source of energy for the Crab nebula [1966]. If optical astronomers had taken this suggestion seriously, they would have discovered the first pulsar flickering in visible light, instead of leaving this discovery to be picked up later by the radio-astronomers.

More important than these individual discoveries was Wheeler's leadership in two major branches of physics that had fallen into neglect and stagnation, general relativity and the foundations of quantum mechanics. Both subjects, after their spectacular successes in the interwar period, were neglected after World War II because they had become purely theoretical, without any fresh input from experiments. Wheeler caused a rebirth of both of them by bringing them back into contact with experiments and observations. He promoted them vigorously, encouraging his numerous students to work on them, and writing review articles and books about them in his inimitable style. As a result of his efforts, both subjects became exciting and fashionable playgrounds for experimenters as well as theorists.

In general relativity, Wheeler's most important insight was to understand the importance of black holes. Before anyone else, he understood that black holes are not merely an odd theoretical consequence of Einstein's theory of gravitation, but they must actually exist and play a vital role in the evolution of the universe. He understood that black holes with masses of a few suns must exist in large numbers as collapsed remnants of massive stars, and black holes with masses of a few million suns may exist as collapsed remnants of nuclei of galaxies. We now know that both kinds of black hole are abundant and dominate many of the violent processes that we see in the sky. The history of the universe would make no sense without black holes. In the foundations of quantum mechanics, Wheeler called attention to the enormous gaps in our understanding and asked fundamental questions that still have to be answered. His questions stimulated a revival of theory and experiment in the phenomena of quantum entanglement. This revival led in turn to the birth of a new science and a new technology, the science of quantum information and the technology of quantum computing.

Throughout his life, Wheeler has oscillated between two styles of writing and thinking, prosaic and poetic. In the fission paper we see the prosaic Wheeler, a master craftsman, using the tools of orthodox physical theory to calculate quantities that can be compared with experiment. The prosaic Wheeler is temperamentally conservative, accepting the existing theories and using them with skill and imagination to solve practical problems. But from time to time we see a different Wheeler, the poetic Wheeler, who asks outrageous questions and takes nothing for granted. The poetic Wheeler writes papers and books with titles such as "Beyond the Black Hole", "Frontiers of Time" and "Law without Law". His message is a call for radical revolution. "As surely as we now know how tangible water forms out of invisible vapor, so surely we shall someday know how the universe comes into being. We will first understand how simple the universe is when we recognize how strange it is. The simplicity of that strangeness, Everest summit, so well directs the eye that the feet can afford to toil up and down many a wrong mountain valley, certain stage by stage to reach someday the goal. Of all strange features of the universe, none are stranger than these: time is transcended, laws are mutable, and observer-participancy matters".

But in science it is not enough to think revolutionary thoughts. If revolutionary thoughts are to be fruitful, they must be solidly grounded in practical experience and professional competence. The people who will lead us into the new world of physics must be conservative revolutionaries like Wheeler, at home in the prosaic world of practical calculation as well as in the poetic world of speculative dreams. The prosaic Wheeler and the poetic Wheeler are equally essential. They are the two complementary characters that together made up the John Wheeler that we knew and loved.

Reference

A convenient reference for all aspects of Wheeler's scientific life is the April 2009 issue of the magazine "Physics Today". This issue contains five historical articles by various authors, including Wheeler himself, with full references to the original documents.

CHAPTER 3.7

CHANDRASEKHAR'S ROLE IN TWENTIETH CENTURY SCIENCE*

1. Basic Science and Derived Science

In 1946, Subrahmanyan Chandrasekhar gave a talk here in Chicago with the title "The Scientist" [Chandrasekhar, 1987]. He was then 35 years old, less than half way through his life, less than a third of the way through his career as a scientist, but he was already reflecting deeply on the meaning and purpose of his work. His talk was one of a series of public lectures organized by Robert Hutchins, then the Chancellor of the University. The list of speakers is impressive, and includes Frank Lloyd Wright, John von Neumann, Arnold Schoenberg and Marc Chagall. That list proves two things. It shows that Hutchins was an impresario with remarkable powers of persuasion, and that he already recognized Chandra as a world-class artist whose medium happened to be theories of the universe rather than music or paint. I call him Chandra because that is the name his friends used for him when he was alive.

Chandra began his talk with a description of two kinds of scientific enquiry. "I want to draw your attention to one broad division of the physical sciences which has to be kept in mind, the division into a basic science and a derived science. You will notice that the division is not between a pure science and an applied science. I shall not be concerned with the latter, as I do not believe that the true values of science are to be found in the applications of science. I shall be concerned only with pure science, and it is the division of this into basic and derived science that I wish to draw to your attention. Broadly speaking, we may say that basic science seeks to analyze the ultimate constitution of matter and the basic concepts of space and

---

*Talk for Chandrasekhar Centennial Symposium, University of Chicago, October 16–17, 2010.

142

time. Derived science, on the other hand, is concerned with the rational ordering of the multifarious aspects of natural phenomena in terms of the basic concepts".

As examples of basic science, Chandra mentions the discovery of the atomic nucleus by Rutherford and the discovery of the neutron by Chadwick. These discoveries were each made by a simple experiment which revealed the existence of a basic building block of the universe. Rutherford discovered the nucleus by shooting alpha-particles at a thin gold foil, and observing that some of the alpha-particles bounced back. Chadwick discovered the neutron by shooting alpha-particles at a beryllium target, and observing that the resulting radiation collided with other nuclei in the way expected for a massive neutral twin of the proton. As examples of derived science, Chandra mentions the discovery by Edmund Halley in 1705 that the comet bearing his name had appeared periodically in the sky at least four times in recorded history with an elliptical orbit described by Newton's law of gravitation, and the discovery by William Herschel in 1803 that the orbits of binary stars are governed by the same law of gravitation operating beyond our solar system. The observations of Halley and Herschel did not reveal new building blocks, but vastly extended the range of phenomena that the basic science of Newton could explain.

Chandra was saying that the basic and the derived discoveries are equally essential to the progress of science. He said, "A very distinguished physicist, apparently feeling sorry for my preoccupation with things astronomical and with an intent to cheering me, said that I should really have been a physicist. To my mind this attitude represents a misunderstanding of the real values of science. And the history of science contradicts it. From the time of Newton to the beginning of this century, the whole science of dynamics and its derivative celestial mechanics have consisted entirely in amplification, in elaboration, and in working out the consequences of the laws of Newton. Halley, Laplace, Lagrange, Hamilton, Jacobi, Poincaré, all of them were content to spend a large part of their scientific efforts in doing exactly this, in the furtherance of a derived science. Indeed, there is a complementary relationship between the basic and the derived aspects of science. The basic concepts gain their validity in proportion to the extent of the domain of natural phenomena which can be analyzed in terms of them. Looked at in this way, science is a perpetual becoming, and it is in sharing its progress in common effort that the values of science are achieved".

Chandra then went on to describe the particular examples of basic and derived science that played the decisive role in his own intellectual development. In the year 1926, when Chandra was fifteen years old but already a physics student at Presidency College in Madras, Fermi and Dirac independently discovered the basic concept of Fermi–Dirac statistics. The basic concept says that, when we have a crowd of electrons distributed over a crowd of quantum states, each quantum state can be occupied by at most one electron, and the probability that a state is occupied is a simple function of the temperature. This basic property of electrons was a cornerstone of the new-born science of quantum mechanics. It solved one of the famous unsolved problems of condensed matter physics, explaining why the specific heats of solid materials decrease with temperature, going rapidly to zero as the temperature goes to zero.

Two years later, in 1928, the famous German professor Arnold Sommerfeld, one of the chief architects of quantum mechanics, came for a visit to Presidency College. Chandra was well prepared. Chandra had read and understood Sommerfeld's classic textbook, "Atomic Structure and Spectral Lines". He boldly introduced himself to Sommerfeld, and Sommerfeld took the time to tell him about the latest work of Fermi and Dirac. Sommerfeld gave this young student the galley-proofs of his paper on the electron theory of metals, a paper which was not yet published and gave the decisive confirmation of Fermi–Dirac statistics. Sommerfeld's paper was a masterpiece of derived science, showing how the basic concept of Fermi and Dirac could explain in detail why metals exist and how they behave. This young Indian student was one of the first people in the world to read it.

Two years after his meeting with Sommerfeld, at the ripe old age of nineteen, Chandra sailed on the steamship Lloyd Triestino to enrol as a graduate student at Cambridge University. He was to work in Cambridge with Ralph Fowler, who had used Fermi–Dirac statistics to explain the properties of white dwarf stars. White dwarfs are stars that have exhausted their supplies of nuclear energy by burning hydrogen to make helium or carbon and oxygen. They collapse gravitationally to a density many thousands of times greater than normal matter, and then slowly cool down by radiating away their residual heat. Fowler achieved another triumph of derived science by calculating the relationship between the density and the mass of a white dwarf star, using the concept of Fermi–Dirac statistics as the basis of the calculation. His calculation agreed well with the scanty observations of white dwarfs available at that time. With the examples of Sommerfeld and Fowler to encourage him, Chandra was sailing to England with the intention of making his own contribution to derived science.

On the ship, Chandra quickly found a way to move forward. The calculations of Sommerfeld and Fowler had assumed that the electrons were non-relativistic particles obeying the laws of Newtonian mechanics. This assumption was certainly valid for Sommerfeld. Electrons in metals at normal densities have velocities that are very small compared with the speed of light. But for Fowler the assumption of Newtonian mechanics was not so safe. Electrons in the central regions of white dwarf stars might be moving fast enough to make relativistic effects important. So Chandra spent his free time on the ship repeating Fowler's calculation of the behavior of a white dwarf star, with the electrons obeying the laws of Einstein's special relativity instead of the laws of Newton. Fowler had calculated a well-defined relationship between the mass and the density of a white dwarf, so that for any given mass there would be a unique central density. For a given chemical composition of the material, the density would be proportional to the square of the mass. This made sense from an intuitive point of view. The more massive the star, the stronger the force of gravity, and the more tightly the star would be squeezed together. The more massive stars would be smaller and fainter, which explained the fact that no white dwarf stars much more massive than the sun had been seen.

To his amazement, Chandra found that the change from Newton to Einstein made a drastic change in the behavior of white dwarf stars. The change from Newton to Einstein makes the matter in the stars more compressible, so that the density becomes greater for a star of given mass. The density does not merely increase faster as the mass increases, but tends to infinity as the mass reaches a finite value which we call the Chandrasekhar limit. With relativistic electrons, there is a unique model of a white dwarf star for every mass below the Chandrasekhar limit, and there are no models with mass greater than the limit. The limiting mass depends on the chemical composition of the star. For stars which have burned up all their hydrogen, the limit is about one and half times the mass of the sun.

Chandra finished his calculation before he reached England, and never had any doubt that his conclusion was correct. When he arrived in England and showed his results to Fowler, Fowler was friendly but unconvinced. Fowler was not willing to sponsor Chandra's paper for publication by the Royal Society in Britain. Chandra did not wait for Fowler's approval, but sent a brief version of the paper to the Astrophysical Journal in America [Chandrasekhar, 1931]. The Astrophysical Journal sent it to Carl Eckart, a famous geophysicist who did not know much about astronomy, to referee. Eckart accepted it, and it was published a year later. Chandra had a cool head. He had no wish to engage in public polemics with the British dignitaries who failed to understand his argument. He published his work quietly in a reputable astronomical journal, and then waited patiently for the next generation of astronomers to recognize its importance. Meanwhile he would remain on friendly terms with Fowler and the rest of the British academic establishment, and he would find other problems of derived science which his mastery of mathematics and physics would allow him to solve.

2. A Violent Universe

There were good reasons for the astronomers to react with skepticism to Chandra's statements in 1930. The implications of his discovery of the limiting mass were totally baffling. All over the sky, we see an abundance of stars cheerfully shining with masses greater than the limit. Chandra's calculation says that when these stars burn up their nuclear fuel, there exist no equilibrium states into which they can cool down. What then can a massive star do when it runs out of fuel? Chandra had no answer to this question, and neither did anyone else when he raised it in 1930.

The answer to Chandra's question was discovered in 1939 by Robert Oppenheimer and his student Hartland Snyder. Oppenheimer and Snyder published their answer in a paper, "On continued gravitational contraction", in the Physical Review [Oppenheimer and Snyder, 1939]. In my opinion this was Oppenheimer's most important contribution to science. Like Chandra's contribution nine years earlier, it was a masterpiece of derived science, taking the basic equations of Einstein and showing that they gave rise to startling and unexpected consequences in the real world of astronomy. The difference between Chandra and Oppenheimer was that Chandra started with Einstein's 1905 theory of special relativity while Oppenheimer started with Einstein's 1915 theory of general relativity. Oppenheimer in 1939 was one of the few physicists who took general relativity seriously. At that time it was an unfashionable subject, of interest mainly to philosophers and mathematicians. Oppenheimer knew how to use it as a working tool, to answer questions about real objects in the sky.

Oppenheimer and Snyder accepted Chandra's conclusion that there exists no static equilibrium state for a cold star with mass larger than the Chandra limit. Therefore the fate of a massive star at the end of its life must be dynamic and not static. They worked out the solution to the equations of general relativity for a massive star collapsing under its own weight. They discovered that there exists a dynamic solution of the equations with the star in a state of permanent free fall. Permanent free fall means that the star continues for ever to fall inward toward its center but never reaches the bottom. General relativity allows this paradoxical behavior because the time measured by an observer outside the star is running faster than the time measured by an observer inside the star. The time measured on the outside goes all the way from now to the end of the universe, while the time measured on the inside runs only for a few days. After the gravitational collapse, the inside observer will see the star still falling freely at high speed, while the outside observer sees it quickly slowing down. The state of permanent free fall is, so far as we know, the actual state of every massive object that has run out of fuel. We now know that such objects are abundant in the universe. We call them black holes.

With seventy years of hindsight, we can see that Chandra's discovery of the limiting mass and the Oppenheimer–Snyder discovery of permanent free fall were major turning-points in the history of science. During the 1930s, between the theoretical insights of Chandra and Oppenheimer, Fritz Zwicky with his systematic observations of supernova explosions was confirming the fact that we live in a violent universe [Zwicky, 1957]. During the same decade, Zwicky discovered the dark matter whose gravitation dominates the dynamics of large-scale structures.

After 1945, radio and X-ray telescopes revealed a universe full of shock-waves and high-temperature plasmas, with outbursts of extreme violence associated in one way or another with black holes.

Every child learning science in school, and every viewer watching popular scientific documentary programs on television, now know that we live in a violent universe. The violent universe has become a part of the prevailing culture. We know that an asteroid collided with the earth sixty-five million years ago and caused the extinction of the dinosaurs. We know that every heavy atom of silver or gold was cooked in the debris of a massive star at the end of its life and thrown out into space by a supernova explosion. We know that life survived on our planet for billions of years because we are living in a quiet corner of a quiet galaxy, far removed from the explosive violence that we see all around us in more turbulent parts of the universe. Astronomy has changed its character totally during the last hundred years. A hundred years ago, the main theme of astronomy was to explore a quiet and unchanging landscape. Today the main theme is to observe and explain the celestial fireworks that are the evidence of violent change. This radical transformation in our picture of the universe began on the good ship Lloyd Triestino when the nineteen-year-old Chandra discovered that there can be no stable equilibrium state for a massive star.

It always seemed strange to me that the work of the three main pioneers of the violent universe, Chandra, Oppenheimer and Zwicky, received so little recognition and acclaim at the time when they were doing it. The neglect of their discoveries was due in part to the fact that all three of them came from outside the astronomical profession. The professional astronomers of the 1930s were conservative in their view of the universe and in their social organization. They saw the universe as a peaceful domain which they knew how to explore with the standard tools of their trade. They were not inclined to take seriously the claims of outsiders with new ideas and new tools. It was easy for the astronomers to ignore the outsiders, because the new discoveries did not fit into the accepted ways of thinking, and the discoverers did not fit into the established astronomical community.

In addition to these reasons for neglect which applied to all three of the pioneers, there were special circumstances which existed for each of them separately. In the case of Chandra, the special circumstances were the personalities of Eddington and Milne, who were the leading astronomers in England when Chandra arrived from India. Eddington and Milne had private theories of stellar structure in which they firmly believed, both inconsistent with Chandra's calculation of a limiting mass. Both of them promptly decided that Chandra's calculation was wrong, and never accepted the physical facts on which the calculation was based. In the case of Zwicky, an even worse situation existed at the California Institute of Technology, where the astronomy department was dominated by Hubble and Baade. Zwicky was doing the first successful photographic sky survey, using a small wide-field camera that could cover the sky a hundred times faster than other telescopes existing at that time. Hubble and Baade were using bigger telescopes with far narrower fields of view, allowing them to see a more limited part of the universe. Zwicky then made an enemy of Baade by accusing him of being a Nazi. As a result, Zwicky's discoveries were largely ignored for the next twenty years. In the case of Oppenheimer, the neglect of his greatest contribution to science was mostly due to an accident of history. His paper with Snyder, establishing in four pages the physical reality of black holes, was published in the Physical Review on September 1, 1939, the same day on which Hitler sent his armies into Poland and began World War II. In addition to the distraction created by Hitler, the same issue of the Physical Review contained the monumental paper by Bohr and Wheeler on the theory of nuclear fission, spelling out, for all who could read between the lines, the possibilities of nuclear power and nuclear weapons [Bohr and Wheeler, 1939]. It is not surprising that the understanding of black holes was pushed aside by the more urgent excitements of war and nuclear energy.

So it happened that each of the three revolutionary pioneers, after a brief period of revolutionary discovery and a brief publication, lost interest in fighting for the revolution. Chandra enjoyed seven peaceful years in Europe, mostly working on the theory of normal stars without revolutionary implications, before moving to America. Zwicky, after finishing the sky survey which revealed the existence of dark matter and several types of supernovae, turned his attention to military problems as World War II was beginning, and became an expert in rocketry. Oppenheimer, after discovering the most important astronomical consequence of general relativity, turned his attention to mundane nuclear explosions and became the director of the Los Alamos laboratory. When I tried in later years to start a conversation with Oppenheimer about the importance of black holes in the evolution of the universe, he was as unwilling to talk about black holes as he was to talk about his work at Los Alamos. Oppenheimer suffered from an extreme form of the prejudice which is prevalent among theoretical physicists, over-valuing pure science and under-valuing derived science. For Oppenheimer, the only activity worthy of the talents of a first-rate scientist was the search for new laws of nature. The study of the consequences of old laws was an activity for graduate students or third-rate hacks. He had no desire to return in later years to the study of black holes, the area in which he had made his most important contribution to science and might have continued to make important contributions in the 1950s. In the 1950s, black holes were an unfashionable subject, and Oppenheimer preferred to follow the latest fashion. Oppenheimer and Zwicky did not, like Chandra, live long enough to see their revolutionary ideas adopted by a younger generation and absorbed into the mainstream of astronomy.

Chandra had to wait fifty three years after his discovery of the limiting mass before his work was recognized by the Nobel Prize committee in Stockholm. In his speech at the banquet in Stockholm, he quoted a passage from Virginia Woolf's novel, "The Waves", which he also used as the epigraph at the beginning of his book, "Ellipsoidal Figures of Equilibrium" [Chandrasekhar, 1968]: "There is a square; there is an oblong. The players take the square and place it upon the oblong. They place it very accurately; they make a perfect dwelling place. Very little is left outside. The structure is now visible; what was inchoate is here stated; we are not so various or so mean; we have made oblongs and stood them upon squares. This is our triumph; this is our consolation". Virginia Woolf expresses in simple words Chandra's unique quality as a scientist. He chooses problems to study without concern for their practical importance. He chooses them for their intrinsic beauty. He devotes to every problem, important or unimportant, the same intensity of creative thought. After each problem is solved, the solution is written down and published with the same meticulous attention to detail. His books are masterpieces of literary style as well as scientific substance. He is, as Robert Hutchins recognized, an artist on the same level as Frank Lloyd Wright and Marc Chagall.

3. The Polarization of Light in the Daytime Sky

I met Chandra for the first time in the summer of 1950 at the University of Michigan. In those days, the University of Michigan continued the tradition which was established in the 1920s, inviting leading physicists from all over the world to take part in a leisurely Summer School, giving lectures and exchanging ideas. Since many of the speakers traveled by ship from Europe, it made sense for them to stay for a month or longer. It was a great opportunity for students like me to listen to the lectures and get to know the famous lecturers. Chandra gave a lecture on the polarization of sunlight in the daytime sky. This was a classical example of derived science, applying to our familiar blue sky the theory that he had worked out in earlier years for the accurate understanding of radiation transport in stellar atmospheres. The atmosphere of a star has much greater optical depth than the atmosphere of the earth, but even in the earth's atmosphere the multiple scattering of sunlight is important. Chandra took delight in calculating the multiple scattering accurately enough to explain the precise measurements of polarization made by observers in the nineteenth century.

Chandra's lecture was a masterpiece of presentation. He took an unimportant episode in the history of science and converted it into a vivid drama. There are in the daytime sky three points where the polarization is zero. The three points were discovered in turn by three famous scientists who gave them their names, the Babinet point, the Brewster point and the Arago point. Brewster was British and the other two were French. Arago was not only a famous astronomer but a brilliant popularizer of science. The Arago point was the last to be discovered and the most difficult for Chandra's theory to explain. If the calculation is not done precisely right, the Arago point will disappear or move to a completely wrong place in the sky. Chandra's talk filled me with the same kind of joy as the lectures which I had heard as a student in Cambridge from the pure mathematician Hardy.

Chandra had known Hardy well during the years he spent as a student in Cambridge, and found him a kindred spirit. Chandra and Hardy had similar views about British imperialism and Indian politics, and they had similar tastes in science. Hardy lectured about the summability of divergent series with the same passion that Chandra devoted to the polarization of daylight. The high point of Hardy's story was the construction by Andrei Kolmogorov of a Fourier series that is almost everywhere divergent. The high point of Chandra's story is the construction by George Stokes of a matrix algebra describing precisely the statistics of elliptically polarized light. Both of them reminded me of Wanda Landowska, a great musician who used to play Bach Fugues on a harpsichord. The polarization of light in the daytime sky is like a Bach fugue, a work of art with deep resonances in a small space.

Another link between Chandra and Hardy was the Indian mathematician Srinivasa Ramanujan. Ramanujan was a mathematical genius who grew up in south India, not far from Chandra's home in Madras but twenty-three years earlier. Ramanujan was invited by Hardy to work with him in Cambridge, and spent the five years 1914 to 1919 in England, becoming a Fellow of Trinity College, Cambridge and a Fellow of the Royal Society, the first Indian to receive either of these honors. His mathematical collaboration with Hardy was the high point in both of their lives. Chandra was nine years old when Ramanujan returned in triumph to India, and the stories that Chandra heard about Ramanujan encouraged Chandra to become a mathematician. Chandra did not know then that Ramanujan was mortally ill with aboebic hepatitis and would die only a year later.

In 1936, after Chandra and Hardy had become friends, Hardy was publishing a book about Ramanujan and needed a good picture of him for the frontispiece [Hardy, 1940b]. Hardy said that the only available picture showed him in academic cap and gown, and made him look ridiculous. Hardy asked Chandra if he could find a better picture the next time he was traveling home to India. Chandra took the trouble to find out where Ramanujan's widow was living, in great poverty and complete obscurity, and he invited her to visit with his family. She still had Ramanujan's passport in her possession, with a much better picture which Hardy used for his book. Just as Hardy always remained proud of having discovered Ramanujan, Chandra remained proud of having discovered Ramanujan's picture. Chandra also made successful efforts to organize an adequate pension for Ramanujan's widow, so that she could live the rest of her life independently and be treated with the respect that was her due.

Hardy and Chandra shared an intense belief in mathematical beauty as the essential quality of good science. That was what attracted both of them to Ramanujan. Both of them wrote eloquently about beauty and about ugliness. Hardy in his little book, "A Mathematician's Apology" [Hardy, 1940a], gave the classic definition of usefulness in science: "A science is said to be useful if its development tends to accentuate the existing inequalities in the distribution of wealth, or more directly promotes the destruction of human life". In Hardy's mind, usefulness was almost synonymous with ugliness. Hardy and Ramanujan published jointly in 1918 a paper, "Asymptotic formulae in combinatory analysis", a monument to Ramanujan's genius and Hardy's good taste, describing their most elegant discovery in the theory of numbers [Hardy and Ramanujan, 1918]. The discovery reveals a deep connection between the most elementary additive properties of ordinary integers and the abstract symmetries of the modular group in the theory of elliptic functions. The paper is written in a literary style worthy of Virginia Woolf. This classic work of Hardy and Ramanujan is another example of derived science according to Chandra's definition. Hardy and Ramanujan did not discover a new mathematical concept or a new law of nature. They discovered a deep connection between old concepts, and the new connection extended the reach of our understanding in many directions.

Throughout his long life, Chandra devoted himself to many different areas of science, and found structures of unexpected mathematical beauty wherever he looked. The polarization of sunlight in the daylight sky, the subject that I heard him describe with such joy at the Michigan Summer School, was one of the smaller items in his cabinet of natural wonders. He was then forty years old, half way through his career. When he was younger, he wrote about natural wonders in the dry professional style customary for articles in scientific journals. Only toward the end of his life, like Hardy, he wrote in a more personal style about the quest for beauty. In 1975 when he was in his sixties, he gave a lecture here in Chicago with the title, "Shakespeare, Newton and Beethoven, or Patterns of Creativity", describing his various encounters with beauty in art and science [Chandrasekhar, 1987]. He took advantage of his status as an honored sage to spread his wisdom over a wide field. At the end of the lecture, he would come back to the unique scientific object that is to him the supreme example of natural beauty: "In my entire scientific life, extending over forty-five years, the most shattering experience has been the realization that an exact solution of Einstein's equations of general relativity, discovered by the New Zealand mathematician Roy Kerr, provides the absolutely exact representation of untold numbers of massive black holes that populate the universe. This shuddering before the beautiful, this incredible fact that a discovery motivated by a search after the beautiful in mathematics should find its exact replica in Nature, persuades me to say that beauty is that to which the human mind responds at its deepest and most profound".

4. Ellipsoids and Black Holes

The pattern of Chandra's life as a scholar was to spend five or ten years on each field of science that he wished to study in depth. He would take a year to master the subject, a few more years to publish a series of journal articles demolishing the problems that he could solve, and then a few more years writing a definitive book, surveying the subject as he left it for his successors. After each book was finished, he left that field alone and looked for the next field to study. This pattern was repeated eight times and was recorded in the dates and titles of his eight books. "An Introduction to the Study of Stellar Structure", 1939 [Chandrasekhar, 1939], summarizes his work on the internal structure of white dwarfs and other types of stars. "Principles of Stellar Dynamics", 1943 [Chandrasekhar, 1943], describes his highly original work on the statistical theory of stellar motions in clusters and in galaxies. "Radiative Transfer", 1950 [Chandrasekhar, 1950], provides the first accurate theory of radiation transport in stellar atmospheres. This is the theory that he applied to the earth's atmosphere in his Michigan lecture. "Hydrodynamic and Hydromagnetic Stability", 1961 [Chandrasekhar, 1961], provided a foundation for the theory of all kinds of astronomical objects, including stars and accretion disks and galaxies, which may become unstable as a result of differential rotation. "Ellipsoidal Figures of Equilibrium", 1968 [Chandrasekhar, 1968], goes back to an old problem, to discover all the possible equilibrium configurations of an incompressible liquid mass rotating in its own gravitational field. This problem was studied by the great mathematicians of the nineteenth century, Jacobi, Dedekind, Dirichlet and Riemann, who were unable to determine which of the various configurations were stable. In the introduction to his book, Chandra remarks, "These questions were to remain unanswered for more than a hundred years. The reason for this total neglect must in part be attributed to a spectacular discovery by Poincaré, which channeled all subsequent investigations along directions which appeared rich with possibilities; but the long quest it entailed turned out in the end to be after a chimera".

After the ellipsoidal figures of equilibrium came a gap of fifteen years before the next book, "The Mathematical Theory of Black Holes", appeared in 1983 [Chandrasekhar, 1983]. These fifteen years were the time when Chandra worked hardest and most intensively on the subject closest to his heart, the precise mathematical description of black holes and their interactions with surrounding fields and particles. His book on black holes was his farewell to technical research, just as Shakespeare's "The Tempest" was his farewell to writing plays. After the black hole book was published, Chandra lectured and wrote about non-technical themes, about the works of Shakespeare, Beethoven and Shelley, and about the relationship between art and science. A collection of his lectures for the general public was published in 1987 with the title, "Truth and Beauty" [Chandrasekhar, 1987]. Finally, during the years of his retirement he spent much of his time working his way through the "Principia Mathematica" of Newton, reconstructing every proposition and every demonstration, translating the geometrical arguments of Newton into the algebraic language familiar to modern scientists. The results of this historical research were published in his last book, "Newton's Principia for the Modern Reader" [Chandrasekhar, 1995], in 1995, shortly before his death. To explain why he wrote this book, he said: "I am convinced that one's knowledge of the Physical Sciences is incomplete without a study of the Principia in the same way that one's knowledge of literature is incomplete without a knowledge of Shakespeare".

To summarize my assessment of Chandra's role in twentieth-century science, I would say that he made two transcendent contributions separated by fifty years. At the age of nineteen he started a revolution in our understanding of the universe by discovering the mass limit for permanently stable stars. This discovery implied, as an inevitable consequence, that all objects with mass greater than the limit must get rid of their excess mass by one means or another. As a result, the universe at large is no longer the quiet domain imagined by Aristotle but a much more dynamic realm of violent instabilities and cosmic cataclysms. At the age of seventy two, Chandra published the book that permanently established the true picture of the final state of massive objects. The true picture is the Kerr solution of the equations of general relativity describing rotating black holes. These are the objects that sit at the centers of galaxies, most of the time quiescent, but from time to time causing the titanic eruptions of energy that we call quasars and gamma-ray bursts, visible to our instruments from all over the universe, nine-tenths of the way back to the beginning of time.

Chandra's work on black holes was the most dramatic example of his commitment to derived science as a tool for understanding nature. Our basic understanding of the nature of space and time rests on two foundations, first the equations of general relativity discovered by Einstein, and second the black hole solutions of these equations discovered by Schwarzschild and Kerr and explored in depth by Chandra. To write down the basic equations is a big step toward understanding, but it is not enough. To reach a real understanding of space and time, it is necessary to construct solutions of the equations and to explore all their unexpected consequences. In the end, Chandra understood more about space and time than Einstein.

References

N. Bohr and J. A. Wheeler [1939] "The mechanism of nuclear fission", *Phys. Rev.* **56**, 426–450.

S. Chandrasekhar [1931] "The maximum mass of ideal white dwarfs", *Astrophys. J.* **74**, 81–82.

S. Chandrasekhar [1939] *An Introduction to the Study of Stellar Structure* (University of Chicago Press, Chicago).

S. Chandrasekhar [1943] *Principles of Stellar Dynamics* (University of Chicago Press, Chicago).

S. Chandrasekhar [1950] *Radiative Transfer* (Clarendon Press, Oxford).

S. Chandrasekhar [1961] *Hydrodynamic and Hydromagnetic Stability* (Clarendon Press, Oxford).

S. Chandrasekhar [1968] *Ellipsoidal Figures of Equilibrium* (Yale University Press, New Haven).

S. Chandrasekhar [1983] *The Mathematical Theory of Black Holes* (Clarendon Press, Oxford).

S. Chandrasekhar [1987] *Truth and Beauty: Aesthetics and Motivations in Science* (University of Chicago Press, Chicago).

S. Chandrasekhar [1995] *Newton's Principia for the Common Reader* (Clarendon Press, Oxford).

G. H. Hardy [1940a] *A Mathematician's Apology* (Cambridge University Press, Cambridge).

G. H. Hardy [1940b] *Ramanujan: Twelve Lectures Suggested by his Life and Work* (Cambridge University Press, Cambridge).

G. H. Hardy and S. Ramanujan [1918] "Asymptotic formulae in combinatory analysis", *Proc. Lond. Math. Soc.* **2**, 75–115.

J. R. Oppenheimer and H. Snyder [1939] "On continued gravitational contraction", *Phys. Rev.* **56**, 455–459.

F. Zwicky [1957] *Morphological Astronomy* (Springer-Verlag, Berlin), pp. 27–30.

CHAPTER 3.8

NICHOLAS KEMMER[*]

Nicholas Kemmer was a theoretical physicist whose most famous contribution to science was the prediction in 1938 of the existence of three kinds of particle, one positive, one negative and one neutral, coupled to protons and neutrons in a symmetrical way so as to produce nuclear forces independent of charge. Three particle species were discovered experimentally ten years later and found to have the nuclear couplings specified by Kemmer. They are the particles now known as pi-mesons or pions. Other sets of three species with the same symmetry were discovered later. The long interval between prediction and verification was caused by World War II, which interrupted the progress of particle physics in general and Kemmer's career in particular. After the war, he devoted his life to teaching rather than research, and became a beloved mentor and friend to several generations of younger physicists. Among the scientists that he launched into successful research careers are Abdus Salam (FRS, Nobel Prize), Paul Matthews FRS, Richard Dalitz FRS, and the present writer. In the American Mathematical Society Mathematical Genealogy Project, which includes physicists as well as mathematicians, Kemmer is academic ancestor to two hundred and seventeen descendants.

In writing this brief account of Kemmer's life and work, I have been greatly helped by the unpublished autobiography which he wrote for the Royal Society. I am indebted to Keith Moore for a copy of the autobiography. I am also indebted to Peter Higgs FRS for contributing a description of Kemmer's adventures as an administrator at Edinburgh University.

---

[*]For Biographical Memoirs of Fellows of the Royal Society, 7 December 1911–21 October 1998, Elected FRS 1956.

1. Life

I divide the memoir into two parts, the life and the work. Kemmer's life began in St. Petersburg in 1911, where he was born into a Russian family belonging to the Baltic German minority of the Russian Empire. Baltic Germans frequently occupied high positions in the Imperial government. Kemmer's father was trained as an engineer and rose rapidly to a high rank in the Ministry of Ways and Communications. In 1916, the blackest year of World War I for both Britain and Russia, Kemmer's father traveled with his family from St. Petersburg to London to purchase British railway rolling stock for the disintegrating Russian army. Luckily for the family and for six-year-old Nicholas, they were safe in London when the Russian revolutions occurred a year later. They stayed in London for five years. Nicholas acquired his interest in science as I did 15 years later, by hanging around the Science Museum and the Natural History Museum in South Kensington. His prep school and my family home were both on Queen's Gate, within easy walking distance from the museums. Besides a taste for science, Nicholas also picked up a liking for England and an accent-free command of our language.

After the Bolshevik victory in Russia, the Kemmers were officially stateless and unable to obtain British nationality. To survive economically, they moved in 1921 to Germany, where Nicholas's father was employed by the British Westinghouse Brake Company. They took German citizenship, and Nicholas switched to his third language for his education. But then another disaster struck. Nicholas's father was a victim of the terrible sleeping-sickness epidemic that ravaged Europe in the early 1920s. He recovered from the encephalitis but became a permanent invalid, with body and mind deteriorating from year to year. This was the same post-encephalitic Parkinsonism that Oliver Sacks later described poignantly in his book "Awakenings". For Nicholas's father there was no awakening. He lived for twenty years with a small pension to support his family, and finally ended his days in a nursing home in Switzerland. Meanwhile Nicholas made his way through the German school system. He concentrated his attention on science, and avoided discussions of literature and history. He concealed his distaste for the fanatical nationalism that was taught in the German schools and embraced by many of his schoolmates. In 1930, he enrolled as a physics student at the University of Göttingen.

To escape the rising tide of Hitler, Kemmer's mother moved from Germany to Switzerland in 1931 and from Switzerland to Spain in 1936. But Kemmer needed to earn a living, and the only place where he had citizenship and could legally work was Germany. He stayed in Germany until 1933, when his teachers in Göttingen, Max Born, Richard Courant and Edmund Landau, were dismissed from their jobs, and the greatest concentration of mathematicians in the world was destroyed. Max Born escaped to Edinburgh, Richard Courant to New York,

Edmund Landau to Cambridge. Kemmer could see that there was no future for him in Germany, and so he joined his mother in Switzerland. There he could not legally work, but he could live and study. He enrolled at the University of Zürich with Gregor Wentzel as his advisor. Within two years he completed a thesis [Kemmer, 1935] and became a Doctor of Philosophy.

Although Kemmer was a student at the University, he quickly established friendly relations with Wolfgang Pauli who was professor at the ETH (Federal Institute of Technology). The ETH was within easy walking distance from the University, and Pauli was a livelier character than Wentzel. Pauli was notoriously rude to people whom he considered to be fools, but he took a liking to Kemmer and treated him with respect. In 1936, Kemmer was in desperate need of a job, and Pauli came to his rescue twice. First Pauli offered him a job as his own assistant. This was a coveted position, given only to people that Pauli considered exceptionally capable. But according to Swiss law, the position could only be given to a foreigner if no equally qualified Swiss candidate was available. The ETH administration told Pauli that a Swiss candidate, Guido Ludwig, was available, and therefore Kemmer was not eligible. Pauli fought hard for Kemmer, but the best deal he could negotiate was to appoint both Kemmer and Ludwig as assistants with the salary divided between them. Kemmer and Ludwig remained friends and published a paper together (4). Meanwhile Kemmer told Pauli that he could not live for long on the half-salary and would have to look for another job. It happened that a research fellowship at Imperial College in London was available, so Kemmer applied for it. He asked Pauli for a letter of recommendation, and Pauli came to his rescue a second time. According to Kemmer, the following conversation ensued, [Enz and von Meyenn, 1988, page 92]. Pauli: "Now, Herr Kemmer, I got this letter from London. Your English is much better than mine. What should I say, 'one of the more promising' or 'one of the most promising'?". Kemmer: "Well, Herr Professor, the latter is much stronger". Pauli: "Oh good, that's the way I'll write it". So Kemmer got the fellowship and moved to London in October 1936. He always said that Pauli, in spite of his reputation, was not cruel.

For Kemmer, the move to London felt like a homecoming. By a happy chance, he was working in the same area of London where he had roamed as a child. He stayed at Imperial College for three years, two years as a Beit Research Fellow and a third year as demonstrator in the mathematics department. The head of the mathematics department was Sydney Chapman, a world-class geophysicist whose primary interest was the aurora borealis. Chapman was famous for bicycling to work every day through the streets of London, and for leading expeditions to Alaska to observe the aurora. He did not pretend to understand the details of Kemmer's research, but welcomed him as a friend and colleague. The London years were the time of Kemmer's peak activity as a research scientist. He remained in close touch with Wentzel and Pauli, and his most important work was done in response to a suggestion of Pauli. Kemmer was still a German citizen when war broke out in 1939. He became an enemy alien and had to appear before a tribunal which would decide whether he should be interned. Chapman went with him to the tribunal and testified that he was harmless and might be useful to the country. As a result, Kemmer was never interned.

In 1940, the British government ordered Kemmer to move to Cambridge. Although he was still an enemy alien, he was directed to join the French atomic energy project led by Hans von Halban and Lew Kowarski. The project had started in Paris under the leadership of Frederic Joliot. When France was over-run in 1940, Joliot stayed in Paris, but von Halban and Kowarski escaped to England with 180 litres of heavy water, at that time almost the whole of the world supply. The French project continued in exile in Cambridge, allowed by the British authorities to share information about nuclear power but not about nuclear weapons. Kemmer stayed with the project for four years in Cambridge and for a final year in Montreal. He never became deeply involved in the engineering work of the project, either in Cambridge or in Canada. His official title was information officer, responsible for keeping track of reports and documents in a bureaucratic organization with arcane rules of secrecy. The project was divided into two factions, one led by von Halban and the other by Kowarsky, who barely communicated with each other. Kemmer was one of the few people who was trusted with information from both sides.

During the wartime years in Cambridge, the physics teaching staff at the University had mostly disappeared into war projects, but there were still students to be taught. Kemmer was glad to take time off from his duties at the atomic energy project, to teach regular courses for undergraduates at the Cavendish Laboratory. He acquired a taste for teaching, and became known to the university authorities as an exceptionally conscientious teacher. Before moving to Montreal, he had become a British subject. When the war ended, he was invited by Cambridge University to return to Cambridge as a University Lecturer, with a Staff Fellowship at Trinity College to make him a full member of the Cambridge academic community. For the first time in his life, he had a regular job and a regular home.

In 1946, I occupied Kemmer's old position as demonstrator in the mathematics department of Imperial College, with Sydney Chapman still in charge. Chapman was as helpful to me as he had been to Kemmer. He saved me from a year of post-war military service by certifying that my job in his department was essential to the nation. I was then a pure mathematician, working on problems in number theory. I told Chapman that I was intending to switch from number theory to physics, and that I hoped to find a mentor who could help me to get started as a physicist. Chapman told me that he knew exactly the right person. The name of the person was Nicholas Kemmer. He had done the same job at Imperial College that I was doing seven years later, he was an expert in theoretical particle physics, and he was just then returning from Canada to Cambridge.

Following Chapman's advice, I contacted Kemmer as soon as I could. I quickly found that he was the teacher I needed. He became a friend as well as a teacher. Our friendship remained alive and well for fifty-two years until his death. In his first year as lecturer, he gave two courses of advanced lectures, one on nuclear physics and one on quantum field theory. Both courses were full of new information, not available anywhere else in England at that time. During his five years as house theoretician to the French atomic energy project, Kemmer had mastered the newly discovered facts and theories of nuclear physics, many of which had not yet been published. He gave the students a thorough understanding of the subject, which I put to good use ten years later when I helped to design a commercial nuclear reactor.

Kemmer's quantum field theory course was like his nuclear physics course, a comprehensive and up-to-date source of information about new ideas. Nothing like it existed anywhere else in England in 1946. Quantum field theory was not a single coherent theory of elementary particles, but a collection of alternative mathematical models, each attempting to combine the ideas of classical relativity with quantum mechanics. The various models were suggested by various theoretical ideas and various fragments of experimental data. Each of them was mathematically incomplete and often inconsistent. Only a few experts knew enough about the mathematical formalisms and the experimental facts to be able to give a comprehensive and comprehensible account of the subject. Kemmer was one of the few. His course was a distillation of the wisdom of continental Europe, at that time still little known in England and America. Quantum field theory had been invented in Europe by Heisenberg, Fermi and Dirac. Although Dirac was English, the theory was for a long time more highly regarded in continental Europe than in England and America. The great experimenter Rutherford expressed the common English view of abstract theories: "The theorists play games with their symbols, but we at the Cavendish Laboratory turn out the real facts of nature".

In 1946, the only existing textbook on quantum field theory was the book [Wentzel, 1943], by Kemmer's teacher Gregor Wentzel, written in Zürich and published in 1943 in Vienna in the middle of the war. Kemmer possessed a copy of Wentzel's book and allowed me to borrow it. It was at that time a treasure without price. I believe there were then only two copies in England. It was later reprinted in America and translated into English. In 1946, few people in England knew of its existence and even fewer considered it important. Kemmer not only lent it to me but also explained why it was important. When I arrived in America as a graduate student one year later, I found myself, thanks to Kemmer, the only person in the Cornell physics department who knew about quantum field theory.

Kemmer's job at Cambridge gave him two heavy burdens. As a University lecturer he had a load of classroom teaching, and as a Trinity College Staff Fellow he had a large number of undergraduates to supervise. I was appalled to see the long hours he had to spend supervising students every day. There was nobody else who supervised students as conscientiously as he did, and so they came to him in big numbers. He never complained, but he paid a heavy price for being such an excellent supervisor. He had no time left over to resume the research career so brilliantly begun ten years before. Kemmer and I were both living in Trinity College. He was a Staff Fellow and was treated by the college as a drudge, while I was a Research Fellow with no duties and complete freedom to do whatever I liked. This was a monstrously unfair division of labour, but Kemmer accepted it without any sign of resentment. He was as generous in spending time with me as he was with his students. He always had time to advise me, to explain the difficult points in Wentzel's book, and to share with me his vision of quantum field theory as the key to a consistent mathematical description of nature. He was the most unselfish scientist I ever knew.

In the spring vacation of 1947, the young physicists of Cambridge scraped together enough money to rent a chalet at Arosa in Switzerland for three weeks. The Swiss authorities at that time were desperate to revive their tourist industry after six years of war. The Germans who were in the past their best customers were down and out, and so the Swiss offered enticing bargains to the British. The British pound was then a non-convertible and internationally worthless currency, but the Swiss changed our pounds into francs at a generous rate of exchange. A dozen of us traveled to Arosa for three weeks of luxurious living and splendid skiing. By a freak of nature, a dust storm from the Sahara had drifted over Switzerland that winter and colored the snow high up on the mountains with bright streaks of orange and pink. Our party included three physicists who later became famous, Nicholas Kemmer, Tommy Gold and Herman Bondi, together with their girl-friends Margaret Wragg, Merle Tuberg and Christine Stockman. The purpose of the trip was not to discuss physics but to celebrate the rites of Spring. The celebration resulted in three weddings soon after we returned to Cambridge. All of us who had shared our lives with Nicholas and Margaret at Arosa had hoped that this would happen. Margaret had never been on skis before, and Nicholas was constantly with her on the slopes giving her help and encouragement. They obviously cared more for each other than for themselves. They were married in the summer of 1947 and remained inseparable until the death of Nicholas fifty-one years later. Together they raised two sons and a daughter, and enjoyed watching eight grandchildren grow up.

During his years as a stateless person and later as an enemy alien in Britain, Kemmer suffered more than his fair share of inconveniences caused by inflexible immigration laws. In 1942, when he became a British subject, he supposed that chapter of his life to be over. But worse was still to come. In 1951, he was invited to be a visiting member at the Institute for Advanced Study in Princeton. He looked forward to spending a year at the Institute, which would have given him a good chance of getting back into productive research. He could not accept the invitation because the US government refused to give him a visa. At that time the US visa authorities were at their most capricious. Kemmer could never find out why his visa was denied. He conjectured that the denial resulted from the fact that he had been a friend of Alan Nunn May, a member of the Montreal atomic energy project who was arrested and convicted of spying for the Soviet Union. In fact, Nunn May had never been working on nuclear weapons, and the information that he gave to the Russians was militarily harmless. Kemmer was known to be unfriendly to the Soviet government, which had declared his father to be a traitor and robbed him of his property. But in the US visa department, logic did not prevail. The denial of a visa in 1951, when Kemmer was young enough to learn as well as to teach, was damaging both to him and to the Institute for Advanced Study. Many years later, he obtained a visa without difficulty and visited America several times.

After seven years of intensive teaching at Cambridge, Kemmer moved to Edinburgh in 1953 to succeed Max Born as Tait Professor of Natural Philosophy. Max Born had always felt isolated in Edinburgh and returned to Germany as soon as he retired. Kemmer made sure he would not be so isolated. He continued to live gregariously, making friends with a heterogeneous crowd of students and colleagues, as he had done at Cambridge. He took pride in the fact that students came to Edinburgh from a wide variety of countries. He served as professor until his retirement in 1979, and then stayed in Edinburgh as Professor Emeritus for the rest of his life. During his years as professor, he devoted much of his time to teaching and mentoring students, and also learned to be an effective administrator. He attracted many capable young scientists to Edinburgh, among them Peter Higgs FRS, who became a close friend and colleague. Higgs first came to Edinburgh in 1954 to spend two years with Kemmer as a post-doctoral student. He returned to Edinburgh permanently as a Lecturer in 1960, became Professor of Theoretical Physics in 1980, and remained there thirty years later. Higgs and Kemmer had similar interests, not only in theoretical physics but also in politics. Both of them were for a while active participants in the Campaign for Nuclear Disarmament.

[The following paragraphs were contributed by Peter Higgs in a personal letter to Dyson]. The status of theoretical physics at Edinburgh University was one of Nick's major preoccupations between 1953 and 1970. When he was appointed in 1953, the theoretical physics group occupied part of the basement of the Natural Philosophy building, but they were not considered to be real physicists by the occupant of the ancient (1583) Chair of Natural Philosophy, Norman Feather (one of Rutherford's disciples). Feather's dictum was "Physics is an experimental subject: there is no such thing as theoretical physics". On his arrival in Edinburgh, Feather had decreed that Max Born and his two colleagues were not suitable people to teach physics students.

Nick's first move was to set up a B.Sc. Honours degree in Mathematical Physics. Meanwhile, a University-owned building in a neighbouring street had become vacant, and in 1955 we moved into it, with the title "Tait Institute of Mathematical Physics". As a tidying up operation, in 1966, Nick persuaded the Senatus to change the title of the Tait Chair from Natural Philosophy to Mathematical Physics. But his long-term aim was to become part of Physics.

In the late 1960s, the British universities were undergoing the Robbins expansion, so new chairs were being created. By 1970, Feather had two colleagues designated Professors of Physics, and the University Grants Committee was pressing him to accept in his department another professorial appointment, preferably a theoretical physicist. Out of the blue, during a discussion of this he turned to Kemmer and said (roughly) "It doesn't make sense to have a theoretical physicist in my department while there is a separate Tait Institute of Mathematical Physics across the street. Why don't we unify our departments to accommodate this new chair?" Nick was quite shattered, because he had been advocating this for years without any success.

The Department of Physics came into existence (replacing Natural Philosophy and Mathematical Physics) in 1971 when the first phase of the James Clerk Maxwell Building was ready for us. The new chair, which had triggered the unification, was never filled, and disappeared during a subsequent funding squeeze. [End of Higgs contribution].

In 1960, Kemmer took the lead in founding and organizing the Scottish Universities Summer School in Physics. This is an annual event, with students and lecturers from many countries meeting to study topics in theoretical physics. Another useful hobby that he pursued in Edinburgh was the translation of Russian books. At that time the leading Russian physics journals were available in English translation, but many important Russian books were not. Since Kemmer was fluent in Russian, and some of the leading Russian physicists wrote excellent books, he was happy to exercise his skill as a translator. He published four translated books, the best known being "The Theory of Space, Time and Gravitation", by Vladimir Fock, a textbook of general relativity, presenting the subject in an unusual and illuminating way [Fock, 1959]. Fock was a highly original physicist, whose contributions to many areas of physics are not well known outside Russia. Another book translated by Kemmer is "What is Relativity?" by Lev Landau and Yuri Rumer [1960]. Landau was a justly famous physicist, Rumer a less famous friend and colleague. Landau and Rumer were arrested on the same day in 1938, victims of Stalin's purges. Landau was released after a year, but Rumer spent a large fraction of his life in the Gulag.

Kemmer received the Hughes Medal from the Royal Society in 1966 and the Planck Medal from the German Physical Society in 1983. He enjoyed traveling to many countries and making new friends. But he did not enjoy sitting in committees. After his retirement as Professor, he was happy to escape from academic committees and spend time cultivating his garden. He was a passionate gardener and knew how to make the best of the Scottish climate.

The last time I saw Nicholas and Margaret was at the celebration of the two-hundredth birthday of George Green at the University of Nottingham in the summer of 1993. This was a happy reunion, in which Julian Schwinger and his wife Clarice also took part. George Green was a self-taught genius who grew up as the child of a miller in Nottingham, spending long nights in the work-room at the top of his father's windmill, where he had the reponsibility for rotating the mill so as to keep the blades properly oriented to the wind. When the wind was steady, he spent the nights reading mathematical books by candlelight. He mastered the theory of partial differential equations, and discovered that the most general solutions of wave-equations are linear combinations of special solutions that are now known as Green's functions. Both Kemmer and Schwinger used Green's Functions extensively in their reconstructions of quantum field theory more than a hundred years later. We were invited to Nottingham to talk about the history of George Green and his functions. Schwinger was the star of the show and gave two brilliant talks. In Schwinger's talks and in Green's papers I could recognize the same functions that I had encountered forty-five years earlier in Wentzel's book and in Kemmer's lectures.

Participants at the Nottingham conference spent an afternoon visiting the Green family windmill, which is still preserved and now used as an educational center with science programs for school-children. Nicholas and Margaret were chatting happily with the children about windmills past and future. Five years after the Nottingham outing, Nicholas died peacefully at his home in Edinburgh.

2. Work

Kemmer's style as a scientist was formed during the two years that he spent as a student in Göttingen, absorbing ideas from the great mathematicians Weyl, Courant and Landau. Göttingen was then the mathematical capital of the world, upholding the great traditions of Gauss and Klein, with the aged but still active Hilbert presiding. Kemmer was lucky to arrive at this mathematical paradise two years before Hitler put an end to it. He knew that he had no time to lose, and plunged immediately into the most advanced courses. What he learned in these courses was mostly formal mathematics, and formal mathematics remained the basis of his thinking for the rest of his life. At the same time, he took a course in experimental physics from Robert Pohl, one of the founding fathers of condensed matter physics, and enjoyed the practical demonstrations. He never intended to be an experimenter himself, but he learned enough about experiments to be able to talk and work with experimenters. He could read the experimental literature and tell the sense from the nonsense.

When Kemmer started as a graduate student in Zürich, Wentzel immediately gave him a research problem on the frontier of existing knowledge, to calculate the behavior of a point electron interacting with the quantized electromagnetic field. This was a formidable problem, and Kemmer did not solve it. It was solved fifteen years later by Tomonaga, Schwinger and Feynman, in three different ways. To solve it required a reconstruction of the theory of the electron as well as of the electromagnetic field. All that Kemmer could do in 1933 was to take the first step toward a solution, calculating the self-energy and the self-force of a point electron with the existing formalism and finding that both were infinite. The appearance of these infinite quantities proved that the existing formalism was inconsistent, and pointed the way toward the discovery of a better formalism fifteen years later. Kemmer's calculation was published in "Annalen der Physik" [Kemmer, 1935], and earned him a doctor's degree in 1935. In the same year, Yukawa in Japan started a new era in particle physics with his proposal that a new particle, the meson, was responsible for nuclear forces, just as the photon was responsible for electromagnetic forces [Yukawa, 1935]. Kemmer, like other physicists at that time, began to think seriously about mesons. Great confusion was caused by the fact that the cosmic-ray particles known at that time did not behave like Yukawa mesons. Yukawa mesons were expected to interact strongly with any kind of matter, while the cosmic-ray particles seemed hardly to interact at all. The confusion was only resolved twelve years later when Cecil Powell and his colleagues at Bristol discovered that there are two kinds of mesons, with the primary pi-mesons decaying into the secondary mu-mesons [Lattes *et al.*, 1947]. Then it became clear that the cosmic-ray particles were mostly mu-mesons while the pi-mesons might be the particles postulated by Yukawa to explain nuclear forces.

In his last year in Zürich, Kemmer made friends with Viktor Weisskopf who had been Pauli's assistant at the ETH. Weisskopf was working on the same problem as Kemmer with a similar lack of success, and so they joined forces. Together they did an improved calculation of the electron self-energy, describing the electron by a field theory including positive as well as negative charges. The self-energy was still infinite, but less infinite than before. This was another useful step toward an improved theory. Kemmer also collaborated with Weisskopf on a calculation of Delbrück scattering. This was the scattering of photons by a static electric field, caused by the quantum fluctuations of electron–positron pairs in the vacuum excited by the field. The scattering process was named after Hans Delbrück, who was then a young physicist at Copenhagen and later became a famous biologist. To calculate it was a strenuous exercise in quantum field theory, stretching the limits of what could be done with the clumsy formalism of quantum electrodynamics as it then existed. The details of the calculation were published in three papers, [Kemmer and Weisskopf, 1936; Kemmer, 1937a; Kemmer and Ludwig, 1937b]. The results were later confirmed by accurate measurements of the scattering of gamma-rays by lead atoms. The collaboration with Weisskopf ended with Kemmer's move to England in 1936.

Kemmer found nobody at Imperial College with whom he wished to collaborate, but he continued to share his ideas by correspondence with Wolfgang Pauli. A large part of Kemmer's work as a research scientist is exceptionally well documented because he discussed it in detail with Pauli and most of their letters are preserved. In the collected correspondence of Pauli edited by Karl von Meyenn [Von Meyenn, 1985, 1993], there are seventeen letters and five postcards from Pauli to Kemmer, nine letters and one postcard from Kemmer to Pauli, written during the four years 1936–1940 when Kemmer was in London and Pauli was in Zürich. They discuss Kemmer's ideas while they are taking shape, and are often more illuminating than the published papers which present only the finished product. As long as Kemmer was in London, he continued to confide in Pauli, and Pauli continued to give him generous help. All the letters from this period are written by hand. They must have cost Pauli many hours of his precious time. In all this correspondence there is no trace of the harsh judgment and rude manner for which Pauli was famous. In 1940, Kemmer moved to Cambridge and Pauli moved to Princeton. There is one more substantial letter about physics, written by Pauli from Princeton in 1941 [von Meyenn, 1993, pp. 95–97]. After that there are only a few brief exchanges, still friendly but not discussing physics.

The most dramatic item in the Pauli–Kemmer correspondence is a postcard from Pauli to Kemmer dated December 15, 1936, soon after Kemmer's arrival in London [Von Meyenn, 1985, p. 491]. Here is an excerpt from the postcard [my translation]. "Take a more careful look at the new American papers on nuclear forces in the November 1 Physical Review. They discuss the possibility that all the non-electromagnetic forces are independent of charge, meaning that the proton–proton and proton–neutron forces are the same. The idea has a certain inner logic. There may be reasonable explanations for this that you could calculate". Pauli knew that Kemmer had mastered the abstract mathematics of quantum field theory, and that he knew little about the phenomenology of nuclear physics. The American papers to which Pauli refers are reporting and interpreting experimental results, and Kemmer would probably not have paid attention to them if Pauli had not sent the postcard.

The most important of the American papers is "The scattering of protons by protons" [Tuve et al., 1936]. This paper is twenty pages long and reports a revolution in experimental particle physics. Tuve and his colleagues at the Carnegie Institution in Washington had taken enormous trouble to measure the differential cross-section for scattering of protons by protons as a function of energy and angle, with a resolution of the order of one percent in both energy and angle, and with an absolute accuracy of the order of one percent. The energy of the incident protons was varied over the range from 600 to 900 kilovolts. To obtain one-percent accuracy of the cross-sections, many thousands of scattering events were measured at each energy and each angle. This was the culmination of an engineering program that had taken them ten years to complete. The source of their protons was a newly constructed Van der Graaff accelerator, which gave more precise control of the particle energy and beam quality than any other accelerator at that time. Their paper ends on a triumphant note: "It thus appears that a real beginning has been made toward an accurate and intimate knowledge of the forces which bind together the primary particles into the heavier nuclei so important in the structure and energetics of the material universe". The measurements were accurate enough to disentangle the nuclear and electromagnetic contributions to the proton–proton interaction. At the highest energy and the largest angle, the measured scattering cross-section was about four times larger than the cross-section resulting from electromagnetic interaction.

The second paper that Pauli called to Kemmer's attention was "Theory of scattering of protons by protons" [Breit et al., 1936a], immediately following the experimental paper in the Physical Review. Gregory Breit, then at the University of Wisconsin in Madison, had been advising and encouraging the Carnegie Institution team for ten years. The Breit paper analyzes the experimental results using accurate Coulomb wave-functions for the protons, and finds the results consistent with a simple short-range attractive nuclear force between protons with oppositely-oriented spins. The main conclusion of the analysis is as follows: "The interaction between protons as derived from the scattering experiments is found to be very nearly equal to that between a proton and a neutron in the corresponding condition of relative spin orientation and angular momentum (singlet S state). The proton–neutron values which come closest to being equal to the proton–proton values are those obtained by Fermi and Amaldi from the scattering and absorption of slow neutrons". Two additional papers in the same issue of the Physical Review [Breit and Feenberg, 1936b; Cassen and Condon, 1936], discuss the implications of the scattering measurements for the binding energies of heavy nuclei, and conclude that all the available evidence is consistent with the hypothesis of charge-independent nuclear forces.

In response to Pauli's suggestion, Kemmer embarked on a two-year program to explore the various possibilities for a quantum field theory describing mesons interacting with protons and neutrons. For each choice of meson theory and meson–nucleon interaction, he calculated the nuclear forces and compared them with experimental results. The details of this work were published in three papers, two by

Kemmer alone and one with Hans Fröhlich and Walter Heitler of Bristol University as collaborators. There were at that time two versions of meson theory, one [Pauli and Weisskopf, 1934] with mesons of spin zero, and one [Proca, 1936] with mesons of spin one. For each choice of meson spin there were two choices of meson–nucleon interaction. In Kemmer's first paper [Kemmer, 1938a] he concluded that only one of the four possibilities was consistent with experiment, and that the mesons must have spin one. In the second paper [Kemmer et al., 1938b] with Fröhlich and Heitler, the spin-one meson field theory was used to calculate the magnetic moments of proton and neutron. The magnetic moments were found to agree in sign and in order of magnitude with the measured values, but the calculation required an arbitrary cut-off of integrals at high frequencies and the agreement with experiment was not impressive. The paper also included a calculation of the proton–proton force going to fourth order in the meson–nucleon interaction. The proton–proton force was found to be repulsive, contrary to the Tuve measurements. The conclusions of [Kemmer et al., 1938b] were discouraging. None of the existing theories showed a convincing agreement with experiment. There was only one ray of hope. The existing theories followed Yukawa in postulating only charged mesons as the carriers of nuclear forces. If neutral mesons also existed, they might give rise to additional nuclear forces that would agree better with experiment. Several other papers [Yukawa et al., 1938; Bhabha, 1938], published at the same time as [Kemmer et al., 1938b], also suggested the introduction of neutral mesons as a possible way forward.

The last of the three Kemmer papers [Kemmer, 1938c] broke the impasse. Kemmer did not merely add neutral mesons to the existing mixture. He understood that a particular combination of neutral with charged mesons would create a new three-fold symmetry of the entire formalism. The new symmetry would be an extension of the old "isotopic spin" symmetry discovered by Heisenberg, but now including mesons as well as protons and neutrons. In a theory with Kemmer's new symmetry, nuclear forces would automatically be charge-symmetric, not only when calculated by perturbation theory but when calculated exactly. The proton–proton and proton–neutron forces would be equal, as Tuve had found. So after two years, Pauli's suggestion to Kemmer bore fruit. Kemmer's symmetric meson theory was the first example of a field theory with an abstract symmetry going beyond the usual symmetries of space and time.

In his paper [Kemmer, 1938c], Kemmer gave careful attention to the question whether the neutral meson should be identical to its antiparticle. Since the positive and negative charged mesons are antiparticles of each other, the threefold symmetry demands that the neutral meson should be antiparticle to itself. Kemmer then concluded the paper by comparing the predicted charge-independent nuclear forces with experiment. He found that with spin-one mesons the agreement was acceptable but not compelling. His model remained for many years a mathematical speculation, notable for its formal beauty rather than for practical use. Eleven years later, when neutral mesons were finally discovered, they turned out to have spin zero and not one. There was, as Kemmer had predicted, a symmetrical triplet of mesons, but they did not explain nuclear forces as he had hoped. As time went on, the picture became more complicated as several kinds of meson were discovered with spin one. A combination of rho-mesons with spin one and pi-mesons with spin zero could explain nuclear forces with some degree of accuracy, but the explanation proved inadequate when experiments measuring deep inelastic scattering became possible. The Kemmer model of the meson fields turned out in the end to explain only a small part of the nuclear forces, but it pushed our thinking in the right direction to find better models. Pauli's basic insight, that a combination of phenomenological nuclear physics with abstract mathematical field theory would be the key to understanding, was correct.

Now, with seventy years of hindsight, we know that protons and neutrons and mesons are little bags of quarks, and that theories representing them as point particles can only be rough approximations. The proton–meson interaction is a rearrangement of five quarks, and the proton–proton interaction is a rearrangement of six quarks. The idea of describing such rearrangements by simple point interactions turned out to be an illusion. Now we have a standard model of particle interactions based on quarks and gauge fields instead of on protons and mesons. The standard model agrees magnificently with experiment. Not much is left of the old meson-theory models with which Kemmer struggled in the 1930s. The one thing that survives from the 1930s is the threefold symmetry which he discovered, now hugely extended and incorporated within a hierarchy of larger symmetries. Kemmer's main contribution to modern physics is not the threefold symmetry itself but the general idea of looking for larger abstract symmetries in nature. The threefold symmetry turned out to be the seed from which a majestic tree of symmetries grew, to become the basis of particle physics in the twenty-first century.

In his last year in London, Kemmer worked on a new formal approach to meson theory based on matrix algebra. Two versions of matrix algebra had been spectacularly successful in describing the behavior of particles with spin one-half. First Pauli had used matrix algebra as the basis for his non-relativistic theory of electron spin, and then Dirac had used another matrix algebra as the basis for his theory of relativistic electrons. The Pauli and Dirac algebras were essential components of the quantum revolution that transformed physics during the 1920s. To Kemmer and Pauli in 1939, it seemed reasonable to hope that similar matrix algebras could be equally successful in providing a new foundation for theories of particles with spin zero and one. Matrix algebras for meson theories had already been suggested by Duffin [1938]. Kemmer investigated these algebras much more thoroughly, so that they are now known as Duffin–Kemmer algebras. The idea was to express the field equations for particles of spin zero and one as matrix equations, just as Dirac had done for particles of spin one-half. Kemmer proved that the most general algebra consistent with the physical requirements would have dimension 126, which happens to be the sum of three squares, $126 = 1^2 + 5^2 + 10^2$. The algebra will then have three irreducible representations, a matrix algebra of rank 10 representing particles with spin one, a matrix algebra of rank 5 representing particles with spin zero, and a trivial algebra of rank 1 representing nothing.

Pauli became enthusiastically involved in these algebraic speculations. During the spring of 1939 he exchanged letters and postcards with Kemmer at a rate of two per week [Von Meyenn, 1985, p. 613]. They wrote long letters filled with equations, punctuated with question-marks and exclamation-marks. Their enthusiasm began to wane in June 1939, when it became clear that an extension of the matrix algebra to describe particles of higher spin, three-halves and two, would not work. Pauli lost interest quickly, and Kemmer more slowly. Their matrix algebra remained a mathematical toy, amusing to play with, but leading to no new physical insight. Kemmer did not bother to write up his results for publication (8) until four years later. For reasons which are still not clear, matrix algebras are enormously useful and fruitful in the theory of particles of spin one-half, but they fail dismally to fulfil Kemmer's hopes for particles of integer spin. Dirac's matrix algebra provides a leitmotiv for one half of nature, the world of protons and electrons, while the other half of nature, the world of photons and gauge fields, marches to a different drum.

Kemmer's last technical research paper (10) was written after his move to Edinburgh, where he continued to think about spin-one fields. During his Edinburgh years, evidence accumulated that spin-one fields play a dominant role in nature. Yang and Mills [1954] discovered their non-Abelian gauge field model, based on a triplet of spin-one fields with the symmetry of the isotopic spin group. Then Glashow published a paper [Glashow, 1959] pointing the way toward the later Glashow–Weinberg–Salam unification of the weak interactions with electromagnetism, basing his argument on another triplet of spin-one fields. Kemmer paid close attention to these developments. He knew more than the authors of the new theories about the intricacies of spin-one fields. He knew that only a restricted class of interactions of spin-one fields was mathematically legitimate. The discoverers, Yang, Mills, Weinberg, Salam, Glashow, did not bother about mathematical legitimacy. They assumed that if spin-one fields existed in nature, then a consistent mathematical formalism must also exist to describe them. Kemmer's paper confirmed that the discoverers were right. Guided by nature and by physical intuition, the discoverers chose interactions that Kemmer showed to be mathematically consistent.

In December 1958, Pauli died of cancer of the pancreas after a short illness. Kemmer published his paper [Kemmer, 1960] in the issue of the journal Helvetica

Physica Acta dedicated to the memory of Pauli. Since Helvetica Physica Acta was not a widely read journal (it is now defunct, having ceased publication in 1998), Kemmer's paper received little attention. Kemmer said that the only reference to it in the literature was a favourable comment by his friend Salam. For Kemmer, the purpose of publication was not to seek public acclaim. He chose to publish his paper in Helvetica Physica Acta because this allowed him to put a final sentence at the end of it: "The writer is sincerely grateful to be able to join in this tribute to the memory of Wolfgang Pauli, one of the greatest physicists of this century, who was also so good a friend and wise a counsellor".

Kemmer was wise enough to be aware that in 1960 he had come to the end of his career as a research scientist. Theoretical physicists usually do their best work as creative scientists before the age of forty, and after that age they need to find another line of work. Kemmer's other line of work was teaching, which kept him busy and happy for the second half of his life. I was lucky to encounter him at a critical moment in both our lives, when he was beginning to teach and I was beginning to learn modern physics. I was only one of the many young people who came to him as students, learned their trade in his classes and tutorials, and ended by becoming his life-long friends.

References to Other Authors

H. Bhabha [1938] "On the theory of heavy electrons and nuclear forces", *Proc. Roy. Soc. A* **166**, 501–528.

G. Breit, E. U. Condon and R. D. Present [1936a] "Theory of scattering of protons by protons", *Phys. Rev.* **50**, 825–845.

G. Breit and E. Feenberg [1936b]. "The possibility of the same form of specific interaction for all nuclear particles", *Phys. Rev.* **50**, 850–856.

B. Cassen and E. U. Condon [1936] "On nuclear forces", *Phys. Rev.* **50**, 846–849.

R. J. Duffin [1938] "On the characteristic matrices of covariant systems", *Phys. Rev.* **54**, 1114.

C. Enz and K. von Meyenn, (ed.) [1988] *Wolfgang Pauli, das Gewissen der Physik* (Friedrich Wieweg u. Sohn, Braunschweig).

S. L. Glashow [1959] "The renormalizability of vector meson interactions", *Nucl. Phys.* **10**, 107–117.

C. M. G. Lattes, G. P. S. Occhialini and C. F. Powell [1947] "Observations on the tracks of slow mesons in photographic emulsions", *Nature* **160**, 453–456.

W. Pauli and V. Weisskopf [1934] "Über die quantisierung der scalaren relativistischen wellengleichung", *Helvetica Phys. Acta* **7**, 709–731.

A. Proca [1936] "Sur la théorie ondulatoire des électrons positifs et négatifs", *J. Phys. Radium* **7**, 347–353.

M. A. Tuve, N. P. Heidenburg and L. R. Hafstad [1936] "The scattering of protons by protons", *Phys. Rev.* **50**, 806–825.

K. Von Meyenn, (ed.) [1985] *Wolfgang Pauli, Wissenschaftlicher Briefwechsel mit Bohr, Einstein, Heisenberg u.a., Band II, 1930–1939* (Springer-Verlag, Berlin).

K. Von Meyenn, (ed.) [1993] *Wolfgang Pauli, Wissenschaftlicher Briefwechsel mit Bohr, Einstein, Heisenberg u.a., Band III, 1940–1949* (Springer-Verlag, Berlin).

G. Wentzel [1943] *Einführung in die Quantentheorie der Wellenfelder* (Franz Deuticke, Wien); [1946] reprinted edition (Edwards Brothers, Ann Arbor).

C. N. Yang and R. Mills [1954] "Conservation of isotopic spin and isotopic gauge invariance", *Phys. Rev.* **96**, 191–195.

H. Yukawa [1935] "On the interactions of elementary particles, I", *Proc. Phys. Math. Soc. Japan* **17**, 48–57.

H. Yukawa, S. Sakata and M. Taketani [1938] "Interactions of elementary particles, III", *Proc. Phys. Math. Soc. Japan* **20**, 319–340.

References

N. Kemmer [1935] "Über die elektromagnetische masse des Dirac–elektrons", *Annalen der Physik* **22**, 674–712.

N. Kemmer and V. Weisskopf [1936] "Deviations from the Maxwell equations resulting from the theory of the positron", *Nature* **137**, 659.

N. Kemmer [1937a] "Über die Lichtstreuung an elektrischen feldern nach der theorie des positrons I", *Helvetica Physica Acta* **10**, 112–122.

N. Kemmer and G. Ludwig [1937b] "Über die Lichtstreuung an elektrischen Feldern nach der theorie des positrons, II", *Helvetica Physica Acta* **10**, 182–184.

N. Kemmer [1938a] "Quantum theory of Einstein–Bose particles and nuclear interaction", *Proc. Roy. Soc. A* **166**, 127–153.

N. Kemmer, H. Fröhlich and W. Heitler [1938b] "On the nuclear forces and the magnetic moments of the neutron and the proton", *Proc. Roy. Soc. A* **166**, 154–177.

N. Kemmer [1938c] "The charge-dependence of nuclear forces", *Proc. Cambridge Phil. Soc.* **34**, 354–364.

N. Kemmer [1943] "The algebra of meson matrices", *Proc. Cambridge Phil. Soc.* **39**, 189–196.

N. Kemmer and V. A. Fock [1959] (Translation). *The Theory of Space, Time and Gravitation* (Pergamon Press, Oxford).

N. Kemmer [1960a]. "On the theory of particles of spin one", *Helvetica Phys. Acta* **33**, 829–838.

N. Kemmer, L. Landau and Y. Rumer [1960] (Translation). *What is Relativity?* Edinburgh, Oliver and Boyd.

SECTION 4

POLITICS AND HISTORY

CHAPTER 4.1

SCIENCE IN TROUBLE

Trouble comes to science on three levels, personal, local and global. The personal behavior of scientists is taken very seriously by our official guardians, the National Academy of Sciences and the departments of government concerned with the funding of science. At a meeting of scientists at Princeton some years ago, various functionaries from the Washington office of the National Academy of Sciences spoke. One of them talked like a Grand Inquisitor. She said her mission was to stamp out Deviant Science. I disagreed sharply with her. I do not find it shocking that some of our best scientists turn out to be cheats or crooks. Creative people in any walk of life have a tendency to be odd. To stamp out Deviant Science means to drive out odd people from our profession. Scientists should be subjected to the same laws as other citizens so far as criminal behavior is concerned. It is a fundamental mistake to pretend that scientists are more virtuous than other people, or to attempt the enforcement of virtue by means of an Academy inquisition.

I was recently invited by the Academy to serve on a committee to investigate the alleged violation of ethical standards by the biologist Robert Gallo. The president of the Academy informed me that service on this committee was an important public duty. Nevertheless, I declined. I did not wish to be a part of any such inquisition. Robert Gallo is a scientific entrepreneur who has made enormous contributions to the understanding of the AIDS virus. The position of Robert Gallo in the world of biology is like the position that Robert Oppenheimer occupied forty years ago in the world of physics. The two Roberts were both successful scientific empire-builders. They both ran big organizations with brilliant flair and with some disregard for bureaucratic rules. They were both accused of deviousness and occasional dishonesty. In the case of Oppenheimer, these accusations were made the basis of the Security Hearing of 1954 which resulted in his condemnation and dismissal from government service. It is now generally agreed that the proceedings against Oppenheimer were unwise and unfair and did serious damage to American science.

The proceedings against Gallo have the same vindictive character as the proceedings against Oppenheimer. Our National Academy of Sciences is now lending its name to such proceedings in the same way as the Atomic Energy Commission did in 1954. In both cases, a great man is being harassed and punished for offenses which are, in comparison with his achievements and his services to society, trivial.

The mathematician Chandler Davis wrote a historical article with the title "The purge", which was published by the American Mathematical Society in 1988 in the book "A Century of Mathematics in America", celebrating the hundredth birthday of the society. The article tells a shameful story of discrimination and persecution inflicted by American academic institutions upon mathematicians who had been accused of disloyalty in the congressional inquisitions of the McCarthy era. The discrimination and persecution continued for several decades. Some of these mathematicians had been trapped by the inquisitors into making false statements about their associations with people who may or may not have been Communists. Others, like Chandler Davis himself, had stoutly refused to make any statements that would incriminate their friends, and were then punished for their silence. Davis was indicted and convicted for Contempt of Congress. The Institute for Advanced Study in Princeton, to its eternal credit, gave Davis a position as a member in 1957 when he was a convicted felon. He worked at the Institute while his conviction was under appeal. The National Science Foundation, to its eternal credit, gave him financial support. After his term as a member of the Institute was over, his appeal was dismissed and he went to jail for six months. After he came out of jail, he was unable to find a job as a mathematician anywhere in the United States. He moved to Canada and became professor of mathematics at the University of Toronto. Others were not so fortunate or so resilient. The physicists David Bohm and Frank Oppenheimer were driven out of academic life in this country and their careers were permanently disrupted.

When one looks at these stories with the benefit of hindsight, one sees that the congressional inquisition was not the worst evil. Politicians generally have short memories. After the physicist Norman Ramsey appeared before the McCarthy committee to defend his friend Wendell Furry who was under attack, Ramsey went home and was surprised to receive a friendly telephone call from Senator McCarthy offering him a job on the committee staff. The inquisitors were mainly interested in attracting public attention to themselves. They were not so much interested in permanently wrecking scientific careers. It would not have been so bad if the result of their inquisitions had been only to send a few people to jail for six months or a year. A year in jail is no joke, but it is not usually fatal. The congressional inquisitors were the lesser evil. The greater evil was the American academic establishment, the university administrators and faculty committees who continued, through venality or cowardice, to discriminate against the inquisitors' victims long after the inquisition was over. The lasting and permanent damage was done by us, by the scientific community to which we belong, not by Senator McCarthy and his bully-boy legal counsel.

This history of the long-range effects of the spy-mania of the 1950s is highly relevant to the present-day hysteria over the alleged misdeeds of the biologists Robert Gallo and David Baltimore. Again we have an inquisitive congressman, John Dingell, who is using his legitimate powers to uncover scandals in the scientific community. He is not a particularly evil character. He is only doing his job as a politician with rather more zeal than the situation requires. There is no reason why we should be afraid of him. If he can uncover some criminal activity with evidence strong enough to send the perpetrator to jail, then the perpetrator should go to jail after being properly tried and sentenced, as Chandler Davis went to jail. We should not contest Congressman Dingell's right to uncover evidence that may send some of our colleagues to jail. That is his legitimate business. But we should strongly oppose the efforts of our National Academy and other academic institutions to be more zealous than Dingell in the extirpation of deviant science. The forces now driving academic institutions to join the bandwagon of moral rectitude are the same forces that drove academic institutions in the 1950s to join the bandwagon of spy-mania. These forces are, now as then, cowardice and venality. If we bravely stand up for our colleagues who are under attack, we risk being attacked ourselves, and, even worse, we risk being deprived of funds on which we have come to rely. Our National Academy of Sciences is not immune to such pressures. Political courage is a rare virtue, not often to be found in National Academies.

David Baltimore is under attack because he put his loyalty to his friend and collaborator Tereza Imanishi-Kari above his loyalty to the rules of scientific rectitude. Chandler Davis went to jail because he put his loyalty to his left-wing friends above his loyalty to the rules of national security. The nature of the ethical dilemma was the same in both cases. Should one sacrifice friends on the altar of general principles? Both Davis and Baltimore showed great courage in their refusal to sacrifice their friends, knowing that the personal cost to themselves would be heavy and that they could expect no support from the American scientific community.

The world of science is like the world of music. Professional music is as full of political intrigue and personal rivalry as professional science. Musical and scientific entrepreneurs are equally obsessed with money and power. The average musician is, like the average scientist, pursuing a difficult vocation under difficult circumstances, sustained by a network of intense professional friendships. Why should we expect scientists to be more virtuous than musicians? Why are we more tolerant of waywardness in musicians than in scientists? It happens that I was intimately exposed to the world of professional music when I was a child in England.

My father was at the center of the musical life of London, and many of the leading musical figures of Europe came by. I said once to my father, "How does it happen that among all these musicians you are the only one who is sane?" He said, "Thank God I am not Beethoven". Many of the visiting celebrities had strange personal habits and violent tempers. Zoltan Kodaly, the leading composer of Hungary, was always on friendly terms with the Hungarian government, whether the government happened to be Fascist or Communist. Richard Strauss was on friendly terms with Hitler. My father did not hold this against them. He was happy that they all kept on coming to London, to keep the international world of music alive as long as possible in those days of turmoil and war. I remember an evening when Kodaly lay silent on the sofa in our living room, his rich brown shoulder-length hair surrounding his handsome head like a halo, a clinical thermometer clutched tightly in his hand. Every few minutes, he put the thermometer in his mouth, pulled it out and looked at it, and said in a deep sepulchral tone, "It is death". The next day, he was back on the podium conducting the British première of his "Galanta Dances". That was the world of professional music as I saw it. Why should we expect the world of professional science to be less weird? Why should we impoverish our science by excluding those controversial spirits and dubious characters that have so greatly enriched our music?

. . . .

Another way that trouble comes to science is on the local level, when people are worried about possible dangers to public health caused by scientific experiments. Public concern about these dangers became acute in the early 1970s after recombinant DNA was discovered and the microbiologist Maxine Singer called attention to possible hazards resulting from its use. Recombinant DNA made it possible in principle to construct new living creatures with genetic characteristics chosen according to the whim of the experimenter. In 1974 a group of leading microbiologists declared a voluntary moratorium, promising not to begin recombinant DNA experiments until the hazards had been carefully examined and a reasonable set of rules to govern such experiments had been publicly accepted. They called for an international conference which was held at Asilomar in 1975 to define the boundaries between safe and dangerous experiments. The National Institutes of Health established rules in 1976 under which experiments could go forward with various levels of containment corresponding to various degrees of viciousness of the creatures under investigation. Similar rules were established by the medical authorities in other countries. This system of regulation of experiments has worked well for the last sixteen years. Scientific progress has been hampered only to a minor extent, and no cases of public injury resulting from experiments have been reported.

Only in two places were the local political authorities unwilling to accept the standards of safety provided by nationally imposed rules. The two places where local authorities decided to take the law into their own hands were Cambridge, Massachusetts and Princeton, New Jersey. Mayor Vellucci of Cambridge had been feuding for many years with Harvard University and happily seized this new opportunity to make trouble. Here is a letter he wrote to the President of the National Academy of Sciences.

"Dear Mr. Handler,

"As Mayor of the City of Cambridge, I would like to respectfully make a request of you. In today's edition of the Boston Herald American, a Hearst Publication, there are two reports which concern me greatly. In Dover, Massachusetts, a "strange, orange-eyed creature" was sighted, and in Hollis, New Hampshire, a man and his two sons were confronted by a "hairy nine foot creature". I would respectfully ask that your prestigious institution investigate these findings. I would hope as well that you might check to see whether or not these "strange creatures", should they in fact exist, are in any way connected to recombinant DNA experiments taking place in the New England area. Thanking you in advance for your cooperation in this matter, I remain

"Very truly yours, Alfred E. Vellucci".

The citizens of Princeton expressed similar worries in more sophisticated language. Mayor Vellucci represented correctly the essence of the problem confronting the biologists both in Cambridge and in Princeton. The problem was distrust. The biologists had somehow to dispel the deep distrust that their activities aroused in the hearts of their neighbors. In both places the municipal authorities appointed a Citizens' Committee composed of local residents, to study the hazards and to advise the authorities as to whether and under what conditions recombinant DNA experiments should be permitted in the town.

The citizens' committee in Princeton consisted of eleven people of which I was one. We were supposed to represent all the important segments of Princeton except for the University. We were not supposed to be experts. The way we were chosen by the Mayor was rather like the way Bill Clinton chose the members of his cabinet. We were six men and five women, nine white and two black, two medical doctors, three writers, three scientists, one school-teacher, one presbyterian minister and one retired housekeeper. Representatives of the University were invited to our meetings but had no voice in our decisions. We started in January 1977 and worked hard for four months, hearing testimony and educating ourselves about the practical details of disease control, bacterial epidemiology, laboratory design and experimental protocol. Our meetings were open to the public. Many citizens came to express their views, and many more came to listen and to learn. On the whole, the committee did an excellent job in providing a neutral ground on which the university biologists and the distrustful citizens could meet and talk and get to know each other and understand each others' concerns. This was the main positive result of our work. We kept the dispute between town and university from descending to the level of legal confrontation. Nobody on either side threatened to sue. We did not bring in lawyers to intensify the existing level of distrust. As a result of our patient work of education, it gradually became obvious to both sides that some political compromise based upon mutual respect could be achieved.

We succeeded in educating ourselves and the public, but we failed to do what the municipal authorities had asked us to do. They asked us to write a report with recommendations for action. When we finished our deliberations in May 1977, we gave them not one report but two, a majority report signed by eight of us and a minority report signed by three. The majority recommended that experiments be permitted to go forward with conditions slightly stricter than the National Institutes of Health required. The minority recommended that experiments requiring a specially built laboratory under the NIH rules be forbidden. According to the minority view, the Borough of Princeton should enforce the local prohibition of dangerous experiments by denying a building permit for the required laboratory. These divided voices were not what the Princeton Borough Council wanted to hear. Since we were unable to produce a unanimous recommendation, the Council was compelled to examine the issues in detail before coming to a decision. Nine months later, the Council decided by a five-to-one vote to follow the recommendations of our majority report. By that time the public concern about recombinant DNA had subsided. The university was able to go ahead with building a laboratory and using it for experiments. So far as Princeton was concerned, the fuss was over.

These events in Princeton happened fifteen years ago. Since that time, the battleground for disputes about recombinant DNA has moved from laboratories to field experiments. It is difficult to argue, after fifteen years of experience without a single casualty attributable to recombinant DNA, that the laboratory experiments are endangering public health. But the practical applications of recombinant DNA in medicine and agriculture require testing in the field. If recombinant DNA is to be used on an industrial scale to improve crop-plants and to suppress agricultural pests, the genetically engineered creatures must be produced in large quantities and let loose in the open air. Proposals to carry out field tests have met bitter opposition, especially in California. Jeremy Rifkin, an environmental activist skilled in legal manoeuvres, harasses the proposers of field experiments with lawsuits and brings their experiments to a halt. In the face of long delays and heavy costs, manufacturers are effectively deterred from pursuing experiments. The peaceful coexistence between science and society achieved by the Asilomar conference of microbiologists in 1975 is in danger of breaking down. The level of distrust between genetic engineers and ordinary citizens is rising. This is not only an American problem. The voices of distrust in Germany and several other European countries are even more strident than they are in California. All over the world, as genetic engineering comes closer to practical application, it runs into a barrier of public distrust. The little storm-in-a-teacup that we experienced in Princeton in 1977 may provide some useful guidance to the world community of scientists in navigating the larger storm that is now brewing. Our citizen's committee learned two lessons that were useful to Princeton in 1977 and may be useful to the big world today.

The first lesson that we learned was the importance of listening. The only effective way to dissipate distrust is for the people who are distrusted to sit down and listen to what their critics have to say. We saw this happen during our long evenings in Princeton Borough Hall. The critics were not won over by university experts giving lectures about the technical safety of biological containment. The critics were won over by personal contact with Edward Cox and Sheldon Judson, the university representatives who listened to the critics' remarks and patiently answered their questions. A citizens' committee is an effective device for allaying distrust because its main business is listening. The National Academy of Sciences is an ineffective device because it only speaks and never listens, telling ordinary citizens in an authoritarian voice what is good for them, hiding its internal doubts and disagreements behind a mask of official consensus. So that is the first lesson that Princeton has to offer to the world. Listen more and talk less. Reach out to the public through citizens' committees and not through National Academies.

The second lesson that our citizens' committee learned is more complicated. It arises from our ultimate failure to agree. We learned that sincere and well-informed people may have fundamentally divergent views about the ethics of science. The official charge to our committee was narrowly focussed on issues of public health. We were told to "study to what extent, if any, research on recombinant DNA within the Borough and Township of Princeton might be harmful to the health, welfare and safety of its citizens". The eight of us who signed the majority report wished to interpret our charge narrowly. We considered that our task was to listen to citizens who were afraid of infection by bugs escaping from laboratories and to evaluate whether their fears were justified. The two medical doctors on our committee told the citizens in no uncertain terms that their fears were exaggerated. The doctors told them that if they wanted to find serious biohazards in Princeton the place to look was not the University but the Princeton Hospital, that the daily procedures in the clinical laboratories of any working hospital were far more risky and far less under control than anything contemplated by the University. After listening to the doctors describing the bacterial fauna at large in the Hospital, it was difficult to regard the small menagery of creatures in the University laboratories as uniquely threatening. Our majority of eight agreed with the doctors that the plans of the University presented no serious danger to public health, and that our final report should confine itself to stating this conclusion. However, we learned in long and friendly discussions that our minority of three, Wallace Alston the presbyterian minister, Susanna Waterman the writer and Emma Epps the retired housekeeper, saw our task differently. The three were unwilling to stay within the narrow interpretation of our charge. Emma Epps had been for many years housekeeper in the home of the great art-historian Erwin Panofsky, and helped to educate the Panofsky twins who afterwards both became famous scientists. Her feelings toward scholarship and science were warm, but her feelings toward Princeton University were bitter. Growing up in Princeton in the days when it was a segregated town, she had to go south for her college education. After she returned and became a leader of the Princeton black community, she had fought and lost many battles against the University. She was not inclined to give the University the benefit of any doubt.

The point of view of our minority was expressed most clearly in a written statement by Susanna Waterman. "We confronted to the best of our abilities the larger ethical, moral and philosophical questions that cannot be ignored, and are of profound importance. 1. Is it possible to separate ethical problems involved in experiment from those involved in application? 2. Is it possible to separate the creation of new forms of plant and microbial life from the creation of new kinds of human beings? 3. Assuming that the genetic manipulation of human beings becomes a practical possibility, should the ethic of the brotherhood of man, or the ethic of the right of the individual to be different, prevail? 4. Does the commercial exploitation of recombinant DNA research introduce different ethical problems from the commercial exploitation of other biological processes? 5. Does the insertion of genes from higher organisms into lower organisms in itself constitute a dangerous breach of evolutionary barriers? 6. Do we have the right to interfere with the processes of natural evolution when we cannot know what it will lead to?" She concludes by saying, "In so far as a society can be measured not by the knowledge that it acquires, but by the values it brings to bear on that knowledge, so this Committee's recommendations reflect not the information that we gathered, but the perceptions and values we brought to bear upon it."

Susanna Waterman's philosophical discourse was intensely irritating to our busy medical doctors, who wanted to get the work of the Committee finished so that they could go back to saving lives in the real world. But I agreed with Susanna that the Committee had a right and a duty to go beyond its charter and examine the deeper long-range problems that biological knowledge raises. After the easy questions of public health have been answered, the difficult long-range questions remain. Our Committee was correctly responding to the anxieties of the public when it devoted a part of its time to the long-range questions. The public is not only concerned about present dangers. The opposition to biological experiments comes at least as much from people who share the worries articulated by Susanna Waterman. If trust between scientists and public is to be restored, the scientific community must be willing to listen to worried citizens and not only to medical doctors. Our committee was successful in creating trust because we listened to Susanna's philosophical questions and showed her that we shared her doubts.

The second lesson that our Committee's experience offers to the larger world is that the good and the evil faces of science should be openly acknowledged and should not be hidden. Science by its nature is unpredictable and therefore risky. Those who speak for the scientific profession should not pretend that the practical consequences of research are either calculable or controllable. The public discussion of science should begin with the recognition that supporters and opponents of research have philosophical beliefs that may be incompatible but are both deserving of respect. We should let the public know that we respect Susanna Waterman and those who share her critical view of science, that we do not claim to have satisfactory answers to her questions, and that the scientific enterprise will always need critical spirits like hers to warn us when we are heading for disaster.

. . . .

The third way trouble comes to science is on the global level. Susanna Waterman asked questions about the long-range effects of recombinant DNA on human society. Similar questions are being asked about the effects of science in general on the lives of ordinary people. These questions were anticipated almost a hundred years ago by the poet Yeats, long before recombinant DNA was dreamed of. Yeats wrote a poem with the title "The Happy Townland":

> There's many a strong farmer whose heart would break in two
> If he could see the townland that we are riding to.
> Boughs have their fruit and blossom at all times of the year,
> Rivers are running over with red beer and brown beer,
> An old man plays the bagpipes in a golden and silver wood,
> Queens, their eyes blue like the ice, are dancing in a crowd.

This little poem presents a rosy picture of the way science was to transform society in the twentieth century, the old farmer lamenting the passing of the old way of life, the young townspeople heedlessly enjoying the blessings of technology. Yeats understood that science produces social change and that some people will be winners and others will be losers. In Yeats's vision of the future, as in the past history of technological change, the losers were mostly old and the winners mostly young. At the beginning of this century when Yeats was writing, old people were building and driving horse-carriages while young people were learning how to build and drive automobiles. So long as the winners are young and the losers are old, the process of technological change is psychologically tolerable. It is seen as part of the normal replacement of one generation by the next. Even if the old farmer is suffering hardship and poverty, he is willing to accept his fate if he knows that his children will be better off than he is. Throughout history, the social upheavals caused by scientific progress have usually been accepted as benign, because they opened new opportunities to the young and only closed opportunities to the old who were soon to disappear as an inevitable consequence of human mortality.

The normal pattern of social change, with young winners and old losers, was abruptly broken by World War I. Especially in England where I grew up in the 1930s, World War I produced a profound public hostility toward science. The greatest horror of the war was not the fact that technology killed millions of people, but the fact that the victims were young while the generals and politicians who organized the technological carnage were old. For the first time, science caused a social upheaval in which the winners were old and the losers young. This inversion of the natural order of winners and losers caused a severe shock to the whole society. It caused a loss of confidence in science and a passionate hatred of technology among many young people of my generation. The mathematician G. H. Hardy spoke for us when he wrote in his "Mathematician's Apology", "A science is said to be useful if its development tends to accentuate the existing inequalities in the distribution of wealth, or more directly promotes the destruction of human life". The tragedies of World War II did not arouse such a revulsion against science, because the sacrifices of the second war were shared more equitably between old and young, between civilians and soldiers.

The war in Vietnam caused a shock to American society in the 1960s similar to the shock caused by World War I in England, with a similar loss of respect for science among the young people who felt themselves to be the losers. The traumas of the Vietnam war are now slowly healing, both here and in Vietnam. We cannot any longer blame the Vietnam war for our persistent social problems. A far more ominous pattern of social change became evident during the 1980s, having nothing to do with past or future wars. During the 1980s, we have seen for the first time a peaceful science and a non-military technology driving a revolution in which the winners are old and the losers young. Instead of Yeats's happy townland of queens dancing in a crowd, we see a townland of poverty and misery with social outcasts roaming the streets, people homeless and disproportionately impoverished. A large fraction of the losers are now young mothers and children, people who were better cared for in the old days when our technologies were less advanced. This state of affairs is ethically intolerable, and if we scientists are honest we must accept a big share of responsibility for allowing it to happen. This is the central ethical imperative that science now has to face. We may say, as Kaiser Wilhelm said at the end of World War I, "Ich hab' es nie gewollt", "I never wanted it to happen", but history will not excuse us for allowing it to happen, just as history has not excused Kaiser Wilhelm.

Why put responsibility upon the scientific community for the decline of urban society and public morality in the United States? Of course we are not alone responsible. But we are more responsible than most of us are willing to admit. We are responsible for the heavy preponderance of toys for the rich over necessities for the poor in the output of our laboratories. We have allowed government and university laboratories to become a welfare program for the middle class while the technical products of our discoveries take away jobs from the poor. We have helped to bring about a widening split between the technically competent and computer-owning rich and the computerless and technically illiterate poor. We have helped to bring into existence a post-industrial society which offers no legitimate means of subsistence to uneducated youth. And at the same time we have subsidized university tuition for children of professors so that the academic profession is gradually converting itself into a hereditary caste. I recently listened to a distinguished academic computer scientist who told us joyfully how electronic databases piped into homes through fiber-optic cables are about to put newspapers out of business. He did not care what this triumph of technological progress would do to the poor citizen who cannot afford fiber-optics and would still like to read a newspaper. I have heard similar boasting from medical scientists about the achievements of high-tech medical specialties that are putting the old-fashioned family doctor out of business. We all know what these triumphs of technology are doing to the poor citizen who cannot afford a visit to a high-tech hospital and would still like to see a doctor.

This indictment of the scientific establishment has not yet been hurled at us from the rooftops by an enraged public. Attacks against science are likely to become more bitter and more widespread in the future, so long as the economic inequities in our society remain sharp and science continues to be predominantly engaged in building toys for the rich. To forestall such attacks, whether or not we feel guilt for the sins of society, the scientific community should invest heavily in projects that benefit all segments of our population. Such projects are not hard to find, and many individual scientists are working on them, working long hours for meagre pay. The two most obvious fields for personal involvement of scientists are education of children and teachers in poor neighborhoods, and staffing of accessible public-health clinics. The physicist Leon Lederman has organized a continuing involvement of scientists from the national Fermilab laboratory in the public schools of Chicago. Such efforts are admirable, but they are very far from being adequate to the size of the problem. What is needed is a major commitment of scientific resources to the development of new technology that will bring our derelict cities and derelict children back to life. If our profession does not put its heart into such a commitment, then we shall deserve the passionate hatred that we shall sooner or later encounter.

Technology has not only done harm to poor people in our own American society. One can draw up an equally damning indictment against scientists for our contributions to the widening split between rich and poor on an international scale, to the worldwide spread of a technology that pauperizes nations and enriches élites. Many individual scientists, here and overseas, have dedicated their lives to repairing the damage that technology has done to poor countries. In the world as a whole, just as in the United States, a far greater commitment of scientific resources is needed in order to create a technology that is friendly to ordinary people wherever they happen to live.

One scientist has told us what it means to carry through an ambitious program of research aimed directly at the relief of human suffering. She has had to fight long battles on many levels, intellectual, technical, financial and personal. She publishes her writings under the name Anna Brito. She is a Portuguese immunologist and poet who worked in a New York hospital studying the biochemistry of leukemia and wrote letters to a friend describing her experiences. Here is an extract from one of her letters, written to her friend June Goodfield and published in June Goodfield's book, "An Imagined World". "I am so tired of so much emotion that I can hardly gather the strength to write. But I must tell you that the little boy, Jesus, with acute T-cell leukemia and iron spots in his red cells, has an iron level in his blood serum of 456 when the normal range is between 48 and 193. I must therefore tell you that from little mice we have traveled all the way to where it matters, to little boys. From little experimental models made of history and past ideas, we have traveled all the way to the beginning of dawn, where from a dark mass the day begins to get shape and promise. And it is a cold hour. And that is what makes it different from love. For love is light and warmth, and understanding is only light. Only much later, and dissociated, is it warmth".

CHAPTER 4.2

PROGRESS IN RELIGION*

First, a big thank you to Sir John Templeton and the administrators of the Templeton Foundation for giving me this undeserved and unexpected honor. Second, a big thank you to the Institute for Advanced Study in Princeton for supporting me as a Professor of Physics while I strayed into other areas remote from physics. Third, a big thank you to the editors and publishers of my books for giving me the chance to communicate with a wider public. Fourth, a big thank you to my wife and family for keeping me from getting a swelled head. And fifth, a big thank you to the Washington National Cathedral for allowing us to use this magnificent building for our ceremonies.

Sir John Templeton has told us clearly the purpose of his awards. They are prizes for Progress in Religion. But it is up to us to figure out what Progress in Religion means. Roughly speaking, there have been two main themes in the lives of the previous prizewinners. The first theme is practical good works, caring for the poor and sick, helping the dying to die with dignity. Outstanding among the doers of good works were Mother Teresa and Dame Cicely Saunders. The second theme is scholarly study and teaching, helping people who are committed to one religion or another to approach God through intellectual understanding, explaining to the uncommitted the logical foundations of belief. Outstanding among the scholarly prizewinners are James McCord and Ian Barbour. I am amazed to find myself in the company of these great spirits, half of them saints and the other half theologians. I am neither a saint nor a theologian.

To me, good works are more important than theology. We all know that religion has been historically, and still is today, a cause of great evil as well as great good in human affairs. We have seen terrible wars and terrible persecutions conducted in the

*Lecture at Washington National Cathedral, May 16, 2000.

name of religion. We have also seen large numbers of people inspired by religion to lives of heroic virtue, bringing education and medical care to the poor, helping to abolish slavery and spread peace among nations. Religion amplifies the good and evil tendencies of individual souls. Religion will always remain a powerful force in the history of our species. To me, the meaning of progress in religion is simply this, that as we move from the past to the future the good works inspired by religion should more and more prevail over the evil.

Even in the gruesome history of the twentieth century, I see some evidence of progress in religion. The two individuals who epitomized the evils of our century, Adolf Hitler and Joseph Stalin, were both avowed atheists. Religion cannot be held responsible for their atrocities. And the three individuals who epitomized the good, Mahatma Gandhi, Martin Luther King and Mother Teresa, were all in their different ways religious. One of the great but less famous heroes of World War II was André Trocmé, the Protestant pastor of the village of Le Chambon sur Lignon in France, which sheltered and saved the lives of five thousand Jews under the noses of the Gestapo. Forty years later Pierre Sauvage, one of the Jews who was saved, recorded the story of the village in a magnificent documentary film with the title, "Weapons of the Spirit". The villagers proved that civil disobedience and passive resistance could be effective weapons, even against Hitler. Their religion gave them the courage and the discipline to stand firm. Progress in religion means that, as time goes on, religion more and more takes the side of the victims against the oppressors.

For Ian Barbour, who won the Templeton Prize last year, religion is an intellectual passion. For me it is simply a part of the human condition. Recently I visited the Imani church in Trenton because my daughter, who is a Presbyterian minister, happened to be preaching there. Imani is an inner-city church with a mostly black congregation and a black minister. The people come to church, not only to worship god, but also to have a good time. The service is informal and the singing is marvelous. While I was there they baptized seven babies, six black and one white. Each baby in turn was not merely shown to the congregation but handed around to be hugged by everybody. Sociological studies have shown that violent crimes occur less frequently in the neighborhood of Imani church than elsewhere in the inner city. After the two-hour service was over, the congregation moved into the adjoining assembly room and ate a substantial lunch. Sharing the food is to me more important than arguing about beliefs. Jesus, according to the gospels, thought so too.

I am content to be one of the multitude of Christians who do not care much about the doctrine of the Trinity or the historical truth of the gospels. Both as a scientist and as a religious person, I am accustomed to living with uncertainty. Science is exciting because it is full of unsolved mysteries, and religion is exciting for the same reason. The greatest unsolved mysteries are the mysteries of our existence as conscious beings in a small corner of a vast universe. Why are we here? Does the universe have a purpose? Whence comes our knowledge of good and evil? These mysteries, and a hundred others like them, are beyond the reach of science. They lie on the other side of the border, within the jurisdiction of religion.

My personal theology is described in the Gifford lectures that I gave at Aberdeen in Scotland in 1985, published under the title, "Infinite in All Directions". Here is a brief summary of my thinking. The universe shows evidence of the operations of mind on three levels. The first level is elementary physical processes, as we see them when we study atoms in the laboratory. The second level is our direct human experience of our own consciousness. The third level is the universe as a whole. Atoms in the laboratory are weird stuff, behaving like active agents rather than inert substances. They make unpredictable choices between alternative possibilities according to the laws of quantum mechanics. It appears that mind, as manifested by the capacity to make choices, is to some extent inherent in every atom. The universe as a whole is also weird, with laws of nature that make it hospitable to the growth of mind. I do not make any clear distinction between mind and God. God is what mind becomes when it has passed beyond the scale of our comprehension. God may be either a world-soul or a collection of world-souls. So I am thinking that atoms and humans and God may have minds that differ in degree but not in kind. We stand, in a manner of speaking, midway between the unpredictability of atoms and the unpredictability of God. Atoms are small pieces of our mental apparatus, and we are small pieces of God's mental apparatus. Our minds may receive inputs equally from atoms and from God. This view of our place in the cosmos may not be true, but it is compatible with the active nature of atoms as revealed in the experiments of modern physics. I do not say that this personal theology is supported or proved by scientific evidence. I only say that it is consistent with scientific evidence.

I do not claim any ability to read God's mind. I am sure of only one thing. When we look at the glory of stars and galaxies in the sky and the glory of forests and flowers in the living world around us, it is evident that God loves diversity. Perhaps the universe is constructed according to a principle of maximum diversity. The principle of maximum diversity says that the laws of nature, and the initial conditions at the beginning of time, are such as to make the universe as interesting as possible. As a result, life is possible but not too easy. Maximum diversity often leads to maximum stress. In the end we survive, but only by the skin of our teeth. This is the confession of faith of a scientific heretic. Perhaps I may claim as evidence for progress in religion the fact that we no longer burn heretics.

That is enough about me. Let me talk now about the great transformations of the world that we are facing in the future. All through our history, we have been changing the world with our technology. Our technology has been of two kinds, green and grey. Green technology is seeds and plants, gardens and vineyards and orchards, domesticated horses, cows and pigs, milk and cheese, leather and wool. Grey technology is bronze and steel, spears and guns, coal, oil and electricity, automobiles, airplanes and rockets, telephones and computers. Civilization began with green technology, with agriculture and animal-breeding, ten thousand years ago. Then, beginning about three thousand years ago, grey technology became dominant, with mining and metallurgy and machinery. For the last five hundred years, grey technology has been racing ahead and has given birth to the modern world of cities and factories and supermarkets.

The dominance of grey technology is now coming to an end. During the last fifty years, we have achieved a fundamental understanding of the processes occurring in living cells. With understanding comes the ability to exploit and control. Out of the knowledge acquired by modern biology, modern biotechnology is growing. The new green technology will give us the power, using only sunlight as a source of energy, and air and water and soil as sources of materials, to manufacture and recycle chemicals of all kinds. Our grey technology of machines and computers will not disappear, but green technology will be moving ahead even faster. Green technology can be cleaner, more flexible and less wasteful, than our existing chemical industries. A great variety of manufactured objects could be grown instead of made. Green technology could supply human needs with far less damage to the natural environment. And green technology could be a great equalizer, bringing wealth to the tropical areas of the world which have most of the sunshine, most of the human population, and most of the poverty.

I am saying that green technology could do all these good things, bringing wealth to the tropics, bringing economic opportunity to the villages, narrowing the gap between rich and poor. I am not saying that green technology will do all these good things. "Could" is not the same as "will". To make these good things happen, we need not only the new technology but the political and economic conditions that will give people all over the world a chance to use it. To make these things happen, we need a powerful push from ethics. We need a consensus of public opinion around the world that the existing gross inequalities in the distribution of wealth are intolerable. In reaching such a consensus, religions must play an essential role. Neither technology alone nor religion alone is powerful enough to bring social justice to human societies, but technology and religion working together might do the job.

We all know that green technology has a dark side, just as grey technology has a dark side. Grey technology brought us hydrogen bombs as well as telephones. Green technology brought us anthrax bombs as well as antibiotics. Besides the dangers of biological weapons, green technology brings other dangers having nothing to do with weapons. The ultimate danger of green technology comes from its power to change the nature of human beings by the application of genetic engineering to human embryos. If we allow a free market in human genes, wealthy parents will be able to buy what they consider superior genes for their babies. This could cause a splitting of humanity into hereditary castes. Within a few generations, the children of rich and poor could become separate species. Humanity would then have regressed all the way back to a society of masters and slaves. No matter how strongly we believe in the virtues of a free market economy, the free market must not extend to human genes.

A few weeks ago I was attending mass in St. Stephen's church in England. In Princeton I am Presbyterian, but in England I am Catholic because I go to mass with my sister. The reading from the gospel of St. Matthew told of the angry Jesus driving the merchants and money-changers out of the temple, knocking over the tables of the money-changers and spilling their coins on the floor. Jesus was not opposed to capitalism and the profit motive, so long as economic activities were carried on outside the temple. In the parable of the talents, he praises the servant who used his master's money to make a profitable investment, and condemns the servant who was too timid to invest. But he draws a clear line at the temple door. Inside the temple, the ground belongs to God and profit-making must stop.

While I was listening to the reading, I was thinking how Jesus' anger might extend to free markets in human bodies and human genes. In the time of Jesus and for many centuries afterwards, there was a free market in human bodies. The institution of slavery was based on the legal right of slave-owners to buy and sell their property in a free market. Only in the nineteenth century did the abolitionist movement, with Quakers and other religious believers in the lead, succeed in establishing the principle that the free market does not extend to human bodies. The human body is God's temple and not a commercial commodity. And now in the twenty-first century, for the sake of equity and human brotherhood, we must maintain the principle that the free market does not extend to human genes. Let us hope that we can reach a consensus on this question without fighting another civil war. Scientists and religious believers and physicians and lawyers must come together with mutual respect, to achieve a consensus and to decide where the line at the door of the temple should be drawn.

Like all the new technologies that have arisen from scientific knowledge, biotechnology is a tool that can be used either for good or for evil purposes. The role of ethics is to strengthen the good and avoid the evil. I see two tremendous goods coming from biotechnology in the next century, first the alleviation of human misery through progress in medicine, and second the transformation of the global economy through green technology spreading wealth more equitably around the world. The two great evils to be avoided are the use of biological weapons and the corruption of human nature by buying and selling genes. I see no scientific reason why we should not achieve the good and avoid the evil. The obstacles to achieving the good are political rather than technical. Unfortunately a large number of people in many countries are strongly opposed to green technology, for reasons having little to do with the real dangers. It is important to treat the opponents with respect, to pay attention to their fears, to go gently into the new world of green technology so that neither human dignity nor religious conviction is violated. If we can go gently, we have a good chance of achieving within a hundred years the goals of ecological sustainability and social justice that green technology brings within our reach.

Now I have five minutes left to give you a message to take home. The message is simple. "God forbid that we should give out a dream of our own imagination for a pattern of the world". This was said by Francis Bacon, one of the founding fathers of modern science, almost four hundred years ago. Bacon was the smartest man of his time, with the possible exception of William Shakespeare. Bacon saw clearly what science could do and what science could not do. He is saying to the philosophers and theologians of his time: Look for God in the facts of nature, not in the theories of Plato and Aristotle. I am saying to modern scientists and theologians: Do not imagine that our latest ideas about the Big Bang or the human genome have solved the mysteries of the universe or the mysteries of life. Here are Bacon's words again: "The subtlety of nature is greater many times over than the subtlety of the senses and understanding". In the last four hundred years, science has fulfilled many of Bacon's dreams, but it still does not come close to capturing the full subtlety of nature. To talk about the end of science is just as foolish as to talk about the end of religion. Science and religion are both still close to their beginnings, with no ends in sight. Science and religion are both destined to grow and change in the millennia that lie ahead of us, perhaps solving some old mysteries, certainly discovering new mysteries of which we yet have no inkling. After sketching his program for the scientific revolution that he foresaw, Bacon ends his account with a prayer: "Humbly we pray that this mind may be steadfast in us, and that through these our hands, and the hands of others to whom thou shalt give the same spirit, thou wilt vouchsafe to endow the human family with new mercies". That is still a good prayer for all of us as we begin the twenty-first century.

Science and religion are two windows that people look through, trying to understand the big universe outside, trying to understand why we are here. The two windows give different views, but they look out at the same universe. Both views are one-sided, neither is complete. Both leave out essential features of the real world. And both are worthy of respect.

Trouble arises when either science or religion claims universal jurisdiction, when either religious dogma or scientific dogma claims to be infallible. Religious creationists and scientific materialists are equally dogmatic and insensitive. By their arrogance they bring both science and religion into disrepute. The media exaggerate their numbers and importance. The media rarely mention the fact that the great majority of religious people belong to moderate denominations that treat science with respect, or the fact that the great majority of scientists treat religion with respect so long as religion does not claim jurisdiction over scientific questions. In the little town of Princeton where I live, we have more than twenty churches and at least one synagogue, providing different forms of worship and belief for different kinds of people. They do more than any other organizations in the town to hold the community together. Within this community of people, held together by religious traditions of human brotherhood and sharing of burdens, a smaller community of professional scientists also flourishes.

I look out from the pampered little community of Princeton, which Einstein described in a letter to a friend in Europe as "a quaint and ceremonious village, peopled by demi-gods on stilts". I look out from this community of bankers and professors to ask, what can we do for the suffering multitudes of humanity in the world outside. The great question for our time is, how to make sure that the continuing scientific revolution brings benefits to everybody rather than widening the gap between rich and poor. To lift up poor countries, and poor people in rich countries, from poverty, to give them a chance of a decent life, technology is not enough. Technology must be guided and driven by ethics if it is to do more than provide new toys for the rich. Scientists and business leaders who care about social justice should join forces with environmental and religious organizations to give political clout to ethics. Science and religion should work together to abolish the gross inequalities that prevail in the modern world. That is my vision, and it is the same vision that inspired Francis Bacon four hundred years ago, when he prayed that through science God would "endow the human family with new mercies".

Chapter 4.3

Tolstoy and Napoleon: Two Styles in History, Education, Science and Ethics[*]

1. War and Peace

My lecture consists of four parts, the first about war and peace, the second about education, the third about science, the fourth returning to war and peace. I chose Tolstoy and Napoleon as the protagonists of my story because they had important things to say and represent sharply different points of view in each of these areas. I begin with war and peace.

Fifty-two years have gone by since the bombing of Hiroshima and Nagasaki. We are fifty years distant from the great and terrible war that ended with those bombings, just as Lev Tolstoy was fifty years distant from the great and terrible war that he recorded in his novel, "War and Peace". We need now to find a historical perspective into which the bombings and the other catastrophic events of fifty years ago can fit. Tolstoy felt the same need. He struggled for five years to find some order in the chaos of the Napoleonic war. "War and Peace" is the record of his struggle, as well as being a record of the war. He did, in the end, develop a coherent view of history. The human story that he tells in the novel is interrupted from time to time by the author, expressing his personal view of the story's larger meaning. He describes the historic calamities of 1812, the death and destruction spread by Napoleon's armies as they moved into Russia, the dispersal and flight of the Russian population, the great battle at Borodino and the burning of Moscow, as events unplanned and unforeseen. According to Tolstoy, the leaders were swept along by forces that they neither understood nor controlled. The nominal leaders on the two sides, Napoleon and Kutuzov, were never really in control. Since Kutuzov was aware of this and Napoleon was not, Kutuzov made fewer mistakes. But in the end the outcome was

---

[*]Lecture given at University of Canterbury, Christchurch, New Zealand, October 10, 2000.

determined by the individual actions of hundreds of thousands of ordinary soldiers, whose decisions, to fight bravely or to run away, nobody could have predicted. "Napoleon", says Tolstoy in his summing up of the lessons of history, "Napoleon is represented to us as the leader in all this movement, just as the figurehead in the prow of a ship to the savage seems the force that guides the ship on its course. Napoleon in his activity all this time was like a child, sitting in a carriage, pulling the straps within it, and fancying he is moving it along".

The recent battle fought over the "Enola Gay" exhibition at the Smithsonian Air and Space Museum in Washington showed how far we still are from achieving a coherent view of World War II. The Smithsonian had planned to show the exhibition in the summer of 1995 to commemorate the bombing of Hiroshima and Nagasaki and the end of the war. The public opposition to the Smithsonian's plans became so bitter that the exhibition was abandoned. My friend Martin Harwit, the Director of the Museum who was responsible for the exhibit, was forced to resign. After the cancellation of the exhibition and the resignation of Harwit, the debate continues. We are struggling, as Tolstoy struggled, to come to terms with a drama too large for any single mind to comprehend. The chief message of my talk today is that Tolstoy can help us in our struggle to understand. We have a lot to learn, and Tolstoy has a lot to teach us.

Tolstoy helped us to understand our predicament fifty years ago, while World War II was raging around us, and not only in retrospect fifty years later. Fifty years ago, Tolstoy and Napoleon were not remote historical figures, but ghosts sitting beside us every day at the breakfast table. I am, in a manner of speaking, only three handshakes away from Napoleon. When I was a child in England, my father used to talk about a memorable trip he had made to Russia in the year 1912 when he was a young man. He was invited to Moscow by some musician friends, and he happened to be there for the centennial celebrations of the defeat of Napoleon in the campaign of 1812. He met a 105-year-old lady who played a prominent part in the celebrations. She remembered, as a five-year-old, standing in the street and giving Napoleon an apple when he made his triumphal entry into the city. The celebrations were mainly concerned, not with Napoleon's triumphal entry into Moscow but with his ignominious departure five weeks later and his disastrous retreat to the west. The story of the 1812 campaign was an important part of English folklore, well known to every English schoolchild. American schoolchildren miss hearing the story because their ancestors happened in 1812 to be fighting on the same side as Napoleon. My father told the story dramatically. He told how England had been engaged in an interminable war against Napoleon for fifteen years. Napoleon had conquered the continent of Europe except for Russia, and England had little hope of defeating him. And then, suddenly and providentially, Napoleon was seized with madness, marched his Grand Army into Russia, and embarked on the sequence of tragic follies that destroyed the Grand Army and saved England. After I had heard this story from my father, I read it in Thomas Hardy's drama, "The Dynasts". I do not know whether anybody nowadays reads "The Dynasts", but it made a deep impression on children of my generation. I remember especially the scene when dawn rises on a group of French soldiers huddled around a dead camp-fire on a snow-swept Russian steppe. The faces of the soldiers have been roasted by the fire while their backs have been frozen by the wind. The Russian general Kutuzov comes riding by on his horse and says, "So perish Russia's enemies". A few years later, as a teenager, I read Tolstoy's more sophisticated version of the same story in "War and Peace".

This story was fresh in my mind one June morning, the morning of June 22, 1941, when I and a bunch of school-boy friends came down to our study-hall to listen to the BBC news on the radio. England was then fighting alone against Hitler, and facing the same discouraging odds that we faced when fighting Napoleon one hundred and thirty years earlier. Hitler had conquered the continent of Europe except for Russia, and it looked as if our war could go on for another fifteen or twenty years if we were lucky enough to survive so long. And then, the radio told us that Hitler had marched his Grand Army into Russia. We went wild with joy. Suddenly, our future was transformed. Instead of a lifetime of war without end, we could look forward to another campaign of 1812. We could reasonably expect to see the end of Hitler within a few years. We could reasonably expect to survive Hitler and emerge into some kind of peace. But the news on the radio still seemed too good to be true. How could it have happened that Hitler caught Napoleon's madness? How could he not know that Russia would be his army's graveyard? Had Hitler never read "War and Peace"? Had none of the German generals read "War and Peace"? It was incredible to us school-boys that the famous German generals knew less about the art of war than we did. And yet, the events of that day made it evident that we had the advantage over the German generals. Evidently, they had not read Tolstoy and we had.

Echoes of these memories of 1941 could be heard in America in the debate over the Smithsonian "Enola Gay" exhibition. The debate was concerned with the question, whether or not the nuclear bombing actually saved lives by bringing the war to a quick end. One of the issues under debate was the number of American soldiers who would have died in the planned invasions of Kyushu and Honshu if Japan had not surrendered in the summer of 1945. On the one side, President Truman and Secretary of War Stimson, in their personal accounts of the decisions for which they bore the primary responsibility, said that the planned invasions of Japan could easily have resulted in a million American and several million Japanese deaths. On the other side, Martin Sherwin and other historians [Sherwin, 1987], who have examined the recently declassified planning documents, point out that the official military estimates of casualties in the planned invasions were far smaller.

The planning documents calculate the number of American deaths to be less than forty thousand for both invasions together. Undoubtedly Stimson and Truman were aware of the low estimates in the planning documents. Some historians say that Stimson and Truman deliberately misled the public when they used much higher estimates of invasion casualties to justify their decision to drop the nuclear bombs. This is a serious accusation, and aroused passionate anger on both sides of the debate. If you believe that the dropping of the bombs was morally unjustified, then you believe that Truman and Stimson were in some sense guilty of a crime, and it is easy then to believe that they were also guilty of dishonesty in their public statements. If you believe that the dropping of the bombs was morally justified, you must also believe that Stimson and Truman were honest. To defend Stimson and Truman against the accusation of dishonesty, you must explain why they did not take the planning estimates of casualties at face value.

I have examined the estimates of casualties in detail [Sherwin, 1987, Appendices U,V,W]. The technical basis of these estimates was simple. The planners assumed a fixed casualty-rate for each army involved in heavy fighting, five per cent casualties per week, of which one per cent would be killed. These numbers were derived from the experience of earlier campaigns in the Pacific islands and in Okinawa. The planners then assumed that the invasions of Japan would involve three quarters of a million men and six weeks of heavy fighting. That was all. The estimates made sense if you believed that the conquest of Japan would be over in a few months. If you believed that the conquest of Japan was an undertaking of extreme hazard and unpredictable duration, the estimates made no sense at all.

I have little doubt that Stimson and Truman were honest. Everything I know about them leads me to that conclusion. I also have little doubt that they were skeptical of any claim that the conquest of Japan could be completed in a few months. They had lived through the agonies of the Italian campaign of 1943–1945, when a small German army conducted a fighting retreat that lasted for two years and inflicted heavy casualties on the invading Allied forces. The geography of Japan is at least as rugged as the geography of Italy. The defending Japanese armies, unlike the Germans in Italy, would have had loyal support from a friendly population. I do not know how Stimson and Truman rated the chances of a successful conquest of Japan. I only know that if I had been standing in their shoes I would have been scared to death. I felt at the time, and I still feel today, that an invasion of Japan in 1945 might have been as total a disaster as Hitler's invasion of Russia four years earlier or Napoleon's invasion in 1812. Just as Napoleon sat in ruined Moscow waiting in vain for an offer of surrender from the Tsar, General MacArthur might have sat in the ashes of Tokyo waiting in vain for an offer of surrender from the Emperor of Japan. Bitter fighting in the Japanese mountains might have continued for twenty years. The conquest of Japan might never have been finished. It was lucky for General MacArthur that the Emperor was staying in Tokyo to welcome him when he arrived.

Both Stimson and Truman had a sense of history and read many books. Probably they had read "War and Peace". Even if they did not read Tolstoy, they had a Tolstoyan understanding of the chanciness of war and peace. They understood that military estimates of casualties at the beginning of an invasion of Japan were worth just as much as German estimates of casualties at the beginning of the invasion of Russia. Hitler, after all, expected the Russian campaign to be over in six months. No army beginning an invasion ever makes plans for a long war. And still, long wars happen. Long wars happen because it is easier to begin wars than to end them. Ending a war is what the "Enola Gay" exhibition was supposed to be about.

The central question, at the heart of the "Enola Gay" controversy, is whether Japan would have surrendered if Hiroshima and Nagasaki had not been destroyed. Nobody will ever know the answer to that question. The best possible opportunity to discover the answer was given to Robert Butow, a historian who happened to be in Japan soon after World War II and wrote a book, "Japan's Decision to Surrender" [Butow, 1954]. Butow had a unique opportunity to examine the official Japanese documents and to interrogate the responsible Japanese leaders while their memories of the concluding weeks of the war were still fresh. Butow asked each of the Japanese witnesses the same question, whether the surrender would have happened without the bombs. Butow's conclusion: the Japanese leaders themselves do not know the answer to that question, and neither do I.

Tolstoy would certainly have agreed with Butow. Tolstoy himself fought with the Russian army defending Sevastopol against French and British invaders in the Crimean War. He saw war as a desperate muddle in which nothing goes according to plan and the historical causes of victory and defeat remain incalculable. Here is Tolstoy's view of the nature of war, expressed in "War and Peace" through the voice of his hero Prince Andrei. Andrei is talking to his friend Pierre on the eve of the battle of Borodino.

"To my mind what is before us tomorrow is this: a hundred thousand Russian and a hundred thousand French troops have met to fight, and the fact is that these two hundred thousand men will fight, and the side that fights most desperately and spares itself least will conquer". "So you think the battle tomorrow will be a victory," said Pierre. "Yes, yes," said Prince Andrei absently. "There's one thing I would do, if I were in power," he began again, "I wouldn't take prisoners. What sense is there in taking prisoners? That's chivalry. The French have destroyed my home and are coming to destroy Moscow; they have outraged and are outraging me at every second. They are my enemies, they are all criminals to my way of thinking. They must be put to death. War is not a polite recreation, but the vilest thing in life, and we ought to understand that and not play at war. We ought to accept it sternly and solemnly as a fearful necessity". Without any great stretch of the imagination, one may hear these words of Prince Andrei as spoken by a thoughtful Japanese soldier preparing to defend his country against the American invaders in the autumn of 1945.

2. Tolstoy on Education

That is enough about war and peace. I will talk now for a few minutes about education. Napoleon and Tolstoy, besides being soldiers, were also educators. And their views about the nature and purpose of education were even more strongly opposed than their views about war and peace. Napoleon understood that public education was important for the modernization of France and the staffing of an Imperial government. He created a system of education in France for these purposes. He imposed similar systems on the countries of Europe that he conquered. The basic pattern of European public education has remained for two hundred years as Napoleon established it. The basic pattern is centralization, standardization and rigorous testing by examination. Education was organized rigidly from the top down. The state authorities, beginning with Napoleon himself, decided what was to be taught, and when, and how. On the whole, the Napoleonic system has done successfully the job for which it was designed. In the countries that adopted it, it produced a population with a high level of literacy and technical competence. It has also acted as a social equalizer, giving talented children from all classes of society an opportunity to become leaders. Napoleon stated explicitly that his aim was to open careers to talents. In many ways, the rigidity of the Napoleonic system works to the advantage of an underprivileged child with talent. The child does not have the option of refusing to join the educated élite. Chiara Nappi, a physicist colleague of mine in Princeton who was educated by the Napoleonic system in Italy and serves on the Princeton school board, has argued convincingly that a small dose of Napoleonic rigidity could be helpful to the underprivileged children of America. She knows from her own experience that a little Napoleonic rigidity can be especially helpful to girls with a talent for science. Girls in Italy do not have the option of dropping out of science. Our American system of education gives children of all ages too many opportunities to drop out. We still have something to learn from Napoleon. But we also have something to learn from Tolstoy.

Tolstoy devoted three years of his life to education. He did not work like Napoleon from the top down. He worked from the bottom up. He organized a school for the children of the peasants on his estate at Yasnaya Polyana. He taught the children himself. He wrote an extensive account of what he had taught them and what he had learned from them. His purpose was to find out, by personal observation at the individual level, how teaching could and should be done.

Before starting his own school, Tolstoy traveled extensively around Europe and observed the Napoleonic system of public education in action. What he saw convinced him that the Napoleonic system was a total failure. Both in Russia and in other European countries, he wrote, "All that the greater part of the people carry away from school is a horror of schooling". He describes the typical child as he saw it in the classroom, "A weary, huddled creature, with an expression of fatigue, terror, and boredom, repeating with the lips alone the words of others in the language of others, a creature whose soul has hidden in its shell like a snail". He contrasts this dismal scene with "the same child at home or in the street, full of the joy of life and love of knowledge, with a smile in its eyes and on its lips, seeking instruction in all things as a joy, expressing its own thought clearly and often forcefully in its own language". He concludes his educational manifesto by saying, "We would not have dared to disturb the calm of theoretical pedagogues and to utter such opinions, which are disgusting to all society, were we to confine ourselves to the reasoning of this article, but we feel that there is a possibility of proving the applicability and legitimacy of these wild beliefs of ours step by step and fact by fact, and to this end alone we shall dedicate our publication".

Unfortunately, Tolstoy had a short attention-span. For three years he gave his school and his pupils his passionate attention, and then he abandoned them. A year later, he wrote in a letter to a friend, "The children come to me in the evenings and bring with them memories for me of the teacher that used to be in me and is there no longer. Now I am a writer with all the strength of my soul". The children sadly reported to Vasily Morozov, who had been Tolstoy's assistant in the school, "We could not get on with Lev Nikolayevich as we used to". All that is left of the school is the memoir of Vasily Morozov and the descriptions that Tolstoy recorded in his journal. Tolstoy's descriptions are vivid and detailed. He was, after all, born to be a writer rather than a teacher. He describes how he taught creative writing to Syomka and Fedka, two of his favorite pupils. He begins by trying to force them to analyze the written text of a story by Gogol. The task of analyzing the language leaves the children puzzled and bored. Tolstoy remarks, "You, the teacher, are insisting on one side of understanding, but the pupil has no need whatsoever of what you want to explain to him". Finally Tolstoy suggests to Fedka that he write a story himself. Tolstoy gives him the theme, a peasant family with a father who leaves home to become a soldier and then returns after six years in the army. Fedka writes the story, and Tolstoy analyzes the text. The process of textual analysis, which had been so sterile when applied to Gogol, comes alive and acquires meaning when it is applied to Fedka's own words. Fedka successfully defends his choice of words against Tolstoy's criticism. In a short time Fedka has become, not merely a gifted storyteller, but a conscious artist aware of his own technical mastery.

Anyone who wishes to learn more about Tolstoy's educational methods and experiences can find them, together with further references to the literature, in a long article by Michael Armstrong [Armstrong, 1983] in the magazine "Outlook". "Outlook" was a quarterly magazine founded by David Hawkins, the philosopher of Los Alamos, and devoted to the education of young children. Unfortunately, like Tolstoy's school and the "Enola Gay" exhibition, it is now defunct. Its demise does not make its ideas and opinions any the less valuable in the world of today.

Napoleon and Tolstoy both proclaimed important truths about education. Napoleon proclaimed that the essence of education is discipline. Tolstoy proclaimed that the essence of education is freedom. These two doctrines seem to be diametrically opposed. How can both of them be true? They can both be true because Napoleon and Tolstoy were concerned with different kinds of children and different meanings of the word education. For Napoleon, education meant the training of an intellectual élite who would find careers in the institutions of a technically organized society. For Tolstoy, education meant opening the eyes of ordinary children to the wonders of the world around them. For Napoleon, education meant cramming docile children with specialized knowledge. For Tolstoy, education meant giving all children, docile or not, direct experience of creative activity in art, science and literature. In the modern world, even more than in Tolstoy's world, both kinds of education are necessary. We need to give every child a taste of the Napoleonic discipline that leads to useful skills and brilliant careers. And we need to give every child, especially the child who is turned off by Napoleonic discipline, a taste of the Tolstoyan freedom that leads to the growth of natural scientific curiosity and artistic expression. Only Napoleon can give us a cadre of technically trained experts to make our society economically competitive. Only Tolstoy can give us a population of culturally and scientifically literate citizens who may possess the wisdom to save our society from Napoleonic follies. Our systems of public education can use all the help they can get, both from Napoleon and from Tolstoy. We have a long way to go.

3. Napoleonic and Tolstoyan Science

I turn now from education to science. In the practice of scientific research, just as in the practice of education, we see two contrasting styles, the Napoleonic and the Tolstoyan. In the last few years I had the opportunity of visiting two extraordinary laboratories, both in Switzerland, the international particle-physics laboratory known as CERN near Geneva, and the IBM research center at Rüschlikon near Zürich. These happen to be the two places where the most spectacular discoveries in the physics of the last twenty years have been made, at CERN in particle-physics and at Rüschlikon in condensed-matter physics. The people at CERN discovered the beautiful world of W and Z particles, the people at Rüschlikon discovered scanning tunneling microscopes and high-temperature superconductors. These two laboratories are good examples to illustrate the two styles of science. CERN with its big machines and its centralized administration belongs firmly to the Napoleonic tradition, even if the new Director General, Christopher Llewellyn-Smith, does not wear the Imperial tiara as flamboyantly as his predecessor Carlo Rubbia. The IBM laboratory at Rüschlikon is just as firmly Tolstoyan, with a social structure resembling an extended family, and nobody giving orders. The two styles are clearly appropriate to the different tasks that the two laboratories are engaged in. To operate successfully a machine of the size and complexity of LEP, the Large Electron-Positron Collider, Napoleonic centralization is unavoidable. Big machines require big egoes to build and operate them. Each experiment at LEP resembles a military campaign, with elaborate logistics and timetables prepared several years in advance. On the other hand, military timetables would have been totally out of place at Rüschlikon, where the major discoveries were unexpected and unplanned. At Rüschlikon, the administration provides excellent equipment for talented scientists to play with, and then gives them freedom to play.

In all areas of science the future will bring opportunities to build new tools and make new discoveries, some requiring Napoleonic organization and discipline, others requiring Tolstoyan chaos and freedom. Both on the national and the international levels, the funding for science is likely to be unstable. Money for science will be increasingly spasmodic and unpredictable. In such an environment, Napoleonic enterprises will be ill adapted, Tolstoyan enterprises will be better adapted to survive. We should be prepared to shift science as far as we can toward a Tolstoyan style of operation. In some sciences such as microbiology and neurobiology, Tolstoyan chaos already prevails to a large extent. The two areas of science for which a shift away from Napoleonic rigidity will be most difficult are particle-physics and space-science. I will say a few words about each of these areas in turn.

When I began life as a particle-physicist fifty years ago, most of the major discoveries were made in Europe by people studying the cosmic rays that bombard the earth from outer space. Particle-physics was then in a Tolstoyan phase. Three young Italians, Conversi, Pancini and Piccioni, working with home-made particle-counters in the chaos of post-war Italy, discovered that the common cosmic-ray meson, later called the muon, had only weak interactions with matter. Cecil Powell, working with microscopes and photographic plates at Bristol, discovered the rarer strongly-interacting cosmic-ray meson which he called the pion. Other strange new particles were discovered by Rochester and Butler using old-fashioned cosmic-ray cloud-chambers in Manchester. The new wave of particle-physics grew in a few short years out of improvised experiments observing whatever particles Nature happened to provide. Meanwhile, the Americans at Berkeley and Cornell and Chicago were preparing to build accelerators that would produce particles in great abundance and put the European cosmic-ray experimenters out of business. Within five years, the accelerators triumphed and particle-physics entered a long Napoleonic phase. For forty years the accelerators grew bigger and more expensive, and the accelerator Napoleons collected their Nobel prizes for leading successively larger and more expensive teams of scientists to victory.

Now particle-physics is struggling to survive the disaster of the Superconducting Supercollider. When I speak of the disaster of the Superconducting Supercollider, I do not mean the cancellation of the Supercollider project four years ago. The disaster occurred five years earlier, when the promoters of the Supercollider led the Congress and the public to believe that we could not do any important particle-physics for less than five billion dollars. This belief was not only false but immensely harmful to the future of science. The harm would have been even greater if the Supercollider project had been allowed to continue. Now that the Supercollider is dead, we have a chance to recover from the disaster. We have a chance to convince the public that there is plenty of good particle-physics to be done with machines costing under a billion dollars.

The time may be ripe for a new Tolstoyan phase in particle-physics. Even the scientists who believed that the Superconducting Supercollider was a splendid idea were aware that the big particle accelerators were running into a law of diminishing returns. Each step upward in the size of accelerators cost more money than earlier steps and took a longer time to produce new discoveries. The ratio of scientific output to financial input was diminishing rapidly as the size increased. At some point not far in the future, the further growth of accelerators was bound to grind to a halt. In the meantime, Nature continues to provide a free supply of cosmic rays, and there are reasons to believe that Nature may also provide a free supply of unknown and exotic particles pervading the universe. Groups of scientists in many countries are building large detectors deep underground, with the aim of detecting neutrinos from the sun and incidentally searching for exotic particles. The underground detectors observe natural events with technically sophisticated methods similar to those used for detecting particles in accelerator experiments. Five modern underground detectors are now in operation. Many more are being built and planned. The new generation of underground detectors has four virtues. First, they are cheaper than accelerators by about a factor of ten. Second, they are flexible and can easily be reprogrammed to search for new phenomena in response to new scientific ideas. Third, they are more likely than accelerator experiments to make totally unexpected or unimagined discoveries. Fourth, they lend themselves better than accelerators to an informal and Tolstoyan style of operation. Accelerators still have virtues that underground detectors lack, high precision of measurement, high abundance of particles, and reproducibility of results. In the future, if we are lucky, radically new techniques of acceleration may allow a radical decrease in size and cost for an accelerator of given energy. Particle-physics will continue to need accelerators as well as underground detectors. But the balance in the future is likely to shift toward underground detectors, and this is a hopeful trend for people like me who are incurable Tolstoyans. I see a bright future for particle-physics after the Napoleonic era ends.

Next a few remarks about space-science. Here, even more than in particle-physics, the Napoleonic style has been dominant for the last thirty years. Missions on a grand scale, such as the Voyager explorations of the outer planets and the Hubble Space Telescope explorations of distant galaxies, have brought back to Earth a wealth of scientific knowledge, and also brought back political glory to the bureaucratic Napoleons of NASA. But within NASA, just as in the community of particle-physicists, winds of change are blowing. Space-scientists are keenly aware that times are changing. Billion-dollar missions are no longer in style. Funding in the future will be chancy. The best chances of flying will go to missions that are small and cheap.

I recently spent some weeks at JPL, the Jet Propulsion Laboratory in California. JPL built and operated the Voyager missions. It is the most independent and the most imaginative part of NASA. I was particularly interested in two planetary missions that the people at JPL hope to fly, Cassini and Neptune Orbiter. Cassini is the last of the grand missions belonging to the Napoleonic tradition. The Cassini spacecraft should be on its way to Saturn by the time I give this lecture, if all goes well. After a long and complicated journey, Cassini will stay in orbit around Saturn and make close encounters with Saturn's rings and satellites, sending back to Earth far more information than the Voyagers could provide during their brief fly-by. Cassini looks like Voyager. It is a massive spacecraft, carrying appendages on which a variety of instruments are deployed. The cost of the mission, including five years of operation in the Saturnian system, is estimated at three and a half billion dollars. From a scientific point of view, Cassini is a superb mission, but from a political point of view it is highly vulnerable. It is a prime example of the sort of mission that reformers in NASA and in Congress would like to abolish.

Neptune Orbiter is not on its way to Neptune. It is only a dream in the minds of a small team of designers. It is not yet an approved proposal, much less an approved mission. The idea of Neptune Orbiter is to do at Neptune, three times as far away as Saturn, the same job that Cassini will do at Saturn, but to reduce the cost by a factor of ten by using radically new technology. Four new technologies are crucial for the reduction of costs. First, shrinkage of instruments and computers without loss of performance. Shrinkage in weight by a factor of forty has already been achieved. Second, solar-electric propulsion, using a xenon ion-jet engine powered by solar energy instead of chemical rockets for the long haul from Earth to Neptune. Third, atmospheric braking in the thin stratosphere of Neptune to reduce the speed of the spacecraft and achieve an orbit around the planet without using large quantities of propellant. Fourth, inflatable structures, deployed in space by blowing up balloons with small quantities of gas at very low pressure. I visited the inflatable structures group at JPL. They have dreams of reducing the costs of all large thin structures such as radio-antennas, solar collectors, optical mirrors and air-brakes by a factor of a hundred. All these new technologies are sprouting at JPL in a Tolstoyan fashion, driven by enthusiasm from the bottom rather than by management from the top. Neptune Orbiter is a daring venture, breaking new ground in many directions. If Neptune Orbiter fails to fly, some other even more daring mission will succeed. In space-science, just as in particle-physics, the collapse of the old Napoleonic order opens new opportunities for adventurous spirits.

4. War and Peace Revisited

After these digressions into education and science, I come back to war and peace, to the ethical dilemma faced by Truman and Stimson in 1945. Was their decision to use the bombs to attack civilian populations right or wrong? There can be no final answer to this question. Looking at the results of the decision at the time it was made, the soldiers then fighting in the Pacific war, and expecting soon to be landing on the beaches of Kyushu, had no doubt that the decision was right. Nothing that has happened since is likely to change those soldiers' minds. They believe that the bombs saved their lives. I myself was not about to be landing on the beaches of Kyushu, but I was expecting in August 1945 to fly to Okinawa with a large force of Royal Air Force bombers to assist in the strategic bombing of Japan. We intended, with this force based in Okinawa, to kill each month, for as long as the war lasted, as many Japanese as died at Hiroshima and Nagasaki. At the time, I felt as the soldiers felt, that Truman's decision was right. The nuclear bombing was brutal and bloody, but no more brutal and bloody than the non-nuclear bombing that Bomber Command had been doing in Germany and was intending to do in Japan. In a single attack on Dresden six months earlier, Bomber Command had killed more people than the nuclear bomb would kill in Hiroshima. If Truman could bring the whole bloody business to an end with two more bombs, then he was clearly justified in using them.

When I look back now at Truman's decision with fifty years of hindsight, I still see him doing the right thing in so far as it was in his power to do the right thing. But I see him now as Tolstoy saw Napoleon, not as a supreme war-lord commanding his forces to move as he pleased, but as a child pulling the strings in a carriage while the carriage rolls along through the night. Truman, like Napoleon, was swept along by forces beyond his control. But Truman was temperamentally closer to Kutuzov than to Napoleon. Truman understood his own limitations and the immensity of the problems that he faced. He understood that the deployment of forces in the Pacific war, including the 509th Composite Group that was deployed on the island of Tinian for the purpose of delivering nuclear weapons onto Japanese targets, was a huge and complex machine programmed to roll inexorably forward to the goal of enforcing a Japanese surrender. He was enveloped in the fog of war, ignorant of the balance of forces within the Japanese government, and receiving no clear message from the nuclear experts who were supposed to advise him about the use of the bombs. He had, and could have had, no compelling reason to halt the machine. Having no compelling reason, he had in fact no power to stop it. The machine rolled on, and Truman in his carriage rolled on behind it. The year 1945 was far too late for Truman or anyone else to reverse decisions that had been made several years earlier, in 1941 or in 1939.

There would have been a chance in 1941 to set history moving along a different path. That was the year when the big decisions were made, to rush full-speed into the creation of a nuclear industry in the United States, to develop and build nuclear weapons as fast as possible, to set up the Los Alamos laboratory to design them and to set up a special military organization to deliver them. The scientists who pushed the development of the bombs in 1941 bear the largest share of responsibility for the dropping of the bombs on Japan in 1945. The blame, if there is to be blame, belongs to the physicists. We know why the physicists in 1941 felt compelled to develop weapons. They were afraid of Hitler. They knew that nuclear fission had been discovered in Germany in 1938 and that the German government had started a secret uranium project soon thereafter. They had reason to believe that Heisenberg and other first-rate German scientists were involved in the secret project. They had great respect for Heisenberg, and equally great distrust. They were desperately afraid that the Germans, having started their project three years sooner than the United States, would succeed in building nuclear weapons first.

The fear was so real that hardly a single physicist who was aware of the possibility of nuclear weapons could resist it. The fear allowed scientists to design bombs with a clear conscience. It would have been impossible for the community of American physicists to say to the world in 1941, "Let Hitler have his nuclear bombs and do his worst with them. We refuse on ethical grounds to have anything to do with such weapons. It will be better for us in the long run to defeat him without using such weapons, even if it takes a little longer and costs us more lives". Hardly anybody in 1941 would have wished to make such a statement. And even if some of the scientists had wished to make it, the statement could not have been made publicly, because all discussion of nuclear matters was hidden behind walls of secrecy. The world in 1941 was divided into armed camps with no possibility of communication between them. Scientists in Britain and America, scientists in

Germany and scientists in the Soviet Union were living in separate black boxes. It was too late in 1941 for the scientists of the world to take a united ethical stand against nuclear weapons. The latest time that such a stand could have been taken was in 1939, when the world was still at peace and secrecy not yet imposed.

In 1939, a great opportunity was missed. Niels Bohr arrived in America in January 1939, bringing from Europe the news of the discovery of fission. Within a few months, Bohr and Wheeler had worked out the theory of fission in America, the possibility of a fission chain reaction had been experimentally confirmed in several countries, and Zeldovich and Khariton had worked out the theory of chain reactions in Russia. All this work was openly discussed and rapidly published. The summer of 1939 was the moment for decisive action to forestall the building of nuclear weapons. Nothing was then officially secret. The leading actors in all countries, Bohr, Einstein, Heisenberg, Kapitsa, Zeldovich, Joliot, Peierls and Oppenheimer, were still free to talk to one another and to decide upon a common course of action. The initiative for such a common course of action would have most naturally come from Bohr and Einstein. They were the two giants who had the moral authority to speak for the conscience of mankind. Both of them were international figures who stood above narrow national loyalties. Both of them were not only great scientists but also political activists, frequently engaged with political and social problems. Why did they not act? Why did they not at least try to achieve a consensus of physicists against nuclear weapons before it was too late? Was their distrust of Heisenberg already in 1939 too strong to be overcome? I do not know the answer to these questions. All I know is that 1939 was the last chance for physicists to establish an ethical tradition against nuclear weapons, similar to the Hippocratic tradition that stopped the leading biologists from promoting biological weapons. The chance was missed, and from that point on the march of history led inexorably to Hiroshima.

The chance that was missed in 1939 was a chance to create an ethic of nonviolence in the newborn world of nuclear engineering. After that came fifty years of history still dominated by wars and other kinds of violence. But I have only a few minutes left, and I turn now from the past to the future. What can we do now, to ensure that nuclear weapons will not be used in the years to come? What can we do to strengthen the ethic of nonviolence that we failed to establish fifty years ago?

Tolstoy gave us a clear answer to these questions. Some years after he had finished "War and Peace", he began to think deeply about ethical questions and underwent a religious conversion. He believed that he had the key to the problems of war and peace. He became the prophet of a new religion, to the great annoyance of his wife and friends. His wife liked him better when he was writing novels than when he was preaching. But he went on preaching for the rest of his life, preaching the ethic of nonviolence. His answer to the problems of war and peace was a total commitment to nonviolence. He preached uncompromising opposition to all weapons, to all military organization, to all coercive action of governments. For Tolstoy, nuclear weapons would have raised no ethical dilemma. He would have regarded nuclear weapons as an ultimate absurdity, demonstrating the folly of military force in a particularly dramatic way.

Religion is an immensely powerful and unpredictable factor in human affairs. Tolstoy's religion never became a serious factor in Europe, and least of all in Russia, either before or after the revolution. The only effective action of citizens against war occurred in 1917, when Lenin encouraged the soldiers of Kerenski's army to desert from the front lines where they were fighting the Germans. But this desertion was not the fulfilment of Tolstoy's dream of a citizenry making war impossible by refusing to co-operate. It was only the opening move in the new war for which Lenin was preparing. The seeds of Tolstoy's gospel of nonviolence fell mostly upon stony soil as they were carried over the world, and nowhere was the soil stonier than in his native Russia.

The great blossoming of nonviolence as a mass political movement was the work of Gandhi in India. Gandhi was a disciple of Tolstoy and corresponded with Tolstoy while he was fighting for the rights of the Indian minority in South Africa. After Gandhi returned to India to organize the campaign for Indian independence, he held his followers for thirty years to a Tolstoyan code of behavior. He proved that satyagraha, soul force, can be an effective substitute for bombs and bullets in the liberation of a people. Satyagraha was an effective weapon in Gandhi's hands because he was, unlike Tolstoy, an astute politician. Gandhi's magic failed at the end of his life, when the campaign against British rule was won and he was trying to bring India to independence as a united country. He had then to deal with quarrels between Hindu and Moslem, deeper and more bitter than the quarrel between British and Indian. Five months after the violent birth of independent India and Pakistan, Gandhi was shot by a Hindu nationalist who considered him insufficiently patriotic. The subsequent history of India proved that the ethic of nonviolence was not strong enough to survive the death of its leader and to withstand the temptations of power.

A flowering of nonviolence on a smaller scale occurred in the civil-rights movement in the United States, with Martin Luther King as the leader. Like Gandhi in India, King used Tolstoyan tactics effectively to fight for the rights of his people. King was assassinated, like Gandhi, with his fight unfinished. But while Gandhi led his people for thirty years, King only had ten. After King's death, the United States reverted further and further into a culture of violence. Never was there a time in the history of the United States when the Tolstoyan ethic of nonviolence was more impotent, when violent crime was more prevalent and guns more popular, than they are today.

Nevertheless, the history of religion is full of surprises, and it is possible that the history of the twenty-first century has a big surprise in store for us. It is possible that the monstrous excesses of violence to be seen today all over the world, the senseless murders in the streets of America, the massacres in Bosnia and in Rwanda, the hundred million uncleared landmines killing children daily in Asia, the absurd accumulations of nuclear weapons that we cannot use and are unable to destroy, all are parts of a pattern that will lead to a radical change in human perceptions. Perhaps it will happen as the biologist Haldane wrote in 1923 [Haldane, 1923]: "The tendency of applied science is to magnify injustices until they become too intolerable to be borne, and then the average man whom all the prophets and poets could not move turns at last and extinguishes the evil at its source". Perhaps the twenty-first century will see a conscious turning of mankind toward the ethic of nonviolence as the only way we can survive. Nonviolence as Tolstoy preached it, nonviolence as Gandhi and King practiced it, nonviolence at all levels, domestic, national, international. A radical turning away from violence in the twenty-first century, like the radical turning away from slavery in the nineteenth. The turning away from violence will not happen in a year or in a decade, but it might happen in a century. This is my hope and my dream.

References

M. Armstrong [1983] "Tolstoy on education: The pedagogy of freedom", *Outlook* **48**, 18–51.

R. J. C. Butow [1954] *Japan's Decision to Surrender* (Stanford University Press).

J. B. S. Haldane [1923] *Daedalus, or Science and the Future* (Kegan Paul, London), p. 85.

G. S. Morson [1987] *Hidden in Plain View: Narrative and Creative Potentials in 'War and Peace'* (Stanford University Press).

M. J. Sherwin [1987] *A World Destroyed: Hiroshima and the Origins of the Arms Race*, Revised edition [Vintage Books, Random House, New York]. The revised edition contains in Appendices U,V,W the recently declassified documents concerning plans and estimated casualties for the invasions of Kyushu and Honshu. These were not available in 1975 when the book was originally published.

CHAPTER 4.4

OF CHILDREN AND GRANDCHILDREN*

Our five-year-old grandson Randall is swinging on a swing in the playground. He is thinking and we do not disturb his thoughts. After a while he announces his verdict, "I think I would rather be an egg than a chicken". Just at that moment a large fire-truck approaches and the fire-crew attaches it to a hydrant. This is a novice crew, with an instructor teaching them how to handle the equipment. They are not yet ready to tackle a real fire. Quickly they roll out the fire-hoses and begin squirting water into the neighboring woods. Randall and his brothers rush to join the action. The opportunity to pursue further the philosophy of eggs and chickens is lost.

A year earlier, Randall is silent in the back of the car while my wife and I are sitting in front. The grown-ups are talking about trivialities, about the groceries we are on our way to buy and the probability of finding a convenient parking-space. When there is a pause in our chatter, we hear the voice of the philosopher in the back of the car. He says with a tone of triumph, "God is dead". We ask what led him to this conclusion and he has a logical answer prepared, "Well, we become spirits when we are dead, and God is a spirit, so God must be dead". Then we arrive at the grocery store and trivial pursuits distract our attention. We do not have time to explore the implications of God's death. Only one thing is clear to us. Randall thought of this by himself. He has not been reading Nietzsche. It is likely that his thought was guided not so much by logical argument as by resentment against God's intrusiveness. He knows that he has a good side and a bad side and that in spite of good intentions the bad side often prevails. He resents the fact that God spies on him all the time. God is invading his privacy, and he needs to keep his thoughts private. It is hard enough to be the oldest of three brothers, without having to deal with God in addition. With a sense of relief and liberation, he declares God dead and regains his privacy.

---

*Contribution to the book *Of Beauty and Consolation* edited by Wim Kayzer.

Some time earlier, Randall is sitting in the bathtub in our home. The bathtub is a good place for thinking. After a long silence, he asks, "Does the devil have a brother?" When my wife replies, "No, I don't believe so", Randall is quiet for a while. Then he says, "That would be double trouble for the world".

Beauty and consolation are the gifts that children bring into the lives of parents and grandparents. People without children find other beauties and other consolations. But children are the most universal sources of beauty and consolation, available to educated and uneducated, to sophisticated and primitive, to rich and poor alike. Human beings of all cultures and races have this in common. Our biological heritage imposed on us an imperative, to love and nurture and educate our children. Raising and educating children are the most important things that our evolution has designed us to do. To raise children is exhausting, to educate them is difficult. Evolution has given us the reward of joy to help us endure the pain of child-rearing and the tedium of education. We would not fulfill our roles as guardians of the young if the young did not fill our hearts with joy. It is no accident that we find their bodies beautiful and their minds miraculous. It is no accident that we find consolation for the sorrows of the world in watching their bodies grow and their minds unfold.

It is a curious paradox that the beauty and consolation provided by children play such a large role in our lives and such a small role in our literature. The literature that we consider great is mostly concerned with dramatic events and extraordinary people. The growth of young children is, by contrast, ordinary. The intellectual awakening of young children is a process as ordinary as the physical awakening of woodland in springtime. Everyone loves the cute remarks of children just as everyone loves the first flowers of spring. Flowers in spring also bring beauty and consolation, but no great literature is written about flowers. Such ordinary things as babies and flowers do not make themes for great books. Although the process by which a child begins to understand and to ask questions is a deep mystery, perhaps the deepest mystery of all that we see around us, we still consider babies ordinary because we are so familiar with them.

There is one genre of literature in which great books about children can be written. This is the genre of books written for children to read. In a book for children, the ordinary becomes extraordinary. The beauty of the ordinary can be honestly portrayed. The mystery of the ordinary is not hidden. Lewis Carroll was one of the greatest writers for children, because he cared passionately for his child-friend Alice and saw the world through her eyes. The world that he saw was absurd and full of paradox because the world that every child sees is absurd and full of paradox. Randall's thought, that he would rather be an egg than a chicken, is a thought that might easily have occurred to Alice. Humpty-Dumpty, the master of logical paradox, was an egg. Carroll understood what every child understands, that words are magic. He understood that eggs are also magic, their smooth round shape hiding the mysteries that give birth to chickens, turtles and dinosaurs. He gave to Humpty-Dumpty the insight that words mean whatever you want them to mean. Such an insight would not have occurred to a chicken.

Another great work of literature written for children is "Charlotte's Web". This goes deeper than the Alice books. It deals directly with death. Charlotte dies. White wrote a masterpiece because he was not afraid to look at death through the mind of a child. At the end of tragedy comes consolation. After death comes rebirth. Charlotte never sees her babies, and her babies never know their mother, but life goes on. A child is able to accept death as a fact of life. Every child more than three years old is aware of death, and most children can speak of it in matter-of-fact words without fear. Two years ago, our little poodle died, and Randall sitting in his high-chair at the supper-table remarked, "My bones are feeling heavier and heavier every day. I think I will be going to Heaven soon". Every child to whom we have read "Charlotte's Web" aloud has loved it. The story is sad but it is also funny and joyful. Children know that life is like that, a mixture of sorrow and joy.

There are also works full of beauty and consolation written for adults about children. Richard Hughes was a master of this genre. His best-known story, "A High Wind in Jamaica" deals with murder. Emily is a resourceful little girl who finds herself accidentally on board a pirate ship in the Caribbean. The pirates are friendly souls who treat her well and mean her no harm. The pirate captain gives her a bed in his own cabin. Then the captain of a Dutch merchant ship is captured and put with Emily in the cabin. In a moment of panic she murders the Dutchman. Arriving back in England, she easily pins the blame for the murder on the pirates. At the court of enquiry, the accused men have no chance of prevailing against her evident innocence. Insouciantly she walks free while they are led away to be hanged. Emily shows us the dark side of childhood, the ruthlessness that nature gave us long ago to help us survive the harsh struggle for existence, the cunning that hides ruthlessness behind an innocent face. Every child has a dark side, though not every child commits murder. Emily is true to her nature. She is lovable, because it is in our nature as human beings to love children even when they do evil. Children must be ruthless and parents must be forgiving, otherwise their genes will not survive.

Another great book about children is William Golding's "Lord of the Flies". Golding shows us an even darker side of childhood. The ruthlessness of a gang is worse than the ruthlessness of a single child. The gang is the natural result of competition for status and power, when children are left by themselves on an island without adults to restrain them. The children need strong leadership to survive in a tropical jungle, and the price of strong leadership is blind loyalty and terror. They are trapped between the silence of the jungle and the brutality of the gang. Children still obey the instincts of tribal cohesion that enabled a social animal to survive the hazards of famines and ice-ages. Golding's children are as individuals no worse than Emily. They are worse only when they are submerged in the gang. Golding's book tells us deep truths about our children and about ourselves. Even when we do evil, we are not all bad. Golding's children, in the depths of their degradation, are lovable too. There is beauty and consolation in tragedy, not only in stories with happy endings.

Tolstoy famously remarked that all happy families are alike but each unhappy family is unhappy in its own way. That is why great books are written about unhappy families. Unhappy families provide tension and drama, the stuff of literature, while happy families provide a card-game for children. It seems to be only in the visual arts that happy families are portrayed as central to the human condition. The holy family, centered on the child Jesus, has inspired thousands of great paintings and sculptures. One does not need to be a Christian to feel the joy of the renaissance painters and the beauty of the masterpieces that they produced. Their religion taught them that this ordinary baby that they painted was the light of the world. Even for those of us to whom the religion is alien, the paintings and the light remain.

CHAPTER 4.5

A FAILURE OF INTELLIGENCE: OPERATIONAL RESEARCH AT RAF BOMBER COMMAND, 1943–1945[*]

I began work at the Operational Research Section (ORS) of RAF Bomber Command on July 25, 1943. The headquarters of Bomber Command was a substantial set of red brick buildings hidden in the middle of a forest on top of a hill in the English county of Buckinghamshire. The main buildings had been built before the war. The ORS was added in 1941 and was housed in a collection of trailers at the back. Trees were growing right up to the windows, so we had little daylight even in summer. The Germans must have known where we were, but they never came to disturb us. I was billeted in the home of the Parsons family in the village of Hughenden. Mrs Parsons was a motherly soul and took good care of me. Once a week she put out her round tin bathtub on her kitchen floor and filled it with hot water for my weekly splash. Each morning I bicycled five miles up the hill to Bomber Command, and each evening I came coasting down. Sometimes, as I was struggling up the hill, an Air Force limousine would zoom by, and I had a quick glimpse of our Commander-in-chief Sir Arthur Harris sitting in the back, on his way to give the orders that sent thousands of boys like me to their deaths. Every day he made the decision, depending on the weather and the state of readiness of the bombers, either to send them out that night or let them rest. Every day he chose the target for tonight.

I arrived at Bomber Command on the day after one of our most successful operations, a full-force night attack on Hamburg. That night the bombers used for the first time the decoy system which we called WINDOW and the Americans called CHAFF. WINDOW consisted of packets of paper strips coated with aluminum paint. One crew member in each bomber was responsible for throwing packets of WINDOW down a chute at a rate of one per minute while flying over Germany. The paper strips floated slowly down through the stream of bombers, each strip being a

[*]*Technology Review* **109**, 62–71 (2006).

resonant antenna tuned to the frequency of the German radars. The purpose was to confuse the radars so that they could not track individual bombers in the clutter of echoes from the WINDOW.

That day when I came to work, the people at the ORS were joyful. I never saw them so joyful again until the day that the war in Europe ended. WINDOW had worked. The bomber losses the night before were only 12 out of 791, one-and-a-half percent, far less than they would have been without WINDOW for a major operation in July when the skies in northern Europe are never dark. The losses at that time were usually about five percent, mostly due to night-fighters which were guided to the bombers by radars on the ground. WINDOW had cut the expected losses by two thirds. Each bomber carried a crew of seven, so WINDOW that night had saved the lives of about 180 of our boys. The first job that my boss Reuben Smeed gave me to do when I arrived was to draw pictures of the cloud of WINDOW trailing through the bomber stream as the night progressed, taking into account the local winds at various altitudes that were measured and reported by the bombers. My pictures would be shown to the aircrew to impress on them how important it was for them to stay within the stream after bombing the target, rather than flying home independently.

Smeed explained to me that the same principles applied to bombers flying at night over Germany as to ships crossing the Atlantic. Ships had to travel in convoys, because the risk of being torpedoed by a U-boat was much greater for a ship traveling alone. For the same reason, bombers had to travel in streams. The risk of being tracked by radar and shot down by a fighter was much greater for a bomber flying alone. But the crews hated to fly in the bomber-stream because they were more afraid of collisions than of fighters. Every time when they flew in the stream, they would see bombers coming close and almost colliding with them, but they almost never saw a fighter. The German night-fighter force was tiny compared with Bomber Command. The pilots were highly skilled and hardly ever got shot down. They carried a firing system called "Schräge Musik" or "Crooked Music", which allowed them to fly underneath a bomber and fire guns vertically upward. The fighter could see the bomber clearly silhouetted against the night sky, while the bomber could not see the fighter. This system efficiently destroyed thousands of bombers, and we did not even know that it existed. This was the greatest failure of our ORS. We learned about Schräge Musik only after the war was over.

The bomber crews judged the risk of a collision to be greater than the risk of being shot down, and tried to keep out of the bomber-stream as much as they could. Smeed said he believed that the crews' judgment was wrong. He believed that the chance of being shot down by a fighter was far greater than the chance of colliding even in the densest part of the stream. But he did not have any solid evidence to back up his belief. He had been too busy with other urgent problems to collect the evidence. He told me that the most useful thing I could do would be to become the Bomber Command expert on collisions. When not otherwise employed, I should collect every scrap of evidence I could find about fatal and non-fatal collisions and put it all together. Then perhaps we could convince the aircrew that they were really safer staying in the bomber-stream.

There were two ways to study collisions, using theory or using observations. I tried both ways. The theoretical way was to use the formula, collision rate for a bomber flying in the stream equals density of bombers multiplied by average relative velocity of two bombers multiplied by mutual presentation area (MPA). The MPA was the area in a plane perpendicular to the relative velocity within which a collision would occur. It was the same thing that atomic and particle physicists call a collision cross-section. It was roughly four times the area of a bomber as seen from above. The formula assumes that two bombers on a collision course do not see each other in time to avoid the collision. For bombers flying at night over Germany, that assumption was probably true.

All three factors in the collision formula were uncertain. The MPA would be larger for an up-and-down collision than for a sideways collision. I assumed that most of the collisions would be up-and-down, with the relative velocity vertical. The relative velocity would depend on how vigorously the bombers were corkscrewing as they flew. Except for the bombing run over the target, they never flew straight and level. A bomber that flew straight and level over Germany would be a sitting duck for antiaircraft guns. The standard manoeuvre to avoid antiaircraft fire was the corkscrew, combining side-to-side with up-and-down weaving. For collisions, it was the up-and-down motion that was most important. From crew reports I estimated up-and-down motions averaging forty miles an hour, uncertain by a factor of two. But the dominant uncertainty in the collision formula was the density of bombers in the stream. I studied the crew reports, which sometimes reported large deviations from the track that they were supposed to fly. For the majority of crews who reported no large deviations, there was no way to tell how close to the assigned track they actually flew. My best estimate of the density of bombers was uncertain by a factor of ten. This made the collision formula practically worthless. The only value of the collision formula was to set an upper bound to the collision rate. If I assumed maximum values for all three factors in the formula, it gave a loss-rate due to collisions of one percent per operation. One percent was much too high to be acceptable, but still less than the overall loss rate of four percent. Even if we squeezed the bomber-stream to the highest possible density, collisions would not be the main cause of losses.

Observational evidence of lethal collisions over Germany was plentiful but unreliable. The crews frequently reported seeing events that looked like collisions, first an explosion in the air, and then two flaming objects falling to the ground. These events were visible from great distances and were often multiply reported.

The crews tended to believe that they were seeing collisions, but there was no way to be sure. Most of the events were probably single bombers, hit by antiaircraft shells or by fighter cannon-fire, breaking in half as they disintegrated.

In the end I found only two sources of evidence that I could trust, bombers that collided over England, and bombers that returned damaged by non-lethal collisions over Germany. The numbers of each kind were small but reliable. The numbers were small enough, so that I could investigate each case individually. The case that I remember best was a collision between two Mosquito bombers over Munich. The Mosquito was a light two-seater bomber that Bomber Command used extensively for small-scale attacks, to confuse the defenses and distract attention from the heavy attacks. Two Mosquitos flew alone from England to Munich and then collided over the target, with only minor damage. It was obvious that the collision could not have been accidental. The two pilots must have seen each other when they got to Munich and started playing games. The Mosquito was fast and manoeuvrable and hardly ever got shot down, so the pilots felt themselves to be invulnerable. I interviewed Pilot Officer Izatt who was one of the two pilots. After I gently questioned him about the Munich operation, he confessed that he and his friend had been enjoying a dog-fight over the target when they bumped into each other. So I crossed the Munich collision off my list. It was not relevant to the statistics of collisions between heavy bombers in the bomber-stream. There remained seven authentic non-lethal collisions between heavy bombers over Germany.

For bombers flying at night over England in training exercises, I knew the numbers of both lethal and non-lethal collisions. The ratio of lethal to non-lethal collisions was three to one. If I assumed that the chance of surviving a collision was the same over Germany as over England, then it was simple to calculate the number of lethal collisions over Germany. There were two reasons why the chances over England and Germany might be different. On the one hand, a badly damaged aircraft over Germany might fail to get home while an aircraft with the same damage over England could make a safe landing. On the other hand, the crew of a damaged aircraft over England might decide to bail out and let the plane crash, while the same crew over Germany would be strongly motivated to bring the plane home. There was no way to calculate the effects of these two differences between England and Germany. Since the two effects would work in opposite directions, I decided to ignore them both. I calculated the number of lethal collisions over Germany to be twenty-one, just three times the number of non-lethal collisions. These numbers referred to major operations over Germany in which about 60000 sorties were flown. So collisions destroyed 42 aircraft in 60000 sorties, a loss-rate of 0.07 percent. This was the best estimate I could make. I could not calculate any reliable limits of error, but I felt confident that the estimate was correct within a factor of two. It was consistent with the less accurate estimate obtained from the theoretical formula, and it strongly confirmed Smeed's belief that collisions were a smaller risk than fighters.

Another job that Smeed gave me was to invent a way of doing estimates of the effectiveness of various counter-measures, using all the evidence from a heterogeneous collection of operations. The first example that I worked on was Monica. Monica was a warning device that gave a high-pitched squeal over the intercom when a bomber was in the beam of a German radar. The squealing became more rapid as the radar signal became stronger. The crews disliked Monica because it was too sensitive and gave too many false alarms. They usually switched it off so that they could talk to each other without interruption. My job was to see from the results of many operations whether Monica actually saved lives. I had to compare the loss-rates of bombers with and without Monica. This was difficult because Monica was distributed unevenly among the squadrons. It was given preferentially to Halifax bombers which usually had higher loss-rates, and less often to Lancaster bombers which usually had lower loss-rates. In addition, Halifaxes were sent preferentially on less dangerous operations and Lancasters on more dangerous operations. To use all the evidence of losses of Halifaxes and Lancasters on a variety of operations, I invented the method which was later reinvented by epidemiologists and given the name "meta-analysis". To put together the evidence from many operations to judge the effectiveness of Monica was just like putting together the evidence from many clinical trials to judge the effectiveness of a drug.

My method of meta-analysis was the following. First I subdivided the data by operation and by type of aircraft. For example, one subdivision would be Halifaxes on Bremen on March 5, another would be Lancasters on Berlin on December 2. In each subdivision, I tabulated the number of aircraft with and without Monica, and the number lost with and without Monica. I also tabulated the number of Monica aircraft expected to be lost if Monica had no effect, and the statistical variance of the number expected. So I had two quantities for each subdivision, the difference observed-minus-expected of losses of Monica aircraft, and the variance of this difference. I assumed that the distributions of losses in the various subdivisions were uncorrelated. So I could simply add up the two quantities, observed-minus-expected losses and variance, over all the subdivisions. The result was a total observed-minus-expected losses and variance for all the Monica aircraft, unbiassed by the different fractions of Monica aircraft in the various subdivisions. If the total of observed-minus-expected losses was significantly negative, it meant that Monica was effective. This was a sensitive test of effectiveness, making use of all the available information. When I did the calculation for Monica, the result was clear. The total of observed-minus-expected losses was slightly positive and less than the square-root of the total variance. Monica was statistically worthless. The crews had been right when they decided to switch it off.

I later applied the same method of analysis to study the question, whether experience helped crews to survive. The Command told the crews that their chances of survival would increase with experience, and the crews believed it. After you have got through the first few operations, they were told, things will get better. To believe this was important for morale, at a time when the fraction of crews surviving to the end of a thirty-operation tour was about twenty five percent. I subdivided the experienced and inexperienced crews on each operation and did the analysis, and again the result was clear. Experience did not reduce loss-rates. The cause of losses, whatever it was, killed novice and expert crews impartially. This result contradicted the official dogma, and the Command never accepted it. I blame the ORS, and I blame myself in particular, for not taking our result seriously enough. The result was clear evidence that the main cause of losses was an attack that gave the experienced crews no chance either to escape or to defend themselves. If we had taken the evidence more seriously, we might have discovered Schräge Musik.

The part of his job that Smeed enjoyed most was interviewing evaders. Evaders were crew members who survived being shot down over German-occupied countries and made their way back to England. About one percent of all those shot down came back. Each week, Smeed would go to London and interview one or two of them. Sometimes he would take me along. We were not supposed to ask them questions about how they got back, but they would sometimes tell us amazing stories without being asked. We were supposed to ask them questions about how they were shot down. They had very little information to give us about that. Most of them said they never saw a fighter and had no warning of an attack. There was just a sudden burst of cannon-fire and the aircraft fell apart around them. Again, we missed an essential clue that might have led us to Schräge Musik.

The one time that I did something useful for Bomber Command in a practical way was in Spring 1944, when Smeed sent me to make accurate measurements of the brightness of the night sky as a function of time, angle and altitude. The measurements would be used by our route-planners to minimize the exposure of bombers to the long summer twilight over Germany. I went to an airfield at the village of Shawbury in Shropshire and flew for several nights in an old Hudson aircraft, unheated and unpressurized. The pilot flew back and forth on a prescribed course at various altitudes, while I took readings of sky brightness through an open window with an antiquated photometer, starting soon after sunset and ending when the sun was eighteen degrees below the horizon. I was surprised to find that I could function quite well without oxygen at 20000 feet altitude. I shared this job with Professor J.F.Cox, a Belgian professor who was caught in England when Hitler overran Belgium in 1940. Cox and I took turns doing the measurements. My flights were uneventful, but on the last of Cox's flights the two engines of the Hudson both failed and the pilot decided to bail out. Cox bailed out successfully and came to earth still carrying the photometer. He broke an ankle but saved the photometer. After the war he became Rector of the Free University in Brussels.

I was nineteen when I came to Bomber Command, fresh from an abbreviated two years as a student at Cambridge University. Members of the ORS were civilians, paid by the Ministry of Aircraft Production and not by the Air Force. The idea was that we should not be afraid to speak our minds to senior officers when giving them advice. The ORS system had been invented by Patrick Blackett to give scientific advice to the Navy. The Navy ORS was extremely effective and made big contributions to winning the war against the U-boats in the Atlantic. Blackett had two enormous advantages. First, he was a world-renowned scientist with a Nobel Prize and a safe job in the academic world, so that he could threaten to resign if his advice was not followed. Second, he had been a Navy officer in World War I and was respected by the admirals that he advised. Basil Dickins, the chief of our ORS at Bomber Command, had neither of these advantages. He was a career civil servant with no independent standing. He could not threaten to resign, and Sir Arthur Harris had no respect for him. His success in his career depended on telling Sir Arthur things that Sir Arthur wanted to hear. So that is what he did. He was the only member of the ORS who met regularly with Sir Arthur. He gave Sir Arthur information rather than advice. He never raised serious questions about Sir Arthur's tactics and strategy.

Our ORS was divided into sections and sub-sections. The sections were ORS1 concerned with bombing effectiveness, ORS2 concerned with bomber losses, ORS3 concerned with history. My boss Reuben Smeed was chief of ORS2. The subsections of ORS2 were ORS2a collecting crew reports and investigating causes of losses, ORS2b studying the effectiveness of electronic counter-measures, ORS2c studying damage to returning bombers, ORS2d doing statistical analysis and other jobs requiring some mathematical skill. I was put into ORS2d. Two other new boys arrived at the same time with me, John Carthy who was in ORS1, and Mike O'Loughlin who shared an office with me in ORS2d. John had been a leading actor in the Cambridge University student theater. Mike had been briefly in the army but was discharged when he was found to be epileptic. John and Mike and I became lifelong friends. John was cheerful, Mike was bitter, and I was somewhere in between. In later life, John was a biologist at London University and Mike taught engineering at the Cambridge Polytechnic. After retiring from the Polytechnic, Mike became an Anglican minister in the parish of Linton near Cambridge.

The ORS consisted of about thirty people, a mixed bunch of civil servants, academic experts and students. Working with us was an equal number of WAAFs, Women's Auxiliary Air Force girls, who wore blue uniforms and were subject to military discipline. The WAAFs were photographic interpreters, calculators, technicians, drivers and secretaries. They did most of the real work of the ORS. They also supplied us with tea and sympathy. They made a depressing situation bearable.

Their leader was Sergeant Asplen, a tall and strikingly beautiful girl whose authority was never questioned. The sergeant kept herself free of romantic entanglements. But two of her charges, a vivacious red-head called Dorothy and a more thoughtful brunette called Betty, became entangled with my friends John and Mike. Romantic entanglements among the troops were not officially discouraged. We celebrated two weddings before the war was over, with Dorothy and Betty discarding their dumpy blue uniforms for an afternoon and appearing resplendent in white silk. The marriages endured, and each afterwards produced four children.

For a week after I arrived at the ORS, the attacks on Hamburg continued. The second on July 27 raised a firestorm which devastated the central part of the city and killed about forty thousand people. We only succeeded twice in raising a firestorm, once in Hamburg and once more in Dresden in 1945. The Germans had good air-raid shelters and warning systems, and did what they were told. As a result of their good civil defense, only a few thousand people were killed in a typical major attack without a firestorm. When there was a firestorm, people were asphyxiated or roasted inside shelters, and the number killed was more than ten times greater. Every time that Bomber Command attacked a city, we were trying to raise a firestorm, but we only succeeded twice and we never knew why. Probably a firestorm could only happen when three things worked together, first a city with a high density of old buildings, second an attack with a high density of incendiary bombs in the central area, and third an unstable state of the atmosphere. When these three things were just right, the flames and the winds combined to make a blazing hurricane. The same thing happened one night in Tokyo in March 1945, and once more at Hiroshima in August. The Tokyo firestorm was the biggest and killed more people than any of the others.

The third Hamburg raid was on the night of July 29 and the fourth on August 2. After the firestorm, the law of diminishing returns was operating. The fourth attack was a fiasco, with high and heavy clouds over the city and bombs scattered over the countryside. But the bomber losses were rising, close to four percent for the third attack and a little over four percent for the fourth. The Germans had learned quickly how to deal with WINDOW. Since they could no longer track individual bombers with radar, they guided the fighters into the bomber-stream and let them find their own targets. Within a month, loss-rates were back at the five percent level, and the life-saving effect of WINDOW was over.

On November 18, 1943, Sir Arthur Harris started the Battle of Berlin. His whole life had been devoted to the proposition that strategic bombing could defeat Germany without the help of land armies, and this was his last chance to prove it. He said publicly that the Battle of Berlin would knock Germany out of the war. The mammoth force of heavy bombers that he commanded had been planned by the British government in 1936 as our primary instrument for defeating Hitler without repeating the trench-warfare horrors of World War I. It was absorbing by itself one quarter of the entire British war effort. In November 1943 it was finally ready to do what it was designed to do, to smash Hitler's empire by demolishing Berlin. The Battle of Berlin started with a success, like the first attack on Hamburg on July 24. We attacked Berlin in good weather with 444 bombers and only 9 were lost. The photo-plots showed that a large fraction of the bombs fell fair and square on the central part of the city. The losses were small, not because of WINDOW but because of clever tactics. Two bomber forces were out that night, one going to Berlin and one to Mannheim. The German controllers were confused and sent most of the fighters to Mannheim.

After that first success, Sir Arthur ordered fifteen more heavy attacks on Berlin, expecting to destroy Berlin at least as thoroughly as he had destroyed Hamburg. All through the winter until March 1944, the bombers were hammering away at Berlin. The November success was never repeated. The weather that winter was worse than usual, covering the city with cloud for weeks on end. Our photo-reconnaissance planes could bring back no pictures to show how poorly we were doing. As the winter went on, the defenses grew stronger, the losses grew heavier, and the scatter of the bombs grew worse. We never raised a firestorm in Berlin. On March 24, in the last of the sixteen attacks, we lost 72 out of 791 bombers, a loss-rate of nine percent, and Sir Arthur admitted defeat. The battle cost us 492 bombers with more than 3000 aircrew. Industrial production in Berlin continued to increase, and the operations of government were never seriously disrupted. There were two main reasons why Berlin won the battle. First, the city is more modern and less dense than Hamburg, spread out over an area as large as London with only half of London's population, so it does not burn well. Second, the repeated attacks along the same routes allowed the fighters to find the bomber-stream earlier and kill bombers more efficiently as time went on.

A week after the final attack on Berlin, we suffered an even more crushing defeat. We attacked Nürnberg with 795 bombers and lost 94, a loss-rate of almost twelve percent. It was then clear to everybody that the losses were unsustainable. Sir Arthur reluctantly abandoned his dream of winning the war all by himself. Bomber Command stopped flying deep into Germany and spent the summer of 1944 giving tactical support to the armies invading France.

One of our group of young students at the ORS was Sebastian Pease, known to his friends as Bas. He had joined the ORS only six months before I did, but he already knew his way around and was at home in that alien world. He was the only one who was actually doing what we were all supposed to do, helping substantially to win the war. The rest of us were sitting at Command Headquarters, depressed and miserable because our losses of aircraft and aircrew were tremendous and we were unable to do much to help. The Command did not like civilians wandering around operational squadrons and collecting information, so we were mostly confined to our gloomy offices at the headquarters. But Bas succeeded in breaking out. He spent most of his time at the squadrons and only came back to headquarters occasionally. Fifty years later, when he was visiting Princeton, he told me what he had been doing.

Bas was able to escape from Command Headquarters to the squadrons because he was the expert in charge of a precise navigation system called G-H. Only a small number of bombers could carry G-H because it required two-way communication with ground stations. These bombers belonged to two special squadrons. 218 Squadron was one of them. The G-H bombers were Stirlings, slow and ponderous machines that were due to be replaced by the smaller and more agile Lancasters. They did not take part in mass bombing operations with the rest of the Command, but did small precise operations on their own with very low losses. Bas spent a lot of time at 218 Squadron and made sure that the G-H crews knew how to use their equipment to bomb accurately. He had a good war, as we used to say in those days. The rest of us were having a bad war.

Some time early in 1944, 218 Squadron stopped bombing and started training for a highly secret operation called Glimmer which Bas helped to plan, to divert German attention from the invasion fleet that was to invade France in June. The operation was carried out on the night of June 5–6. The G-H bombers flew around in tight circles while moving slowly out over the English Channel, so that they appeared to the German radars to be a fleet of ships. They flew very low and dropped packages of WINDOW, while boats underneath carried specially designed radar transponders. While the real invasion fleet was moving out towards Normandy, the fake invasion fleet composed of G-H bombers was moving out toward the Pas de Calais two hundred miles to the East. The spoof was successful, and the strong German forces in the Pas de Calais did not move to Normandy in time to stop the invasion. While Bas was training the crews, he said nothing about it to his friends at ORS. We knew only that he was out at the squadrons doing something useful. Even after Glimmer was over and the invasion had succeeded, Bas never spoke about it. My boss Reuben Smeed was a man of considerable wisdom. One day at Bomber Command, he said, "In this business you have a choice. Either you get something done, or you get the credit for it, but not both". Smeed's dictum applies to many other things besides Operational Research. Bas was a fine example of Smeed's dictum. He made his choice, and he got something done. In later life he became a famous plasma physicist and ran the Joint European Torus, the main fusion program of the European Union.

After the war, Smeed became Professor of Traffic Studies at University College London. He was director of the Traffic Studies Group, which applied the methods of operational research to traffic problems all over the world. His group designed intelligent control systems to operate traffic lights so as to optimize the flow of traffic through cities. Smeed had a fatalistic view of traffic flow. He said that the average speed of traffic in central London will always be nine miles per hour, because this is the minimum speed that people will tolerate. The intelligent traffic lights increase the number of cars on the roads but do not increase the speed. As soon as the traffic flows faster, more drivers come to slow it down. It will be interesting to see whether the average speed will remain nine miles per hour, now that you have to pay a serious toll to drive in central London.

Smeed also had a fatalistic view of traffic accidents. He collected statistics of traffic deaths from many countries, all the way back to the invention of the automobile. He found that under an enormous range of conditions the number of deaths per year in a country is given by a simple formula: number of deaths equals 0.0003 times the two-thirds power of the number of people times the one-third power of the number of cars. This formula is known as Smeed's Law. He published it in 1949, and it is still valid fifty-seven years later. It is of course not exact, but it holds within a factor of two for almost all countries at almost all times. It is remarkable that the number of deaths does not depend strongly on the size of the country, the quality of the roads, the rules and regulations governing traffic, or the safety equipment installed in cars. Smeed interpreted his law as a law of human nature. The number of deaths is determined mainly by psychological factors which are independent of material circumstances. People will drive recklessly until the number of deaths reaches the maximum that they can tolerate. When the number exceeds the tolerable limit, they drive more carefully. Smeed's Law merely defines the number of deaths that we find psychologically tolerable. One of the consequences of Smeed's Law is surprising. If you are a driver, your chance of being killed declines rapidly as the number of cars on the roads increases. That is the main reason why the death-rate per driver is much higher in poor than in rich countries. During the next ten years, as the number of cars in China doubles, the death-rate per driver will go down sharply.

The last year of the war was quiet at ORS Bomber Command. We knew that the war was coming to an end and that nothing we could do would make much difference. With or without our help, Bomber Command was doing well. In the fall of 1944, when the Germans were driven out of France, it finally became possible for our bombers to make accurate and devastating night attacks on oil refineries and synthetic oil production plants. Accurate bombing became possible for two reasons. First, the loss of France made the fighter defenses much less effective. Second, a new method of organizing attacks was invented by 5 Group, the most independent of the Bomber Command groups. The new method was pioneered by 617 Squadron, one of the 5 Group squadrons, which carried out the famous attack on the Ruhr dams in March 1943. The good idea, as usually happens in large organizations, grew from the bottom up rather than from the top down. The new way of attacking oil targets had a Master Bomber flying at low altitude over the target in a Mosquito, directing the attack by radio in plain language. The Master Bomber would first mark the target accurately with target indicator flares, and then tell the heavy bombers overhead precisely where to aim. A deputy Master Bomber in another Mosquito was ready to take over in case the first one was shot down. 5 Group carried out many such precision attacks with great success and low losses, while the other groups flew to other places and distracted the fighter defenses. In the last winter of the war, the German army and air force finally began to run out of oil. Bomber Command could justly claim to have helped the allied armies who were fighting their way into Germany from East and West.

While the attacks on oil plants were helping to win the war, Sir Arthur continued to order major attacks on cities, including the attack on Dresden on the night of February 13, 1945. The Dresden attack became famous because it caused a firestorm and killed a large number of civilians, many of them refugees fleeing from the Russian armies that were over-running Pomerania and Silesia. It caused some people in Britain to question the morality of continuing the wholesale slaughter of civilian populations when the war was almost over. Some of us were sickened by Sir Arthur's unrelenting ferocity. But our feelings of revulsion after the Dresden attack were not widely shared. The British public at that time still had bitter memories of World War I, when German armies brought untold misery and destruction to other people's countries, but German civilians never suffered the horrors of war in their own homes. The British mostly supported Sir Arthur's ruthless bombing of cities, not because they believed that it was militarily necessary, but because it was teaching the German civilians a good lesson. This time the German civilians were finally feeling the pain of war on their own skins.

I remember arguing about the morality of city-bombing with the wife of a senior Air Force officer, after we heard the results of the Dresden attack. She was a well-educated and intelligent woman who worked part-time for the ORS. I asked her whether she really believed that it was right to kill German women and babies in large numbers at that late stage of the war. She answered, "Oh yes. It is good to kill the babies especially. I am not thinking of this war but of the next one, twenty years from now. The next time the Germans start a war and we have to fight them, those babies will be the soldiers". After fighting Germans for ten years, four in the first war and six in the second, we had become almost as bloody-minded as Sir Arthur.

At last, at the end of April 1945, the order went out to the squadrons to stop offensive operations. Then the order went out to fill the bomb-bays of our bombers with food packages to be delivered to the starving population of the Netherlands. I happened to be at one of the 3 Group bases at the time, and watched the crews happily taking off on their last mission of the war, not to kill people but to feed them.

CHAPTER 4.6

THE INDIVIDUAL OR THE GROUP? A QUESTION THAT ARISES IN SCIENCE, LAW AND LANGUAGE*

1. Introduction

First of all, my thanks to Judge Kern and Judge Cotter and anyone else who was responsible for inviting me to meet with this distinguished group. I spoke with you a year ago about global warming and various other controversial questions that appear frequently in the daily news. But I do not expect you to remember what I said then. You do not need to worry, there will not be a quiz. Today I will talk about another controversial question, a question that does not often appear in the daily news but influences our thinking and our behavior on a longer time-scale.

The question, whether the individual or the group is more important, is one of the basic problems that we have to deal with as human beings and as members of a civilized society. It goes to the root of our ethics, our laws and our politics. A little less than a year ago, in August 2007 to be precise, this question hit me from three different directions in one week. The first hit came from Caroline Humphrey, a distinguished anthropologist at Cambridge University, who has worked extensively in Mongolia and Central Asia and happens to be fluent in Russian. The second hit came from the lawyer Lawrence Latto in Washington, former editor of the Columbia Law Review, an expert on American constitutional law. The third hit came from Richard Dawkins, now professor of public communication of science at Oxford University. Dawkins has recently been in the limelight as author of a new book, "The God Delusion". I disagree strongly with his claim that anyone who wishes to become a scientist must also become an atheist, but that is not the subject of my talk today. My subject is the individual and the group. Humphrey is an expert on

*Medina Seminar given to Judiciary Leadership Development Council Princeton University, June 16, 2008.

language, Latto on law, and Dawkins on science. I found it striking that the same question, the clash in human society between the interests of the individual and the interests of the group, should arise in three widely separated contexts. I will describe the three hits in turn and then put them together and arrive at some conclusions.

I am well aware that it is foolish of me to talk about a question of law to an audience of judges. Each of you knows a hundred times more about constitutional law than I do. For this reason I hesitated to include the second section of this talk where I discuss the views of Lawrence Latto. Many of you may be familiar with Latto's argument, and you may have good reasons for dismissing it as legally unsound. I decided nevertheless to include the legal section of the talk, since it provides an essential link between the theoretical worlds of language and science and the practical world of law. Without this legal section, the talk would lose its main point. I must therefore ask you to excuse me for venturing into your territory for a few minutes. The third section of the talk, in which I return to my own territory of science, will be the longest and the most substantial. I now begin the first section, which is concerned with language.

2. The First Hit: Language

Caroline Humphrey gave a talk at a meeting of the American Philosophical Society with the title, "Alternative Freedoms", in April 2005, and her text was published in the Proceedings of the Society, volume 151, pages 1–10, in March 2007. Her concern was the illusion prevalent in America that the American idea of freedom has a meaning that can be readily understood and shared by people all over the world. If our idea of freedom is to be shared by people belonging to other cultures, then our word for freedom must be translatable into their languages. If translation of the word fails, then understanding of the idea must also fail. Humphrey examines in detail the translation of the word "freedom" into Russian, and shows us how and why the translation fails.

There are three Russian words that are commonly used to translate the English word "Freedom". They are "Svoboda", "Mir" and "Volya". Each has different overtones of meaning and history that are foreign to our way of thinking. "In medieval times", Humphrey writes, "svoboda, which is based on the root 'svoi' (self, ours), seems to have meant something rather different, that is, the security and well-being that result from living amongst one's own people. Svoboda first of all was the agglomeration of practices of our own way of life, most fundamentally contrasted with those of alien people and enemies. It suggests an image of a social kind of freedom, one that was not centred on the singular individual".

Humphrey then discusses the later history of the word "Svoboda". "The officially defined svoboda in Soviet times came to have a double sense. First of all independence, not being ruled by foreigners with an alien set of values. And second, privileged political status in a situation where a part of society was unfree". Varlam Shalamov was one of the unfree. He was a prisoner in the Gulag who afterwards published his memoirs. When he was under investigation at the Butyrka prison in the 1960s, he enquired about his rights under the constitution. "It doesn't apply to you", the investigator replied, "Your constitution is the criminal code". The adjectival form "svobodny" meaning "free" was used in contexts where some were free and some were not. So the notion of "our own kind of freedom" came into the language, implying that Russians can be free in ways that others cannot. In Soviet official language, the freedoms of socialist life were proclaimed as superior to the illusory freedoms of the West. To enjoy Soviet freedoms, it was necessary to be fully immersed in Soviet society. Svoboda was not an escape from the constraints of Soviet rule, but a willing acceptance of constraints in return for privileges.

Humphrey continues with "Mir". "Mir has the meaning of the universe, all humanity, the world, or any given world, and in the past it also referred to the rural commune, the social world of the peasant. Mir points to the well-being naturally present between all persons, communities and their environment. This is the idyllic image of the universalised community, which ignores its fatal downside, namely, that if individuals subordinate themselves to such a totality they may be easily manipulated by any government claiming to represent it. I hope this helps explain the deeply non-intuitive fact (to us) that there are Russian villagers today who identify freedom, precisely with Stalinism. Freedom here melds svoboda with mir, producing an emotion of security, warmth and expansiveness that is still remembered by older people today". Humphrey quotes from Margaret Paxson, who recorded the memories of people in the village of Solovyovo, "During the time of Stalin, we lived better than we live now. Everyone was free. There was everything everywhere. How we lived better then! How we were joyous! One wrong word and they could take you away in the middle of the night. We lived in friendship and generosity! Remember how they took away that woman, for one little rhyme, she never came back". In the view of these villagers, fear was compatible with freedom. The night arrests were unfair, but they helped to bond the community together. After Stalin died and the bonds of fear were loosened, the community had fallen apart in instability, slackness, disorder.

Lastly comes "Volya". "Many people hold that the true Russian word for freedom is volya. Volya means 'will' as well as individual, personal freedom. Volya is sensation, emotion and action. But volya too has dark shadows. In political life there can be a volya that indicates a despotic freedom of action. The psychic consequences of unrestrained volya are likely to be destructive to the very person holding it". At the time of the revolution the Bolsheviks highjacked volya to their own purposes with countless slogans, for example, "We are striving to submit the economy to our volya!" or "The Party knows no limits to its volya!". Volya came to mean collective political power as well as individual freedom. Humphrey quotes from a novel describing a woman who was a party boss in the days of Stalin. For her, volya meant total freedom to act as she saw fit for the good of society. After Stalin dies and Khrushchev changes the party line, she refuses to conform and the Party calls her to account. Having lost her volya, she collapses and is left miserably alone. Humphrey sums up the three notions of freedom in Soviet Russia as follows: "What I have tried to describe so far is three different concepts of freedom, how they were related to one another in the Soviet context, and how each of them came at a heavy cost, of distance from reality, of fear, of anger or isolation".

Humphrey then turns from the Soviet period to modern times. The image that she uses to illustrate Russian notions of freedom today is the Chukotka region of far eastern Siberia, which her student Niobe Thompson has studied in depth. Chukotka today is divided into two worlds, the rich, warm, insulated world of oil-fields and luxury hotels, and the cold, dark, dangerous world of the Arctic wilderness outside. Inside the fence, freedom is svoboda, the freedom of a tightly organized Mafia to run its own business. Outside, freedom is volya, the freedom of groups of pioneers to roam and take their chances in an endless expanse of frozen lakes and forests. "Volya has become the sphere for personal ethics, and svoboda that of efficacious yet immoral action. The new svoboda furthermore is associated with privilege and foreignness, indeed with the humiliation of what Russians see as imperialism and global power. Meanwhile, Mir has evaporated, perhaps, though, to morph into the freedom of pure space (prostor), nature, the environment".

None of the three words carries the same resonances as our word "freedom". The nuances of language reveal profound differences between the English and Russian cultures. The main difference lies in the relative importance of the individual and the group. Russian culture missed the Renaissance which turned Western Europe toward the cult of the individual. During the time of the Renaissance, Russian culture was dominated by the fight against Tartar hegemony, a fight that demanded loyalty to the group and self-sacrifice of the individual. To defeat the Tartars, Russians had to learn to think like Tartars. All the way to the twenty-first century, Russian Tsars, and their successors Lenin and Stalin and Putin, were consolidating their personal power like Tartar Khans, while English Kings were at the same time yielding their power to assemblages of merchants and lawyers. For England and America, freedom means strict limitation of the power of government to control the individual. For Russia, freedom means belonging to a community with a strong leader who will keep the oligarchs in their place.

When we look beyond Europe and Russia to the rest of the world, translation of our words and our ideas becomes even more difficult. Moslem culture and Sino-Japanese culture are even more group-centered than Russian culture. Our individualistic concept of freedom seems natural only to a minority of people. Since we are a minority, it is important for us to understand how the rest of the world thinks. We may even have something to learn from the majority.

3. The Second Hit: Law

The second hit was an article with the title, "Has the Supreme Court lost its way?", published in the September 2007 issue of the magazine, "Washington Lawyer" by Lawrence Latto, a lawyer who runs a private practice in Washington. Here I must warn you again that I am a layman trespassing on legal territory where I have no professional competence. Latto analyzes the history and the consequences of a famous case, "Regents of the University of California versus Bakke", decided by the Supreme Court of the United States in 1977 with a vote of five to four, the nine judges explaining their reasons in six separate opinions. Allan Bakke was a white student who applied for admission to the medical school of the University of California at Davis. Bakke was denied admission. Of the hundred places for new students, the University had set sixteen aside for members of four specified minority groups. Bakke sued the University, claiming that his exclusion was a denial of his right to equal protection of the laws. He claimed that the University had violated the Civil Rights Act of 1964, which prohibited discrimination on the basis of race by any program or activity receiving Federal financial assistance. The case came first to the state Supreme Court of California, which decided in favor of Bakke. The University appealed the decision to the United States Supreme Court, which again decided in favor of Bakke. Bakke was admitted to the medical school, and the quota of sixteen places for minority students was abolished. During the subsequent thirty years, the Bakke decision has served as a precedent for many decisions in similar cases when public school or public university administrators tried to maintain special programs for racial minority groups. At issue in all these cases is the clash between two rights, the right of an individual to equal protection of the laws, and the right of a group that is disadvantaged as a result of past discrimination to obtain an equal share of educational opportunities. The Bakke decision, and the later decisions based on the Bakke precedent, said that the right of the individual must prevail over the right of the group. These decisions have reinforced the tendency of Americans to think of America as a society of individuals rather than a society of communities.

Lawrence Latto is saying that the Bakke decision was legally and morally wrong, and that the Supreme Court has lost its way by following the Bakke decision as a precedent. To support his view of the matter, Latto quotes President Lyndon Johnson, "You do not take a person who, for years, has been hobbled by chains and liberate him, bring him up to the starting line of a race and then say, 'You are free to compete with all the others', and still justly believe that you have been completely fair. Thus it is not enough just to open the gates of opportunity. All our citizens must have the ability to walk through those gates". Latto allows that Johnson "probably was less qualified by temperament and ability than anyone to be a judge", but still maintains that Johnson was right. For Johnson the basic issue was fairness. In a society that has historically treated minority groups unfairly, equal rights for the individual are not enough. Minority groups think of themselves as groups rather than as individuals. To them, fairness means fairness to the group. To the black community, fairness means that the community should have a share of professionally trained doctors proportional to their population. It is not enough for the individual black student to have an equal chance to compete with better-prepared whites.

The main part of Latto's essay is a detailed analysis of the legal opinions that the judges wrote to justify their decisions. The judges were preoccupied with fine points of constitutional law rather than with trying to achieve fairness. The main issue as they saw it was which of two criteria should govern the constitutionality of state programs providing special assistance to minority citizens. The two criteria are called rationality and strict scrutiny. Rationality means that the program is constitutional if it is reasonably designed to achieve its aim of helping disadvantaged minorities. Strict scrutiny means that the judges have the duty to scrutinize the program rigorously and declare it unconstitutional if they find that it is not "narrowly tailored" to achieve a "compelling governmental interest". The majority of five judges believed in strict scrutiny, the minority of four believed in rationality, and this ideological division determined the outcome of the case.

Latto describes how the doctrines of rationality and strict scrutiny originated in a lengthy and abstruse opinion written in 1937 by Justice Stone. Justice Stone wrote the opinion for a unanimous Supreme Court, upholding a judgment that condemned the Carolene Products Company, a dairy in Georgia, for violating the Filled Milk Act of 1923, which prohibited the interstate sale of milk containing any fat other than butter fat. Justice Stone used the criterion of rationality to uphold the Filled Milk Act, which was all that he needed to settle the case against Carolene Products. But then he went on to write an even lengthier footnote to his opinion, observing that the criterion of rationality might not always be sufficient. He remarked that a more exacting judicial scrutiny might be required for a review of statutes directed at particular religious or racial minorities. "Prejudice against discrete and insular minorities may be a special condition, which tends seriously to curtail the operation of those political processes ordinarily to be relied upon to protect minorities, and which may call for a correspondingly more searching judicial inquiry". Latto concludes his summary of Stone's opinion by saying, "The footnote is written in the terse, precise, and somewhat turgid fashion of a judge of the very highest rank".

Five of the justices who deliberated upon the Bakke case forty years later decided that this was a case to which the Stone footnote was clearly relevant. They wrote several opinions explaining how they applied their strict scrutiny to the affirmative action program of the University of California and found it unconstitutional, either because it was not "narrowly tailored" enough, or because it was not "responsive to a compelling governmental interest". Their judgment was a historical turning-point, providing a legal precedent for overturning affirmative action programs in schools and universities all over the country. As the years went by, the ascendancy of individual rights over group rights became more and more firmly established.

Latto devotes two of the seven pages of his essay to the opinion written by Justice Powell to explain his view of the Bakke case. Powell effectively decided the case, after the other eight justices had expressed various opinions that split them four to four. Powell first expressed agreement with his four colleagues who voted to apply strict scrutiny as the criterion for assessing the University of California admissions process. The university had named four separate purposes that were served by preferential admission of minority groups. The most urgent of these purposes was to increase the supply of trained doctors who would be likely to practice in minority neighborhoods. Powell asked for statistical proof that minority doctors would in fact practice in minority neighborhoods. The mere statement that this was likely was not sufficient. No statistical proof was forthcoming, and so Powell declared this purpose hypothetical. Being hypothetical, it could not provide a legal basis for action. Another purpose for preferential admission of minorities was to enhance the quality of education for all students, majority and minority. Experienced educators had written that a more diverse student body results in improved education for all. Powell accepted this purpose as valid, although it might have been considered equally hypothetical. He then went on to reject it as a justification for the admissions process, on the grounds that the process was not narrowly tailored to achieve it. Continuing in this way, applying strict scrutiny to the purposes of the admissions process one by one, Powell succeeded in eliminating all of them. The admissions process that excluded Bakke was therefore unconstitutional.

After telling the story of the judges and their opinions, Latto asks the question, "Why did they make this bad decision?" and answers it, "Probably because of the arrogance that too often occurs in persons with unlimited and unaccountable power. The threshold question was, who should be the primary decision maker, the Congress, the state legislatures and their agencies, or the Court? Naturally, they believed that they were better equipped to do the job. Also it increased, rather than diminished, their role. But they missed the point and, being so entranced by the intricacies of their craft at which they were so skilled, they found all the wrong reasons for making this choice." Latto then quotes another opinion from the past, written by Oliver Wendell Holmes: "It must be remembered that legislatures are ultimate guardians of the liberties and welfare of the people in quite as great a degree as the courts. Had the Court listened to these words and considered thoughtfully its proper role under our governmental structure, it would not have insisted that it have the primary role in second-guessing a state agency's decision."

Latto ends his account with a gentle reprimand directed against the Supreme Court. "The Court's decision in 'Brown versus Board of Education', that segregation in public schools was unconstitutional, was met with vigorous resistance that lasted for decades but did not prevent much progress from being made toward a country with more freedom and opportunity for all. It is still a work in progress. It is strange that the Supreme Court, which we view as the protector of our liberties and even of certain natural rights, should now erect a barrier against more rapid progress toward that end". Latto is here using the words "freedom" and "liberties" to mean the rights of minority groups to equal education and equal opportunities. The Supreme Court and the majority of Americans use the same words with a different meaning, paying attention to the rights of individuals only.

There is a striking difference in law between the United States on the one hand, and other countries such as Canada, France and Germany, which also belong to the Western cultural tradition, on the other. In Canada, France and Germany there are strong laws prohibiting "hate speech". In those countries, people are put in jail for speech inciting hatred against minority groups in the society. In the United States, hate speech is legal provided that it does not threaten immediate bodily harm to the people we hate. The first amendment to the US constitution guarantees freedom of speech to every individual. In the USA the right of the individual to speak freely prevails over the right of the minority community to be treated with respect. In most other Western countries, the right of the community prevails. United States judges could, if they wished, protect minority communities against hate speech while leaving the first amendment substantially intact.

4. The Third Hit: Science

The third hit was my collision with Richard Dawkins. I came into collision with Dawkins, not physically but electronically. The hit happened at a gathering of scientists invited by the impresario John Brockman to stay overnight at Eastover, his beautiful farm in rural Connecticut. John Brockman is a literary agent who presides over a group of people called the Reality Club, with a website called "Edge", where he invites scientists and writers to engage in discussions about questions of the day. The meeting at Eastover was a rare event at which devotees of the website could meet face to face. I was delighted to find there two famous biological entrepreneurs, Craig Venter who is world champion reader of genomes, and George Church who is world champion writer of genomes. Craig Venter has spent the last year cruising around the world in a boat equipped with sequencing equipment, fishing out of the ocean samples of thousands of species of marine organisms previously unknown to science, and sequencing their DNA. George Church has spent the last year perfecting his techniques for synthesizing long stretches of DNA embodying a given sequence of base-pairs. Venter is able to translate any naturally occurring piece of DNA into a piece of computer-code specifying the sequence of its base-pairs. Church is able to translate any computer-code sequence back into a piece of synthetic DNA with the specified base-pairs accurately placed in the right order.

Working together, these two technologies will soon give us the ability to design and construct living creatures for fun and for profit. They give new meaning to the phrase, "Genetically Modified Organisms". Most people are afraid of genetically modified organisms, because they imagine such organisms being the property of big corporations such as Monsanto. There are good reasons for distrusting Monsanto. Monsanto has made a habit of inserting genes for poisonous pesticides into crop-plants such as cotton and soybeans. But genetically modified organisms may not belong much longer to the big corporations. I am making a prediction, that biological technology will soon be domesticated, just as computer technology was domesticated during the last fifty years. I see the technologies of Craig Venter and George Church rapidly becoming user-friendly and cheap enough to be accessible to ordinary farmers and gardeners and animal-breeders. I am saying that these technologies are instruments of liberation rather than oppression, giving small farmers and breeders the chance to be creative and also to be economically productive. This was the background out of which our conversation at Eastover arose.

I was talking to the crowd at Eastover about the past and future of life on this planet. Following the ideas of the great biologist Carl Woese, I divided the past history of life into three periods. The first period I call the Pre-Darwinian era, when all life consisted of primitive cells sharing genetic information freely, with no division into separate species. Biologists call this free sharing "horizontal gene transfer", since it is unlike the normal vertical transfer of genes from mother to daughter. The second period I call the Darwinian Interlude, when all life was divided into species and each species refused to share genetic information with others. The third period I call the Post-Darwinian Era, the era in which we are now living.

"The Darwinian interlude", I said, "lasted for two or three billion years. It probably slowed down the pace of evolution considerably. The basic biochemical machinery of life had evolved rapidly during the few hundreds of millions of years of the pre-Darwinian era, and changed very little in the next two billion years of microbial evolution. Darwinian evolution means the competition for survival of non-interbreeding species. Darwinian evolution is slow because individual species, once established, evolve very little. With rare exceptions, Darwinian evolution requires established species to become extinct so that new species can replace them.

"Now, after three billion years, the Darwinian interlude is over. It was an interlude between two periods of horizontal gene transfer. The epoch of Darwinian evolution based on competition between species ended about ten thousand years ago, when a single species, Homo Sapiens, began to dominate and reorganize the biosphere. Since that time, cultural evolution has replaced biological evolution as the main driving force of change. Cultural evolution is not Darwinian. Cultures spread by horizontal transfer of ideas more than by genetic inheritance. Cultural evolution is running a thousand times faster than Darwinian evolution, taking us into a new era of cultural interdependence which we call globalization. And now, as Homo Sapiens domesticates the new biotechnology, we are reviving the ancient pre-Darwinian practice of horizontal gene transfer, moving genes easily from microbes to plants and animals, blurring the boundaries between species. We are moving rapidly into the post-Darwinian era, when species will no longer exist, and the rules of Open Source sharing will be extended from the exchange of software to the exchange of genes. Then the evolution of life will once again be communal, as it was in the good old days before separate species and intellectual property were invented".

Dawkins was not physically present at the Eastover meeting. He was sitting in Oxford, linked to us electronically and paying close attention to our discussions. As soon as I finished speaking, Dawkins went into attack mode. He sent a text message to Brockman which Brockman read aloud to the assembled company. "This statement of Dyson constitutes a classic schoolboy howler", said Dawkins speaking through the mouth of Brockman, "a catastrophic misunderstanding of Darwinian evolution. Darwinian evolution, both as Darwin understood it, and as we understand it today in rather different language, is not based on the competition for survival of species. It is based on competition for survival within species. Darwin would have said competition between individuals within every species. I would say competition between genes within gene pools. The difference between those two ways of putting it is small compared with Dyson's howler". The word "howler" is British slang for a grossly stupid mistake. Brockman went on reading Dawkins' text message, "In the arms race between predators and prey, or parasites and hosts, the competition that drives evolution is all going on within species. Individual foxes don't compete with rabbits, they compete with other individual foxes within their own species to be the ones that catch the rabbits".

I responded to Dawkins the next day. "It is absurd to think that group selection is less important than individual selection. Consider for example Dodo A and Dodo B, competing for mates and progeny in the dodo population on the island of Mauritius, a hundred years before the arrival of humans. Dodo A competes much better and has greater fitness, as measured by individual selection. Dodo A mates more often and has many more grandchildren than Dodo B. Two hundred years later, the species is extinct and the fitnesses of A and B are both reduced to zero. Selection operating at the species level has trumped selection at the individual level. Selection at the species level wiped out the progeny of both A and B because the species neglected to maintain the ability to fly, which was essential to survival when human predators appeared on the island. This situation is not peculiar to dodos. It arises throughout the course of evolution, whenever environmental changes cause species to become extinct".

After this lengthy description of Dawkins's hit and my response, let me explain why this question of the relative importance of individual selection and group selection is crucial to the evolution of our own species. Our species is unique as the inventor of civilization. How could this have happened? Civilization means the peaceful cooperation of societies much larger than the families or tribes that are held together by genetic relatedness. Peaceful cooperation requires that individuals behave altruistically, helping others at some cost to themselves. Our species did not invent altruism. Many other species, for example ants, bees, wolves and baboons, behave altruistically. But these other species are organized in genetically related groups, and are altruistic only within the group. Humans invented the extension of altruism beyond the family and the tribe. This extension is called by biologists "the evolution of altruism". There are strongly divergent opinions as to how it could have happened, with Dawkins and me on opposite sides of the argument.

Two facts are clear. First, group selection favors the growth of big societies with altruistic rules of behavior. Big societies with strong internal discipline tend to prevail in the struggle for survival, either by wiping out small societies or by absorbing them. Second, individual selection within a big society does not favor altruism. Within a big group with altruistic rules, individual selection favors the cheater who breaks the rules. Cheating becomes easier when the group becomes larger. This is the paradox which makes the evolution of altruism hard to understand. Individual selection by itself will lead to a society of cheaters and a breakdown of altruism.

One way to resolve the paradox is to combine altruism with vengeance. Vengeance means a positive delight associated with the act of punishing cheaters. Vengeance is the dark side of altruism, but seems to be necessary for altruistic societies to evolve beyond a certain size. We know that vengeance is deeply rooted in our nature. Vengeance is especially enjoyable, as Gilbert's Mikado observed, when the punishment fits the crime. I remember as a child in England enjoying the celebration of the punishment of Guy Fawkes on November 5 each year far more than I enjoyed the celebration of the birth of Jesus on December 25. Guy Fawkes was the traitor who was caught with thirty-six barrels of gunpowder in a cellar underneath the Houses of Parliament, preparing to blow up the King and Parliament together in the Gunpowder Plot of November 5, 1605. He was publicly burnt alive in a particularly gruesome fashion. As children, we sang the bellicose second verse of our National Anthem, "Oh Lord our God arise, Scatter his enemies,

And make them fall. Confound their politics, Frustrate their knavish tricks, On Thee our hopes we fix" and so on, with far more gusto than the insipid first and third verses. Delight in punishing cheaters is innate in our species, and has made us what we are, an intimate mixture of altruism and vengeance, love and hate. Altruism means to love our neighbors as ourselves. Vengeance means to take delight in punishing sinners. Humans have evolved to slip easily into either role, good Samaritan or lord high executioner.

There is a close analogy between the organization of human beings into societies and the organization of cells into human beings. A human being is a community of cells. Our somatic cells, all the cells that are not germ cells, are totally altruistic. A somatic cell gives up its individual freedom and lives to serve the community. Even in this perfectly altruistic community, cheating is a problem. Occasionally a somatic cell cheats and begins to live and reproduce itself independently. The offspring of the cheater become a cancer which destroys the community. But the community has devised an extreme form of punishment to discourage cheating. The punishment is called apoptosis or programmed cell death. Every cell carries a supply of lethal chemicals that will kill it, as soon as they are released. If a cell shows any signs of being a cheater, the community gives it a signal that releases the chemicals and causes it to die. Cancers only arise when a cell is a cheater and the punishment system happens to be disabled. A similar punishment system based on compulsory suicide was used in ancient Athens to punish the philosopher Socrates. The more altruistic the society, the more vindictive the punishment.

Sociologists have invented an ingenious game called the Ultimatum Game, to test quantitatively the vindictiveness of humans. The game is very simple and has been played by members of primitive tribes in many countries as well as by educated people in modern cities. In each game there are two players, A and B. A substantial sum of money, at least a day's wages in the local currency, is placed on a table, with the following rules explained in advance and agreed to by the players. Player A must divide the sum into two parts, one for A and the other for B. B then has the choice of saying yes or no to this division. If B says yes, then both A and B collect their shares as determined by A. If B says no, then the money is removed and both players get nothing. The point of the game is that B has nothing to gain from saying no, except the joy of punishing A for not being more generous. If both players were only interested in maximizing their gains, then A would offer very little to B and B would still say yes. But in fact, the result of the game is very different. Usually A offers to B between one quarter and one half of the total, and usually B says yes. In the unusual case when A offers less than one quarter, B usually says no. This same pattern of results is seen in many different cultures all over the world. If A offers less than a quarter, he is seen as a cheater, and B values the joy of punishing a cheater more highly than the joy of collecting a prize.

This harmless and illuminating game leads to the following hypothesis about the evolution of altruism. Altruism and vengeance evolved together in human societies. Within each society, individual selection favored the survival of altruists because cheaters were severely punished. In the competition between societies, group selection favored the survival of vengeance because societies that did not punish cheaters could not work or fight effectively. Individual selection and group selection were both essential to making us what we are. Individual selection enforced by punishment gave us altruism, and group selection enforced by warfare and competition for resources gave us vengeance. This hypothesis does not tell us the whole truth about the evolution of human nature, but it must be at least a part of the truth. Richard Dawkins' version of evolution, with selection operating only between individuals, could hardly have caused us to evolve from single cells to multicellular tadpoles, and certainly not from clans of apes to nations of civilized humans.

5. Conclusions

Let me briefly summarize what we have learned from these three writers coming at us from different directions. Caroline Humphrey the anthropologist tells us that our western concept of freedom as belonging to the individual rather than the group is not widely shared. Even in the west it is a recent development, originating in the Renaissance and unknown to the ancient Greeks and Romans. Athens in the fifth century BC considered itself to be a shining example of a free society, but still condemned Socrates to death for corrupting young people with ideas that did not conform to community norms. Lawrence Latto the lawyer tells us that the United States Supreme Court has led our country astray by giving human rights of individuals priority over human rights of disadvantaged groups. The legal sanctification of individual rights by the Supreme Court is even more recent than the Renaissance. The decisive tilt toward individual rights happened only thirty years ago, when the Bakke case gave judges an opportunity to launch a backlash against the civil rights legislation of the 1960s. Richard Dawkins the biologist scolds me sharply for my statement that group selection is at least as important as individual selection in the evolution of humans and other creatures. He maintains dogmatically the doctrine that natural selection acts only on individuals or on individual genes. Dawkins' doctrine is also of recent origin. The notion of group selection was widely accepted by evolutionary biologists, including Darwin, in the nineteenth and twentieth centuries. The subtitle of Darwin's great work, "On the Origin of Species by Means of Natural Selection, or, the Preservation of Favoured Races in the Struggle for Life", shows that Darwin thought of selection as acting on races rather than on individuals. Dawkins is historically as well as scientifically wrong when he asserts that Darwinian evolution depends on selection between individuals only.

Having heard the three messages, I agree with Humphrey and Latto and disagree with Dawkins. I see the roots of freedom and human rights and altruistic behavior growing historically within tight-knit communities and continuing today to belong to the communities. Individual rights and individual freedom are precious, but they are a recent and precarious addition to the old communitarian tradition. The road to a freer and more peaceful world lies through compromise, with equal respect paid to individual and community values. Anthropologists and lawyers and biologists should all be ready to help us make such compromises.

References

J. Brockman [2007] "Life: what a concept! An edge special event at Eastover Farm", on web-site (www.edge.org).

R. Dawkins [1976] *The Selfish Gene* (Oxford University Press, Oxford).

F. J. Dyson [2007] *A Many-colored Glass: Reflections on the Place of Life in the Universe* (University of Virginia Press, Charlottesville).

C. Humphrey [2007] "Alternative freedoms", *Proc. American Philos. Soc.* **151**, 1–10.

L. J. Latto [2007] "Has the Supreme Court lost its way?" *Washington Lawyer*, September 2007, pp. 35–41.

CHAPTER 4.7

NUKES AND GENOMES*

1. Abolition of Nuclear Weapons

First I say thank you to Edie Munk and the other organizers of the Helen Edison lectures for inviting me to give this talk. Half of the talk will be about the abolition of nuclear weapons and the other half about the domestication of biotechnology. These are two of the great problems that we shall be struggling with for the next fifty years. Twenty-five years ago I preached a sermon in the Unitarian Church at Davis about nuclear weapons. I said the number one danger to ourselves and to the world was nuclear weapons, but there were solid grounds for hope that we could get rid of them. Both these statements are still true today. The world has changed drastically in the intervening twenty-five years. Some of the changes were for the better and some for the worse. The best change, one that I had not dreamed would be possible, was the peaceful collapse of the Soviet Union. The worst change, one that I had also not dreamed of, was the launching by the United States of a preventive war. As a result of these changes, the way people think about nuclear weapons has changed, but the essential dangers of nuclear weapons and the possible remedies have hardly changed at all.

Now people are mostly worrying about nuclear weapons belonging to Iran, North Korea or Pakistan, countries that they call "rogue states", or nuclear weapons belonging to terrorist groups like Al Qaeda. They call this the problem of nuclear proliferation. It is a real problem, and it has been a real problem for fifty years. But it is not a problem that we can solve by ourselves. The main problem for us, the problem that is in our power to solve, is our own weapons. We have about ten thousand nuclear weapons, enough to wipe out a large fraction of the world population. The Russians have a roughly equal number. These vast accumulations

---

*Helen Edison Lecture given at Scripps Aquarium, UCSD, July 21, 2009.

of weapons are far more dangerous to the world as a whole than the small numbers available to Iran or Pakistan. People complain that the Russians are sloppy in taking care of their nukes, but I will never forget the time I walked into a room in one of our nuclear weapon storage sites and found forty-one hydrogen bombs lying around on the floor, not even tied down. I counted them carefully, and made sure there were forty-one. I wondered if anyone would have noticed if one or two had been missing. There is some sloppiness on our side too.

There are two ways to talk about nuclear weapons. You can talk about ethics and morality, saying that nuclear weapons are uniquely evil because they are weapons of genocide, an offense against God, and we have a moral and religious duty to get rid of them. Or you can talk about hard-boiled military requirements, saying that nuclear weapons are ineffective from a practical point of view. For this talk I chose the second alternative. I shall skip the moral argument, assuming that you all have heard it and more or less agree that nuclear weapons are evil. I will concentrate on the practical argument, trying to convince you that nuclear weapons are useless for the business of winning wars. There is nothing sensible that we can do with our own nuclear weapons to stop Iran, North Korea or Pakistan from having nuclear weapons too. Our own nukes are useless for any sane military purpose. The basic problem, when we are trying to use nuclear weapons to win a war against a poor country, is that we have a multitude of good military targets while they have very few. With nuclear weapons we can kill large numbers of people and make sure that the survivors view us with enduring hatred, but that does not mean that we have won the war.

I conclude that one of the primary aims of our foreign policy should be to get rid of nuclear weapons altogether. We must understand that the words "get rid" are misleading. We can never know for sure that our enemies or our friends are not hiding a stockpile of weapons somewhere. Nuclear warheads are notoriously easy to hide. When we say "get rid of nuclear weapons", we mean that they are legally prohibited, in the same way that biological weapons are legally prohibited. This means that all remaining weapons must be clandestine, without any large and conspicuously deployed delivery systems. And it means that we know for sure that our own weapons are gone. In my opinion the removal of our own weapons would make the world safer, even if other countries keep some of theirs. The most tempting targets for a surprise attack are nuclear-armed aircraft carriers, and these targets mostly belong to us. By getting rid of such targets, we substantially reduce the chances of a nuclear war beginning on the high seas or in the Persian Gulf.

There are two ways to get rid of weapons, either by unilateral action or by multilateral negotiation. Both ways have been tried, sometimes successfully, during the last fifty years. I will briefly describe four historical examples. The first example happened in 1963 when I was working at the US Arms Control and Disarmament

Agency. The nuclear arms race was then racing toward bigger and bigger hydrogen bombs. The Soviet Union was leading with a sixty-five-megaton bomb, advertized as a prototype for a hundred-megaton bomb. We were afraid the next step of the race would be a gigaton mine, too big to be carried on airplanes or missiles. Gigaton mines would be deployed in big underwater containers or unmanned submarines and would destroy coastal cities with giant tsunamis. Even the most bloodthirsty Air Force generals and Navy admirals did not want them. They knew that the United States had many coastal city targets and the Soviet Union had few. Kennedy and Khrushchev negotiated the atmospheric test-ban treaty that put a stop to this madness. All future nuclear tests had to be underground. The practical limit to the yield of an underground test was ten megatons. After this, the arms-race ran quickly in the opposite direction, toward smaller bombs with smaller yields. Some years later, the threshold test-ban treaty prohibited all tests with yields greater than a hundred and fifty kilotons. But Kennedy and Khrushchev missed the opportunity to negotiate a comprehensive test-ban treaty that would have slowed down the nuclear arms-race much more.

My second example is the abolition of biological weapons by President Nixon in 1969. This was a unilateral executive action, taken quietly by Nixon without any fanfare. No international negotiation or Senate ratification process was needed. Opponents of the decision had no opportunity to raise political objections or introduce procedural delays. Nixon simply declared that the entire US biological weapons program would be terminated and the stockpiles destroyed. This happened because the Harvard biologist Matthew Meselson owned a summer place on Cape Cod next door to Henry Kissinger who was Nixon's National Security Advisor. Meselson persuaded Kissinger that it was time to get rid of biological weapons, and Kissinger persuaded Nixon. At a congressional hearing, Meselson asked the army generals who were in charge of the biological weapons program, "What are your plans for using these weapons?", and the generals had no answer. The generals had to admit that, even if we were attacked with biological weapons, they did not have any realistic plan to use our own biological weapons in response. From a purely military standpoint, our own weapons were useless. Three years after Nixon's unilateral action, he negotiated the 1972 international convention making biological weapons illegal, and the Soviet Union signed the convention. The convention was unverifiable and the Russians cheated on a massive scale. Still we were better off with the convention than without it. The Soviet program remained covert, with no open deployment of biological weapons. The threat of terrorist biological weapons remains, but the threat would be worse if we still had our own biological weapon stockpiles for the terrorists to steal.

My third example of an attempt to get rid of weapons is the one that failed. In 1986, Ronald Reagan and Mikhail Gorbachev held a summit meeting in Reykjavik to negotiate an Arms Control treaty. Reagan was a passionate abolitionist who wanted to get rid of nuclear weapons altogether, and Gorbachev had similar feelings. The two of them escaped from their official advisors and talked privately. They came close to agreeing to abolish all their nuclear weapons of all kinds. There were two reasons why they failed to agree. First, both of them had conservative advisors who were terrified of any drastic change in the status quo. Second, Reagan was deeply attached to his Star Wars missile defense program and refused to give it up, while Gorbachev was afraid that the Star Wars system could be converted to an offensive first strike mission. Gorbachev's fears were exaggerated but not unreasonable. Reagan's stubbornness cost him the chance to change the course of history.

The fourth historical example of getting rid of weapons was brilliantly successful. It happened in 1991 when George Bush Senior was president. Two years before, Gorbachev had allowed Germany to reunify and demolish the Berlin Wall, and the Cold War had effectively ended. Bush decided the time had come to get rid of all the tactical nuke systems in the US army and surface navy. This meant that roughly half of our total deployed weapons were removed in one afternoon by unilateral action. It was the biggest act of nuclear disarmament in history. A few years before this happened, I visited the missile cruiser Princeton in Long Beach harbor. The Princeton is named for the town where I live. It then carried 98 Tomahawk cruise missiles in two big boxes, 49 with nuclear warheads and 49 non-nuclear. The captain had to be careful to remember which was which. It was an accident waiting to happen, an easy way for a nuclear war to start at sea. The army tactical nukes were equally dangerous, deployed all over the world in exposed places. Now they are all gone. One of the most dangerous documents in the bad old days was Army Field Manual 101-31-1 (1963 edition, unclassified) used for training troops to fight tactical nuclear wars. I have a copy in my office. This document was available to every military intelligence service in the world. It told the world how the US army was training our troops to use tactical nukes on the battlefield. Every army in the world that wanted to keep up with the Joneses should have tactical nukes and tactical nuke training manuals too. Fortunately, the US army is no longer distributing that field manual. Unfortunately, some tactical nukes remain in the US Air Force. The army and the surface navy are now both happy to be non-nuclear. They can do their jobs much better without the encumbrance of taking care of nuclear weapons. Nobody now wants to put the weapons back where they were. Bush was careful to time the announcement of his action to coincide with the settlement of a big lawsuit against the tobacco industry. The media concentrated their attention on the tobacco settlement and the nuclear disarmament slipped by unnoticed. Some time later, Gorbachev responded with a similar withdrawal of Soviet tactical nukes. The withdrawal was maintained by Russia and Ukraine after the Soviet collapse.

These four examples convince me that unilateral action is usually more effective than multilateral negotiation as a way of moving forward to drastic disarmament. Both ways should certainly be tried, and both are needed. The most recent move toward nuclear abolition was started in 2006 by Max Kampleman, who was with Reagan at Reykjavik and served as Reagan's negotiator of Arms Control agreements. Kampleman joined with other elder statesmen, including Henry Kissinger, William Perry, Sam Nunn, and George Schultz who was Reagan's secretary of state, to publish a public statement calling for worldwide nuclear abolition as the goal of US foreign policy. They proposed to revive the Reykjavik discussion with Russia and then bring in other countries to reach a multilateral abolition agreement. In my opinion, they put too much emphasis on verification and enforcement. It would be better to begin with unilateral moves without enforcement. A world without major open deployments of nuclear weapons would be much safer, even if Israel and Iran keep some stockpiles hidden away. There is no reasonable way to enforce an agreement if Israel and Iran do not wish to join. Every country should have the right not to join, or to withdraw after six months notice. A withdrawal clause is standard in all arms-control agreements, for good reasons.

The main obstacle to overcome, if we are trying to convince the American public to get rid of our nuclear weapons, is the deeply held belief that our nuclear weapons give us some kind of security. This belief is supported by the myth that our nuclear weapons at Hiroshima and Nagasaki brought World War II to an end. Recent studies by the historian Hasegawa and others, including my Princeton friend Ward Wilson, have convinced me that the myth is false and needs to be demolished. Wilson has a more general historical argument, supporting the view that nukes are militarily ineffective. Whether the weapons were nuclear or not, killing civilians has never been an effective way to win wars. One of the extreme examples of civilian killing was the war of 1865–1870 in Paraguay. More than half of the population of Paraguay was killed, but Paraguay did not surrender to the alliance of Brazil, Argentina and Uruguay. In the European part of World War II, the bombing of cities in England and Germany did not induce any talk of surrender, but on the contrary induced a mood of solidarity in the civilian population and a determination to fight on to the bitter end. As we shall see, this was also true in Japan.

2. Why did Japan Surrender?

Here is the argument of Ward Wilson, convincing me that, contrary to the general belief of Americans, the nuclear bombs at Hiroshima and Nagasaki did not end World War II. There are two separate questions that are often mixed up. (a) Was the American decision to use the bombs right or wrong? (b) Was the Japanese decision to surrender caused by the bombs or not? I am only discussing question (b). The answer to (a) depends on quite different considerations, especially on the ignorance of the US authorities about events in Japan.

Wilson makes six points to answer no to question (b). His points come from recent books by historian Hasegawa and others.

Point 1. Members of the Supreme Council, which customarily met with the Emperor to take important decisions, learned of the nuclear bombing of Hiroshima on the morning of August 6, 1945. Although Foreign Minister Togo asked for a meeting, no meeting was held for three days.

Point 2. A surviving diary records a conversation of Navy Minister Yonai, who was a member of the Supreme Council, with his deputy on August 8. The Hiroshima bombing is mentioned only incidentally. More attention is given to the fact that the rice ration in Tokyo is to be reduced by ten percent.

Point 3. On the morning of August 9, Soviet troops invaded Manchuria. Six hours after hearing this news, the Supreme Council was in session. News of the Nagasaki bombing, which happened the same morning, only reached the Council after the session started.

Point 4. The August 9 session of the Supreme Council resulted in the decision by the Emperor to surrender. The timing of the session shows clearly that it was responding to the Soviet invasion rather than to the nuclear bombs.

Point 5. The Emperor, in his rescript addressed to the Japanese armed forces ordering them to surrender, does not mention the nuclear bombs but emphasizes the historical analogy between the situation in 1945 and the situation at the end of the Sino–Japanese war in 1895. In 1895, Japan had defeated China and occupied Manchuria. Then European powers led by Russia intervened and moved into Manchuria. The Russians occupied the naval base at Port Arthur, which was the most important military objective of the Japanese invasion. The great emperor Meiji, who had modernized Japan, accepted a humiliating peace, dictated by the European powers. He agreed to withdraw all Japanese forces from Manchuria. Meiji was the grandfather of Hirohito and was held in awe by everyone in Japan, especially by the military leaders. By making a dishonorable peace with the Europeans, Meiji had kept the Russians out of Japan. Ten years later in 1905, Japan defeated Russia and drove the Russians out of Port Arthur. The language of Emperor Hirohito's rescript shows that he had this analogy in his mind when he made the decision to surrender to the Americans. His mind was primarily concerned with history and not with technology. The decisive events were not the bombings but the Russian declaration of war and the Russian invasion of Manchuria.

Point 6. The Japanese leaders had two good reasons for lying when they later said that the nuclear bombs had caused the surrender. The first reason was explained afterwards by Lord Privy Seal Kido, another member of the Supreme Council: "If military leaders could convince themselves that they were defeated by the power of science but not by lack of spiritual power or strategic errors, they could save face to some extent". The second reason was that they were telling the Americans what the Americans wanted to hear, and the Americans did not want to hear that the Soviet invasion of Manchuria brought the war to an end.

That is the end of Ward Wilson's six points. To me they are convincing. Like everyone else who was in England or America in 1945, I shared the common belief that the nuclear bombs ended the war. Wilson's arguments caused me to change my mind. But it will not be easy to persuade the American public that our brave soldiers and brilliant scientists did not win the war against Japan by themselves. After that myth is dispelled, there are three other myths that need to be demolished. First, there is the myth that, if Hitler had acquired nuclear weapons before we did, he could have used them to conquer the world. Much more likely, if Hitler had had nuclear weapons, our soldiers would have got to Berlin a year sooner. If London and Moscow had been nuked, the soldiers invading Germany from east and west would have had a powerful incentive to move faster. Second, there is the myth that the invention of the hydrogen bomb changed the nature of nuclear warfare. In fact, if you look at the weapons we now have in the stockpile, they are almost exactly the same as they would have been if the H-bomb had never been invented. Third, there is the myth that international agreements to abolish weapons without perfect verification are worthless. In fact, many international agreements are unverified and even violated and are still useful. A good example is the Rush–Bagot agreement of 1817, ratified by the United States Senate in 1818, which strictly limited naval forces on the Great Lakes and helped to keep the US–Canada border peaceful. The agreement was massively violated by both sides but remains in force today. These three myths are all false. After they are demolished, dramatic moves toward a world without nuclear weapons may become possible. But for this to happen, peace-loving citizens and hard-boiled presidents and soldiers must work together.

Historically, all the big dramatic moves to get rid of weapons have been made by right-wing Republican Presidents, Nixon, Bush Senior and Reagan. It is easy to see why. If you are not a Republican, the Republicans will accuse you of being soft. They could not accuse Nixon, Bush or Reagan of being soft. This presents us with a dilemma. If we want to get rid of nuclear weapons quickly, perhaps we should have voted for McCain. But I still think we have a better chance with Obama.

3. The Age of Wonder

I have been reading a newly published book, "The Age of Wonder", by Richard Holmes, an English author who specializes in romantic biographies (Harper Press,

2008 London). "The Age of Wonder" is pop history at its best, racy, readable and well documented. It gives us a picture of the leading personalities of the Romantic Age which lasted roughly sixty years from 1770 to 1830. The remarkable feature of that age is that the intellectual leaders were a mixture of scientists and poets, with considerable overlap and close friendships between them. The physician Erasmus Darwin published his speculations about evolution in a poem, "The Botanic Garden" in 1791. The chemist Humphrey Davy wrote a large number of serious poems and published many of them. He was a personal friend of the poets Wordsworth and Coleridge. Davy began his scientific career at the age of twenty when he was appointed Superintendent of the Pneumatic Medical Institute in Bristol. The Pneumatic Institute was a clinic where patients were treated for ailments of all kinds by inhailing various gases. Among the gases available for inhailing was nitrous oxide. Davy experimented enthusiastically with nitrous oxide, using himself and his friends, including Coleridge, as subjects. After one of these sessions, he wrote, "I have felt a more high degree of pleasure from breathing nitrous oxide than I ever felt from any cause whatever — a thrilling all over me most exquisitely pleasurable. I said to myself I was born to benefit the world by my great talents".

Davy was so successful in Bristol that he was invited at the age of twenty three by Count Rumford to become director of the Chemical Laboratory at the Royal Institution in London. There he switched from physiology to chemistry and discovered the new elements sodium and potassium. Later he invented the Davy safety lamp which made it possible for coal-miners to work underground without killing themselves in methane explosions. The lamp made him even more famous. Coleridge invited him to move north and establish a chemical laboratory in the Lake District where Coleridge and Wordsworth lived. Coleridge wrote "I shall attack Chemistry like a Shark". Davy wisely stayed in London, where he became President of the Royal Society and chief panjandrum of British science. The poet Byron knew him and gave him a couple of lines in his poem "Don Juan".

> This is the patent age of new inventions
> For killing bodies, and for saving souls,
> All propagated with the best intentions:
> Sir Humphrey Davy's lantern, by which coals
> Are safely mined for, in the mode he mentions;
> Timbuktoo travels: voyages to the Poles;
> Are always to benefit mankind — as true,
> Perhaps, as killing them at Waterloo.

William Lawrence, a famous physician, published a popular book, "Natural History of Man", a scientific account of human anatomy and physiology, based on current researches in field anthropology and in the dissecting room. Lawrence fiercely attacked the doctrine of Vitalism that was then fashionable. According to the Vitalists, there exists a Life Force which animates living creatures and makes them fundamentally different from dead matter. Lawrence was a materialist, and believed in no such force. The poet Shelley was a patient and friend of Lawrence, and saw him frequently during the year 1817, when Shelley's young wife Mary, then nineteen years old, was at work on her novel "Frankenstein". The battle between vitalists and materialists probably gave her the idea for the novel. The novel portrays Frankenstein creating his monster silently by candlelight, using the delicate dissecting tools of a surgeon, and portrays the monster as an articulate philosopher lamenting his loneliness in poignantly poetic language. Six years later, the novel was turned into a play, "Presumption, or the Fate of Frankenstein", which was a huge success in London, Bristol, Paris and New York. The play turned Mary Shelley's intellectual drama upside-down. It became a combination of horror-story with black comedy, and that is the way it has remained ever since, on the stage and in the movies. In the play, the monster is created by zapping dead flesh with sparks from a huge electrical machine, and the creature emerges as a dumb and misshapen caricature of a human, the epitome of brutal malevolence. And then comes the surprise. Mary Shelley went to see the play and loved it. She wrote in a letter to a friend, "Lo and behold, I found myself famous! Mr. Cooke played the monster's part extremely well — all he does was well imagined and executed — it appears to excite a breathless excitement in the audience — in the early performances all the ladies fainted and hubbub ensued". Mary Shelley was seduced by the magic of show business in 1823, just as easily as young writers are seduced by the magic of show business today.

The gist of Richard Holmes' story is that the scientists and poets of the romantic age started their lives as brilliant, undisciplined, credulous and adventurous amateurs. They blundered into science or literature in pursuit of ideas which were often absurd. They became sober professionals only after they had achieved success. Another outstanding example was the astronomer William Herschel, who discovered the planet Uranus. Herschel was a professional musician and an amateur astronomer. He dropped music and took to astronomy because he believed that all the heavenly bodies were inhabited by intelligent creatures and that the round objects which he saw on the moon were cities built by the Moonies. He continued throughout his life to publish wild speculations, many of which turned out later to be correct. He understood that when he looked at remote objects he was looking not only into deep space but into deep time. He correctly identified many of the nebulous objects in the sky as external galaxies like our own Milky Way, and calculated that he was seeing them as they existed at least two million years in the past. He proved by observation that the universe was not only immensely large but immensely old.

In this talk I am proposing the idea that a new Romantic Age is at hand, extending over the first half of the twenty-first century, with the technological billionaires of today playing roles similar to the enlightened aristocrats of the eighteenth century. I have been lucky enough to know personally some of the leading characters of the modern Romantic Age, Craig Venter, George Church, Kary Mullis and Dean Kamen. Craig Venter is world champion reader of genomes, George Church is world champion writer of genomes, Kary Mullis is the surfer who taught the world how to copy genomes. Dean Kamen is the engineer who invented the Segway and taught the world how to make artificial hands that really work. I also knew one of their kindred spirits who lived fifty years earlier and died tragically young, the mathematician John von Neumann. The new Romantic Age is centered on biology and graphic art, as the old one was centered on chemistry and poetry. Richard Holmes chose as the central figure and impresario of the old Romantic age Sir Joseph Banks, a wealthy botanist who went native in Tahiti for a few months when Tahiti was still an earthly paradise. Banks was friend and counselor to most of the leading characters. Impresarios for the new Romantic Age are the literary agent John Brockman and his artist wife Katinka Matson. John and Katinka operate the Edge website where scientists and artists are encouraged to argue and pontificate.

4. Freedom of Inquiry

The possible dangers of the new biotechnology did not pass unnoticed when it began. Thirty years ago, the invention of gene-splicing, allowing biologists to move genes easily from one species of animal, plant or microbe to another, caused widespread public alarm. The new technology might have created immediate public health hazards, for example, if genes for virulence should be transferred from disease germs to bacteria like E. Coli that normally inhabit the human gut. It might also create long-term dangers to the planetary ecology by upsetting the balance of nature in unpredictable ways. The town of Princeton where I live appointed a citizens' committee to advise the municipal authorities concerning the regulation of gene-splicing experiments. The immediate question was whether to allow Princeton University to build a laboratory where such experiments could be done. But the committee also devoted much of its time to discussing broader issues and longer-range dangers. As a member of the committee, I wrote a statement of my view of the broader issues:

"The members of this committee have assumed a responsibility as guardians of the public health and safety of Princeton. In exercizing our responsibility, we face a moral dilemma similar to that faced by every parent who watches a son or a daughter climbing a tree. The consequences of a fall may be tragic beyond imagining, a broken neck followed by a lifetime of physical helplessness, or a damaged brain and a stunted spirit. But the child shouts, "Can't you see how careful I'm being?" or "But Jane's mother lets her climb as high as she likes". A wise parent is not afraid to say no and enforce parental authority when the child is behaving recklessly. A wise parent also knows that, no matter how terrible the consequences of an accident may be, we cannot always be saying no. If we forbid tree-climbing arbitrarily and absolutely, we not only destroy a part of the joy of childhood, but also destroy the respect upon which our authority as parents ultimately depends. So this committee, as guardian of the public safety, recommends that the Princeton community not be afraid to assert its authority to put a stop to any experiments which the community considers reckless. But we recommend also that this authority be sparingly exercized. Where there is no evidence of negligence or recklessness, the community should let experiments go forward, accepting the risks which are inseparable from living in an unstable and rapidly changing world. We are not saying that scientists who do experiments are children and that we committee-members are grown-ups possessed of superior wisdom. The whole human society on earth is the child, science is the tree, and the wise upbringing of our species into a new era of global interdependence is an inescapable responsibility which we all share".

Thirty years later, this statement still describes our situation surprisingly well. For thirty years, gene-splicing experiments have gone forward under guidelines which require strict confinement of dangerous organisms, and nobody has been killed or hurt. The gene-splicing technology, combined with other new technologies such as rapid sequencing of genomes and in vitro fertilization of embryos, has led to enormous advances in our understanding of biological processes. The long-range dangers of biotechnology to human health and to the environment remain real but incalculable. In the long run, the dangers do not come from any particular technology or from any particular organism. The dangers come from knowledge. The dangers are incalculable because we can never know in advance what new hazards knowledge will bring.

Dangers can come not only from biology but also from physics. The threat of nuclear terrorist bombs is at least as serious as the threat of biological terrorist attacks. As a physicist, I am well aware that physicists in the past were responsible for inventing more devilish devices than biologists. But there is a big difference between the nuclear and biological dangers. Nuclear science is a dead subject, while biology is alive and growing fast. The time when nuclear scientists invented new kinds of nuclear weapons ended around 1970. The last attempts of physicists to bring new horrors into the world by creative use of science concerned the so-called "nuclear hand-grenade". The nuclear hand-grenade was supposed to be a pure fusion device not requiring fissionable material. The idea was to use chemical high explosive to drive a small implosion which would ignite a tiny quantity of fusion fuel. The neutrons from the fusion reaction would give a lethal dose of radiation to anybody within a few hundred yards from the explosion. Groups of physicists in Russia and in the United States were enthusiastically playing with such devices in the 1960s. Fortunately, all their designs failed, and further study of the ignition process showed that a portable nuclear hand-grenade is impossible. If the designs had succeeded, guerrilla fighters all over the world might now be armed with nuclear hand-grenades instead of AK-47s, and every local terrorist threat might be nuclear. The world has become a little safer, since we now know for sure that any portable nuclear explosive must be a fission bomb.

If a group of terrorists is planning to explode a nuclear bomb today, they need to be competent engineers, and they need to acquire fissionable materials and designs. They do not need to know the latest science. New discoveries in nuclear science will not make the task of the terrorists easier. Restrictions on freedom of enquiry in nuclear science will not make the world less dangerous. It is only the older generation of physicists who bear the responsibility for devilish devices that they invented with unnecessary enthusiasm in times past.

That is all I have to say about physics. I now return to biology, the field in which new knowledge is undeniably creating new dangers. A good example to illustrate the way in which new biological knowledge arises is the recent work of David Haussler and his colleagues at the University of California Santa Cruz, published in the online edition of Nature, August 16, 2006. They discovered a small patch of DNA in the vertebrate genome which has been strictly conserved in the genomes of chicken, mouse, rat and chimpanzee, but strongly modified in humans. The patch is called HAR1, short for Human Accelerated Region 1. It evolved hardly at all in three hundred million years from the common ancestor of birds and mammals to the common ancestor of chimpanzees and humans, and then evolved rapidly in six million years from the common ancestor of chimpanzees and humans to modern humans. In the last six million years, eighteen mutations became fixed in this patch of the human germ-line. Some major change must have occurred in the developmental program that this patch helps to regulate. Two other facts are known about HAR1. First, HAR1 is part of a gene coding for an RNA molecule and not for a protein. Second, the gene is expressed in the developing cortex of the human embryo brain from seven to nineteen weeks after conception. This is the place and time at which the detailed architecture of the cortex is organized. Haussman's team later found another similar patch of DNA in the vertebrate genome which they call HAR2. It is active in the developing wrist of the human embryo hand. The brain and the hand are the two organs that most sharply differentiate humans from our vertebrate cousins.

The discovery of HAR1 and HAR2 is probably an event of seminal importance, comparable with the discovery of nuclear fission in 1938 or the discovery of the double helix in 1953. It opens the door to two new sciences, the study of RNA as an agent of embryonic development, and the study of human evolution at the molecular level. These two new sciences will profoundly change the possible applications of biological knowledge for good or evil. None of the new applications could have been predicted before the discovery of HAR1 and HAR2 was made. No committee scanning research proposals in human genetics would have picked out Haussler's program as particularly dangerous. The only way to stop dangerous knowledge from arising would be to stop research in human genetics altogether.

There are two main ways in which biological knowledge can be dangerous. Knowledge of microbes and of biologically important molecules can lead us to new lethal agents and new ways of infecting populations. Knowledge of human biology and evolution can expose new vulnerabilities for weapon-builders to exploit. In addition to these dangers of malicious abuse of biological knowledge, there are also dangers of well-meaning abuse, such as parents trying to give their babies competitive advantages by genetic manipulation of embryos. The long-range dangers of biological knowledge are real and serious.

The question then is how to respond to the dangers. There are two possible responses. The first is to try to stop the increase of dangerous knowledge by restricting and controlling science. The second is to allow freedom of scientific enquiry and impose restrictions only on dangerous applications of knowledge. The second response will certainly not eliminate the danger. The history of biological weapons in the twentieth century shows how difficult it is to control dangerous applications after knowledge has been acquired. But I believe that the first response will be even less effective. The first response fails, because the attempt to suppress scientific enquiry puts incompetent people in positions of authority and increases the likelihood of disastrous mistakes.

If we are to establish a system of censorship, to decide which scientific enquiries are safe to pursue, the crucial question is, who will be the censors. We can be sure that the job of a censor will not attract first-rate scientific minds. The job will be intensely political, and is likely to attract people with a political axe to grind. In the year 1644, the poet John Milton made a famous speech to the English parliament opposing the censorship of books. The argument that Milton used applies equally well to the censorship of scientific enquiry. "He who is made judge", said Milton, "to sit upon the birth or death of books, whether they may be wafted into this world or not, had need to be a man above the common measure, both studious, learned, and judicious . . . If he be of such worth as behoves him, there cannot be a more tedious and unpleasing journey-work, a greater loss of time levied upon his head, than to be made the perpetual reader of unchosen books and pamphlets . . . Seeing, therefore, those who now possess the employment, by all evident signs wish themselves well rid of it . . . , we may easily foresee what kind of licensers we are to expect hereafter, either ignorant, imperious, and remiss, or basely pecuniary".

This last phrase of Milton identifies precisely the two kinds of people who became candidates for the job of scientific censor in more recent times. "Ignorant, imperious and remiss" describes the communist apparatchiks of Russia in the time of Lysenko. "Basely pecuniary" describes the capitalist lobbyists who swarm around the chambers of government today in Washington. Science will always have to defend itself against enemies of freedom on two sides, against ideological enemies on one side and commercial enemies on the other. The ideological enemies are not only Christian fundamentalists on the right but dogmatic Marxists and environmentalists on the left. The commercial enemies are not only monopolistic corporations interested in profits but corrupt politicians interested in power. The choice that we have to make is not between scientific freedom and science governed by a wise group of philosopher-kings. The choice is between scientific freedom and science governed by political hacks of one kind or another.

To stop scientists from thinking dangerous thoughts will not give us protection against the abuse of science, and the dangers of abuse will not go away. Disease germs designed for epidemiological efficiency, directed against human populations, animals or crop-plants, can be enormously destructive. The science required to create such horrors already exists. What then should we do to protect ourselves? There is no magic remedy that can make the world safe, but there are two major steps that we could take toward a safer world. Step one is to establish effective and comprehensive public health services in countries such as the United States which lack such services. Step two is to remove so far as possible all barriers to communication between scientists and public health experts of all countries. If we talk to our potential enemies, we have a better chance to find out who they are and what they might be doing. The international community of scientists can do a better job of collecting intelligence about its members than any of our secret services. When we are dealing with enemies of the human race hidden in ordinary houses and laboratories, a more open world is a safer world. Openness rather than secrecy is our best defense.

CHAPTER 4.8

NOAH'S ARK EGGS AND VIVIPAROUS PLANTS*

Science-fiction stories about starships usually depict the universe as a collection of stars and planetary systems separated by vast stretches of empty space. The space between stars is imagined to be filled with dilute interstellar gas and nothing else. The real universe is much more interesting. The real universe contains a multitude of objects of various sizes, giving interstellar travelers places to stop and visit friends and collect fresh supplies between the stars. We know almost nothing about these objects except for the fact that they exist. We know that in the space around our own planetary system there are two populations of comets, known as the Kuiper Belt and the Oort Cloud. The Kuiper Belt is the source of short-period comets, and the Oort Cloud is the source of long-period comets. We know that they exist because the comets which we see coming close to the sun are visibly disintegrating and cannot survive for a long time. The tail which makes a comet beautiful is proof of its mortality. Meteor showers are the debris marking the graves of dying comets. To keep new comets appearing at the observed rate, the source populations must be large, of the order of billions of comets of each kind. A few of the biggest and closest objects in the Kuiper Belt population can be directly observed, orbiting the sun with orbits concentrated around the plane of the planets. The brightest and most famous of these objects is Pluto. The Oort Cloud is invisible from the earth. It is a spherical population of objects at much greater distances from the sun, loosely attached to the sun by weak gravitational forces.

There is no reason to believe that the space between the Oort Cloud and the nearest stars is empty. We know that a large fraction of all stars are born with planetary systems. It is also likely that large numbers of planets are born unattached to stars. Furthermore, we know that the normal processes of formation and evolution

---

*Foreword to *Starship Anthology*, edited by Jim Benford.

of planetary systems result in ejection of planets and comets from the systems. As a result of these processes, the universe probably contains more unattached planets than stars, and billions of times more unattached comets. The space between our solar system and the nearest stars is probably infested with unattached planets and far more numerous unattached comets. In addition, there may be other objects of intermediate kinds which we have not yet observed, from snow-balls to black-dwarf stars. It is conceivable that some of the intermediate objects might be alive, a population of mythological monsters making their home in space.

The existence of abundant way-stations between the stars is likely to have a decisive influence on the development of starships. We shall not jump in one huge step from planetary to interstellar voyages. We shall be exploring one group of objects after another, first the Kuiper Belt, then the Oort Cloud, then a string of further-out oases in the desert of space, before we finally come to Proxima Centauri. In the history of mankind on this planet, there were two very different kinds of explorers who learned to navigate the oceans. There were the European navigators who sailed from fixed bases in Europe to destinations in America and Asia and came home to Europe with loot from their trade and conquest. Columbus was typical of these explorers, making three voyages back and forth across the Atlantic. But before the Europeans, there were Polynesian navigators who built canoes to sail long distances on the Pacific and populated the Pacific islands from Asia to Hawaii and New Zealand. The Polynesians did not have home bases in Asia and America, and they were not interested in sailing all the way across the ocean. They made their voyages from island to island, stopping to make a new home when they found a new island suitable for raising their crops, pigs and children.

The Polynesians were navigating the Pacific for a thousand years before the Europeans crossed the Atlantic. Island-hopping came first, intercontinental voyages later. It is likely that the future of our traveling beyond the Solar System will follow the same pattern. The evolution of starships, like the evolution of Polynesian canoes and European galleons, will proceed by a process of trial and error. Unattached comets and planets will be like the islands in the Pacific Ocean. We will begin like the Polynesian navigators, modestly. Developing starships one step at a time, we can learn from our mistakes how to do the job right. Perhaps, after a thousand years, we will be ready to build grand super-highways conveying traffic along non-stop routes from star to star.

Two things are needed to make starships fly, a place to go and a way to get there. The first problem is mainly a problem of biology, the second a problem of engineering. Let us look at biology first. To have a place to go, we must learn how to grow complete eco-systems at remote places in the universe. It is not enough to have hotels for humans. We must establish permanent ecological communities including microbes, plants and animals, all adapted to survive in the local environment.

The populations of the various species must be balanced so as to take care of each others' needs as well as ours. Permanent human settlement away from the earth only makes sense if it is part of a bigger enterprise, the permanent expansion of life as a whole. The best way to build human habitats is to prepare the ground by building robust local ecologies. After life has established itself with grass and trees, herbivores and carnivores, bacteria and viruses, humans can arrive and build homes in a friendly environment. There is no future for humans tramping around in clumsy space-suits on lifeless landscapes of dust and ice.

The recent revolution in molecular biology has given us new tools for seeding the universe with life. We have learned to read and write the language of the genome, to sequence the DNA that tells a microscopic egg how to grow into a chicken or a human, to synthesize the DNA that tells a bacterium how to stay alive. We have sequenced the genomes of several thousand species. The speed of sequencing and of synthesis of genomes is increasing rapidly, and the costs are decreasing equally rapidly. If the increase in speed and the decrease in costs continue, it will take only about twenty years for us to sequence genomes of all the species that exist on our planet. The genetic information describing the entire biosphere of the planet will be available for our use. The total quantity of this information is remarkably small. Measured in the units that are customary in computer engineering, the information content of the biosphere genome amounts to about one petabyte, or ten to the power sixteen bits. This is a far smaller amount of information than the databases used by enterprises such as Google. The biosphere genome could be embodied in about a microgram of DNA, or in a small room full of computer memory-disks.

Looking ahead fifty or a hundred years, we shall be learning how to use genetic information creatively. We shall then be in a position to design biosphere populations adapted to survive and prosper in various environments on various planets, satellites, asteroids and comets. For each location we could design a biosphere genome, and for each biosphere genome we could design an egg out of which an entire biosphere could grow. The egg might weigh a few kilograms and look from the outside like an ostrich egg. It would be a miniature Noah's ark, containing thousands or millions of microscopic eggs programmed to grow into the various species of a biosphere. It would also contain nutrients and life-support to enable the growth of the biosphere to get started. The first species to emerge from a Noah's ark egg would be warm-blooded plants designed to collect energy from sunlight and keep themselves warm in a cold environment. Warm-blooded plants would then provide warmth and shelter for other creatures to enjoy. In this way, life could be seeded in great abundance and variety in all kinds of places, traveling on small spacecraft carrying payloads of a few kilograms. Since life is inherently an unpredictable phenomenon, many of the biospheres would fail and die. Those that survived would evolve in unpredictable ways. Their evolution would continue for ever, with or without human intervention.

We would be the midwives, bringing life to birth all over the universe, as far as our Noah's ark eggs could travel.

The second problem, the problem of engineering, is to build machines that can take us from here to there. To have space travel over long distances at reasonable prices, we must build a public highway system so that the costs of the initial investment can be shared by a multitude of users. A public highway system in space will require terminals using sunlight or starlight to generate high-energy beams along which spacecraft can fly. The beams may be laser-beams or microwave-beams or pellet streams. The massive energy-generating machinery at the terminals remains fixed. The spacecraft are small and light, and pick up energy from the beams as they fly along. Unlike chemical or nuclear rockets, they do not carry their own fuel. For the system to operate efficiently, the volume of traffic must be big enough to use up the energy of the beams. Spacecraft must be flying along the beams almost all the time. As with all public highway systems, the system can only grow as fast as the volume of traffic. The cost of travel will be high at the beginning and will become low when every terminal is crowded with passengers waiting for a launch.

In every public transport system, things work better if we build separate vehicles for passengers and freight. On the roads, cars for passengers and trucks for freight. On the railroads, fast short trains for passengers and slow long trains for freight. The Space Shuttle was a system designed to put passengers and freight on the same vehicle, and that was one of the reasons why it failed. It was supposed to be cheap, safe and reliable, with frequent flights and a high volume of traffic, and it turned out to be expensive, unsafe and unreliable. The public highways of the future will be like roads and railroads and not like the Shuttle. But the relation between passengers and freight in the future will be the opposite of what it was in the past. In the past, humans were small and light, freight was big and heavy. Cars were small and agile, trucks were big and clumsy. In space today, this relation between human passengers and freight is already inverted. Because of the miniaturization of instruments and communication systems, unmanned spacecraft have become smaller and lighter than manned spacecraft. Payloads of unmanned missions have remained roughly constant while their performance and capability have improved by leaps and bounds. Payloads of manned missions have remained larger while politicians fail to decide what they are supposed to do.

In the future, when missions go beyond the solar system, the difference between passengers and freight will become greater. Freight will no longer be bulk materials such as fuel and water. Freight will be information, embodied in ultralight computer memory or in DNA. Freight will be several orders of magnitude lighter than human passengers. Payloads of unmanned missions may be measured in grams, while payloads of manned missions will always be measured in tons. As a result, the public highway system will consist of two parts, a heavy-duty system transporting human passengers between a small number of metropolitan human habitats, and a light-freight system transporting packages of information along a wider network of routes to more distant destinations. A typical light-freight mission might be like the Starwisp proposed by Bob Forward. The Starwisp is an ultralight sail made of fine wire mesh, driven through space by a high-power beam of microwaves. The wire mesh is not only the vehicle but also the payload, carrying sensors to explore the environment and transmitters to send information collected by the sensors to humans far away. Starwisp could also be a vehicle for carrying Noah's ark eggs to bring life to remote places. It is likely that the travel-times of voyages will become longer than a human life-time. After life has spread that far, it will no longer make sense for humans to travel with it. Instead of imprisoning human travelers for a lifetime in a spacecraft, it would make more sense to load the spacecraft with a few human eggs, which could grow into humans at the destination. In the end, we would populate the galaxy by broadcasting the information required for growing humans, rather than by carrying deep-frozen human bodies for thousands of years.

When we are thinking about the spread of life into the universe, the most important fact to remember is that almost all the real-estate in the universe is on small objects. Real-estate means surface area. The universe contains objects of all sizes. Most of the mass and volume belong to big objects such as stars and planets. Most of the area belongs to small objects such as asteroids and comets. Most of the life will have to find its home on small objects. The majority of small objects have three qualities which make them unfriendly to life. They are far from the sun or other stars, they have no atmosphere, and they are cold. In spite of those disadvantages, they can be seeded with life. They can support biospheres as diverse and as beautiful as ours.

The key technology for bringing life to small cold objects in space is the cultivation of warm-blooded plants. Warm-blooded plants are more essential to the ecology of cold places than warm-blooded animals are to the ecology of our warm planet. Life on earth might have evolved happily without birds and mammals, but life in a cold place could never get started without warm-blooded plants. Two external structures make warm-blooded plants possible, a greenhouse and a mirror. The greenhouse is an insulating shell protecting the warm interior from the cold outside, with a semi-transparent window allowing sunlight or starlight to come in but preventing heat radiation from going out. The mirror is an optical reflector or system of reflectors in the cold region outside the greenhouse, concentrating sunlight or starlight from a wide area onto the window. Inside the greenhouse are the normal structures of a terrestrial plant, leaves using the energy of incoming light for photosynthesis, and roots reaching down into the icy ground to find nutrient minerals. Since there is no atmosphere to supply the plant with carbon dioxide, the roots must find mineral sources of carbon and oxygen to stay alive. We see in the light emitted from comets, as they come close to the sun, that these icy objects contain plenty of carbon and oxygen as well as nitrogen and other elements essential to life.

The embryonic warm-blooded plant must grow the greenhouse and the mirror around itself while still protected within the greenhouse of its parent. The seeds must develop into viable plants before they are dispersed into the cold environment. These plants must be viviparous as well as warm-blooded. It seems to be only an accident of evolution on our own planet that animals learned to be viviparous and warm-blooded while plants did not.

The optical concentration that the mirror must provide will depend on the distance of the plant from the sun or star providing the energy. Roughly speaking, the optical concentration must increase with the square of the distance from the source. For example, if the plant is on the surface of Enceladus, a satellite of Saturn at ten times the Earth's distance from the sun, the intensity of sunlight is one-hundredth of the intensity on Earth, and the optical concentration must be by a factor of a hundred. If the plant is in the Kuiper Belt at a hundred times the Earth's distance, sunlight is reduced by a factor of ten thousand and the mirror must concentrate by a factor of ten thousand. Existing biological structures can do much better than that. The human eye is not an extreme example of optical precision, but it can concentrate incoming light onto a spot on the retina by a factor larger than a million. That is why staring at the sun is bad for the health of the eye. A mirror as precise as a human eye would be good enough to keep a plant warm at a distance ten times further from the Sun than the Kuiper Belt. Eagles and hawks have better eyes than we do, and a simple amateur telescope costing less than a hundred dollars is better still. There is no law of physics that would prevent a warm-blooded plant from growing a mirror to concentrate enough starlight to survive anywhere in our galaxy. The main difficulty in achieving a high concentration of starlight is that the mirror must track the source accurately as the object carrying the plant rotates. The plant must be like a sun-flower, tracking the sun as it moves across the sky. If high accuracy is needed, the plant must grow an eye to see where it is pointing.

These speculations about viviparous plants and Noah's ark eggs and life spreading through the galaxy are my personal fantasies. They are only one possible way for the future to go. The real future is unpredictable. It will be rich in surprises that we have not imagined. All that we can say with some confidence is that biotechnology will dominate the future. The awesome power of nature, to evolve unlimited diversity of ways of living, will be in our hands. It is for us to choose how to use this power, for good or for evil.

Section 5

Technical Papers

CHAPTER 5.1

THE COULOMB FLUID AND THE FIFTH PAINLEVÉ TRANSCENDENT[*]

Certain Fredholm determinants arising in the theory of random matrices and Coulomb gases were discovered [Jimbo *et al.*, 1980] to be special cases of the Fifth Painlevé Transcendent. Recently Mehta found some new exact identities satisfied by these determinants. The present paper proceeds in three stages. First, Mehta's proofs of his identities are presented in a simpler and more transparent fashion. Second, a classical thermodynamic model of a Coulomb fluid is solved exactly, and the solution is expressed in terms of elliptic integrals. Third, the exact Mehta identities and the Coulomb fluid solutions are woven together in order to obtain a systematic expansion of the Fredholm determinants. The main new result is the calculation of oscillating factors in the Fredholm determinants. The oscillating factors turn out to be Jacobian elliptic functions. They can be understood physically as arising from the crystalline long-range order of the Coulomb gas or of the eigenvalues of a large random matrix.

Dedication

This chapter is dedicated to Frank Yang in honor of his 70th birthday. Forty years ago, Frank published a paper [Yang, 1952] in which he calculated exactly the spontaneous magnetization of an Ising ferromagnet in two dimensions. The result amazed us because of its beautiful simplicity; Frank's calculation amazed us because of its beautiful complexity. The calculation was a virtuoso exercise in the theory of Jacobian elliptic functions. The result was a simple algebraic expression from which all traces of elliptic functions had disappeared. After working through this astonishing display of mathematical fireworks, I expressed some disappointment that Frank had chosen such an unimportant problem on which to lavish his skill. I remarked, with the arrogance of youth, that if Frank could ever do a beautiful job like this on a problem of major importance, then he would really amount to something as a scientist. After that, I had the pleasure of watching Frank grow over my head into a scientist of world stature. Only two years later, he found and solved

---

[*]Contribution to a book celebrating the 70th birthday of Frank Yang, edited by Professor Shing-Tung Yau, published in 1993.

a problem of importance commensurate with his gifts. With his 1954 discovery of non-Abelian gauge theory, he laid the foundations on which, after many competing attempts had ended in failure, successful unified theories of particle interactions were finally built. With his work on gauge fields and particle symmetries, he showed that it is possible to combine a deep understanding of nature with a mastery of beautiful mathematics.

For forty years since the paper on the Ising ferromagnet was published, Frank has continued to do beautiful work on both important and unimportant problems. Between his major contributions to the central areas of particle physics and many-body theory, he indulges his taste for classical mathematics by continuing to solve unimportant problems. He still loves the intricate baroque music of elliptic functions, the music that he played in his 1952 paper. I offer this chapter to celebrate his birthday, in the hope that he will enjoy the same music in the context of another unimportant problem. Imitation is the sincerest form of flattery, even if it comes forty years late.

1. Mehta's Identities and Coulomb Gases

The fifth Painlevé transcendent [Painlevé, 1902] is a name for a class of analytic functions that satisfy nonlinear differential equations. The fifth transcendent includes as a special case [Jimbo *et al.*, 1980], the Fredholm determinants describing the distribution of level-spacings in the theory of random matrices [Mehta, 1991; Mahoux and Mehta, 1992]. The determinants are

$$\Delta(z,t) = \det(1 - zK), \quad \Delta_R(z,t) = \det(1 - zK)^R = \exp \mathrm{Tr}(R \log(1 - zK)),$$
(1.1)

where R and K are operators defined by the kernels

$$R = \delta(x + y), \quad K = (1/2)P_t \bar{e} P_1 e = P_t(\sin[\pi(x - y)]/[\pi(x - y)]), \quad (1.2)$$

P_t is the projection-operator onto the interval $[-t, t]$, and

$$e = \exp(i\pi xy), \quad \bar{e} = \exp(-i\pi xy). \quad (1.3)$$

Following [Mehta, 1992], we introduce the derivatives

$$A = -(1/2)(d \log \Delta/dt) = (u|(G - 1)|u), \quad B = -(1/2)(d \log \Delta_R/dt) = (\hat{u}|G|u),$$
(1.4)

$$G = (1 - zK)^{-1}, \quad (1.5)$$

with the states u and $\hat{u}$ defined by

$$|u) = \delta(x - t), \quad |\hat{u}) = R|u) = \delta(x + t). \quad (1.6)$$

The adjoint state $(a|$ and the transposition operation O^T are defined without complex conjugation, so that $(a|O^T|b) = (b|O|a)$. Mehta also uses the equivalent kernel

$$L = (1/2)eP_t\bar{e}P_1 = (\sin[\pi t(x-y)]/[\pi(x-y)])P_1, \quad Le = eK, \qquad (1.7)$$

so that

$$\Delta(z,t) = \det(1-zL), \quad \Delta_R(z,t) = \det(1-zL)^R. \qquad (1.8)$$

If the dilatation operator c is defined by

$$c = t^{1/2}\delta(x-ty), \quad c^T c = cc^T = 1, \qquad (1.9)$$

and the states v and $\hat{v}$ by

$$|v) = \delta(x-1) = t^{1/2}c^T|u), \quad |\hat{v}) = \delta(x+1) = t^{1/2}c^T|\hat{u}), \qquad (1.10)$$

then (1.2) and (1.7) imply

$$K = cL^T c^T, \quad L = c^T K^T c, \qquad (1.11)$$

and (1.4) becomes

$$tA = (v|(H-1)|v), \quad tB = (\hat{v}|H|v), \quad H = (1-zL)^{-1}. \qquad (1.12)$$

Lastly, Mehta defines the complex matrix element

$$S = (v|He|u) = (v|eG|u), \quad S^* = (v|He|\hat{u}) = (v|eG|\hat{u}), \qquad (1.13)$$

and proves five elegant nonlinear equations connecting the three functions A, B and S, namely

$$dA/dt = 2B^2, \qquad (1.14)$$
$$d(tA)/dt = z|S|^2, \qquad (1.15)$$
$$d(tB)/dt = z\Re(S^2), \qquad (1.16)$$
$$2\pi tB = z\Im(S^2), \qquad (1.17)$$
$$dS/dt = i\pi S + 2S^*B. \qquad (1.18)$$

Of these equations, only (1.14) was previously known, proved by Gaudin many years ago (see Appendix A.16 of [Mehta, 1991]). The four new equations bring new tools to the study of the Painlevé transcendent.

Since Mehta's proofs of (1.14)–(1.18) are written in a cumbersome notation that obscures their beauty, I reproduce them here translated into a simpler algebraic language. First, differentiation of H using (1.7) and (1.12) gives

$$dH/dt = zH(dL/dt)H = (z/2)HeT\bar{e}H, \quad T = |u)(u| + |\hat{u})(\hat{u}|. \tag{1.19}$$

Inserting (1.19) into (1.12), using the definition (1.13) of S, gives immediately (1.15) and (1.16). To prove (1.14) and (1.18) takes only a little more algebra. Define the differentiation-kernel D by

$$D = \delta'(x - y). \tag{1.20}$$

Acting on the states $|u)$ and $e|v)$, D gives

$$D|u) = -d|u)/dt, \quad (u|D = d(u|/dt, \quad (v|eD = -i\pi(v|e. \tag{1.21}$$

Differentiating (1.4) and (1.13) and using (1.21), we obtain

$$dA/dt = (u|M|u), \quad dS/dt - i\pi S = (v|eM|u), \quad M = (dG/dt) + [D, G]. \tag{1.22}$$

Now (1.2) gives

$$dK/dt = TK, \quad [D, K] = -UK, \quad U = |u)(u| - |\hat{u})(\hat{u}|, \tag{1.23}$$

which gives with (1.22)

$$M = zG((dK/dt) + [D, K])G = zG(T - U)KG = 2G|\hat{u})(\hat{u}|(G - 1). \tag{1.24}$$

When (1.24) is substituted into (1.22), the result is (1.14) and (1.18). To prove (1.17) we introduce the kernel

$$X = x\delta(x - y), \tag{1.25}$$

so that (1.4) gives

$$2tB = -(\hat{u}|[X, G]|u) = -z(\hat{u}|G[X, K]G|u). \tag{1.26}$$

But (1.2) and (1.3) imply

$$2\pi i[X, K] = 2iP_t \sin[\pi(x - y)] = P_t e[|v)(\hat{v}| - |\hat{v})(v|]e, \tag{1.27}$$

so that (1.26) becomes

$$4\pi itB = z[(\hat{u}|G^T e|\hat{v})(v|eG|u) - ((\hat{u}|G^T e|v)(\hat{v}|eG|u)] = z[S^2 - (S^*)^2], \tag{1.28}$$

and (1.17) is proved.

The theory of random matrices is in turn connected with the thermodynamics of a Coulomb gas [Dyson, 1962a], composed of unit point charges at positions x_j on the infinite line $(-\infty < x < \infty)$, with a repulsive interaction-energy

$$W_{jk} = -\log|x_j - x_k| \tag{1.29}$$

between each pair of charges. The probability-distribution of eigenvalues of a large random matrix is identical to the distribution of charges in the Coulomb gas at a particular temperature. We denote by

$$E_\beta(s, n) \tag{1.30}$$

the probability that any fixed interval of length s will contain precisely n charges, when the gas is in thermal equilibrium at temperature $T = \beta^{-1}$ and at an average density of one charge per unit length. The choice $\beta = 2$ is appropriate for eigenvalues of random complex Hermitian matrices. The Coulomb gas and the Fredholm determinant (1.1) are connected [Basor et al., 1992] by the relation

$$E_2(n, 2t) = ((-1)^n/(n!))(\partial^n \Delta(z, t)/\partial z^n)|_{z=1}. \tag{1.31}$$

This implies that

$$\Delta(z, t) = \sum_{n=0}^{\infty} (1 - z)^n E_2(n, 2t) = \langle (1 - z)^n \rangle \tag{1.32}$$

is the expectation-value of $(1 - z)^n$, where n is the number of charges in an interval of length $2t$ for a Coulomb gas with $\beta = 2$.

Another way of writing (1.32) is

$$\Delta(z, t) = (Z_2(v, s)/Z_2(0, s)), \tag{1.33}$$

where

$$Z_\beta(v, s) = \sum \exp(-\beta(W + vn)) \tag{1.34}$$

is the partition-function of the Coulomb gas with the external potential v applied to the charges in a fixed interval of length s, and

$$1 - z = \exp(-\beta v). \tag{1.35}$$

In (1.34) the sum represents an infinite-dimensional integral over all configurations of the gas, and W represents the sum of all the Coulomb interactions (1.29). We do not attempt here to give (1.34) a rigorous meaning, since we are using it only for heuristic purposes. Our aim is to find approximations to $Z_\beta(v, s)$ which are valid when s is large. In the regime of large s, the discrete Coulomb gas may be approximated by a continuous Coulomb fluid [Dyson, 1962b]. Then (1.34) is approximated by

$$Z_\beta(v, s) = \exp(-\beta\Phi), \qquad (1.36)$$

$$\beta\Phi = \min\left(\beta(W + vn) + ((\beta/2) - 1)S\right), \qquad (1.37)$$

where Φ, W and S are the free energy, the Coulomb energy and the entropy, and the minimum is taken over all configurations of the fluid. We shall make a further approximation, based on the fact that the energy W is of order s^2 while the entropy S is only of order s for large s. We replace (1.37) by

$$\beta\Phi = \beta(\min(W + vn)) + ((\beta/2) - 1)S, \qquad (1.38)$$

so that the configuration of the fluid is found by minimizing the energy at zero temperature, and the entropy term is merely subtracted from the zero-temperature energy. The error in replacing (1.37) by (1.38) is likely to be of order unity or $\log s$ at most, and is no greater than the error arising from the replacement of a discrete gas by a continuous fluid. In particular, when $\beta = 2$, the entropy term is missing in (1.37), and (1.38) is exact. Section 2 will be concerned with the precise definition of the Coulomb fluid and with the calculation of (1.38).

2. The Coulomb Fluid

We consider a one-dimensional Coulomb fluid defined by a non-negative charge density $\rho(x)$ on the infinite line $(-\infty < x < \infty)$. The line is supposed to carry a uniform fixed neutralizing charge density $(\rho_0(x) = -1)$. The potential at any point x is

$$V(x) = -\int \hat{\rho}(y) \log|x - y| dy, \qquad \hat{\rho}(y) = \rho(y) - 1. \qquad (2.1)$$

The electrostatic energy of the fluid is

$$W = -(1/2) \iint \hat{\rho}(x)\hat{\rho}(y) \log|x - y| dx\, dy = (1/2) \int \hat{\rho}(x)V(x)dx. \qquad (2.2)$$

We allow as density-functions the class of non-negative $\rho(x)$ for which the integrals (2.1) and (2.2) are finite. This means that point charges (delta-function singularities) are excluded. The functions that we are interested in have

$$\rho(x) = 1 + O(x^{-2}) \quad \text{as } x \to \pm\infty. \qquad (2.3)$$

We fix attention on the interval I of length s defined by

$$-t < x < t, \quad t = (s/2). \tag{2.4}$$

We consider two alternative conditions to fix the state of the fluid. Let the total charge in I be

$$N(\rho) = \int_{-t}^{t} \rho(x)dx. \tag{2.5}$$

For any given positive n, we may define the ground-state as the function ρ that minimizes W subject to the condition

$$N(\rho) = n. \tag{2.6}$$

For any given v, we may define the ground-state as the function ρ that minimizes the quantity

$$W_v = W + vN, \tag{2.7}$$

while N is free to vary. The two definitions are equivalent. The ground-state of the fluid with a fixed total charge n in I is the same as the ground-state with a fixed external potential v applied to the charge in I in addition to the self-generated potential $V(x)$.

The electric field generated by the charge $\rho(x)$ is

$$E(x) = -(dV/dx) = \int \hat{\rho}(y)(x - y)^{-1}dy. \tag{2.8}$$

Minimization of W subject to (2.6), or of W_v, gives in both cases a condition satisfied by the minimizing ρ,

$$E(x) = 0 \quad \text{whenever } \rho(x) \neq 0. \tag{2.9}$$

The whole line $(-\infty < x < \infty)$ must therefore be covered by intervals in which $E(x) = 0$ and intervals in which $\rho(x) = 0$. On the outermost intervals where (2.3) holds, we must have $V(x) = 0$. When (2.7) is minimized, we must also have

$$V(x) + v = 0 \quad \text{whenever } \rho(x) \neq 0, \quad x \text{ in } I. \tag{2.10}$$

According to (2.8), $(1/\pi)E(x)$ and $\hat{\rho}(x)$ are Hilbert transforms of each other, which implies that

$$F(x) = \rho(x) + (i/\pi)E(x) \tag{2.11}$$

is the boundary value for real x of a function $F(z)$ analytic in the upper half-plane $(\Im z > 0)$. The condition (2.9) implies that $F(x)$ is alternately purely real and purely imaginary. At each of the transition points where it changes from real to imaginary,

$F(x)$ must have a branch-point with a square-root singularity. Apart from these square-root singularities, no other singularities are allowed. Further, two of the branch-points must be at the ends of the interval I. These conditions together with (2.3) fix the form of $\rho(x)$ and $E(x)$ uniquely. There must exist a positive a such that

$$\rho(x) = [(x^2 - a^2)/(x^2 - t^2)]^{1/2}, \quad E(x) = 0,$$

$$|x| < \min(a, t) \quad \text{or} \quad |x| > \max(a, t), \tag{2.12}$$

$$\rho(x) = 0, \quad E(x) = \pi\,\text{sign}(x)\text{sign}(a - t)[(x^2 - a^2)/(t^2 - x^2)]^{1/2},$$

$$\min(a, t) < |x| < \max(a, t). \tag{2.13}$$

There are five intervals, with $E(x)$ zero in intervals 1, 3 and 5, and $\rho(x)$ zero in intervals 2 and 4. The four branch-points are at

$$x = \pm t, \quad x = \pm a. \tag{2.14}$$

The charge distribution (2.12) is the same, whether we are minimizing W for fixed n or minimizing W_v for fixed v. In the first case, a is determined by (2.6) and v is determined by (2.10). In the second case, a is determined by (2.10) and n is determined by (2.6). There is a one-to-one relation between n and v, n being the charge on I induced by the external potential v, and $(-v)$ being the internal potential on I when the charge on I is n.

There are two regimes corresponding to the two possible signs of v. The first regime is

$$v > 0, \quad n < s, \quad a < t, \tag{2.15}$$

which means that the potential is driving charge out from I and two gaps with zero charge are formed inside I adjacent to the end-points. The second regime is

$$v < 0, \quad n > s, \quad a > t, \tag{2.16}$$

which means that the potential is attracting charge into I and two gaps with zero charge are formed outside I adjacent to the end-points. In the limiting case

$$v = 0, \quad n = s, \quad a = t, \tag{2.17}$$

the gaps disappear and the charge-density $\rho(x) = 1$ everywhere. Another limiting case occurs when

$$v = \pi t, \quad n = 0, \quad a = 0, \tag{2.18}$$

which means that all the charge is driven out of I. No states exist for v larger than πt. If we try to continue the solution (2.12) to negative values of a^2, the function (2.11) is no longer analytic in the upper half-plane.

The entropy of the fluid is defined by

$$S = -\int \rho(x) \log \rho(x) dx. \tag{2.19}$$

The analytic properties of $\rho(x)$ allow us to calculate S exactly. The function $\log F(z)$ is analytic in the upper half-plane and tends to zero like $|z|^{-2}$ at infinity. Therefore

$$\int F(x) \log F(x) dx = 0, \tag{2.20}$$

when the logarithm of $F(x)$ is defined as a boundary value coming from the upper half-plane. According to (2.11) and (2.13), the boundary value of the logarithm is

$$\log F(x) = \log \rho(x) \quad \text{when } E(x) = 0, \tag{2.21}$$

$$\log F(x) = \log (|E(x)|/\pi) + (i\pi/2)\text{sign}(E(x)) \quad \text{when } \rho(x) = 0. \tag{2.22}$$

The real part of (2.20) then becomes

$$S + (1/2) \int |E(x)| dx = 0. \tag{2.23}$$

But (2.8) and (2.10) imply

$$\int |E(x)| dx = 2|V(0) - V(\infty)| = 2|v|, \tag{2.24}$$

and so (2.23) reduces to

$$S = -|v|, \tag{2.25}$$

a remarkably simple relation between entropy and potential. The entropy is always negative because we normalized it to be zero for the fluid with uniform density $\rho(x) = 1$.

The relation between v and the total energy W is not so simple but involves elliptic integrals. We now proceed to the calculation of W. According to (2.2), (2.8) and (2.11),

$$W = (1/2) \int \rho(x)V(x)dx - (1/2) \int V(x)dx = (1/2)nV(0) - (1/2) \int xE(x)dx, \tag{2.26}$$

$$= -(nv/2) + (i\pi/2) \int x(F(x) - 1)dx. \tag{2.27}$$

The last integral is easily evaluated by moving the path of integration to a large semicircle in the upper half-plane. We thus find

$$W = -(nv/2) + (\pi^2/4)(t^2 - a^2), \tag{2.28}$$

$$W_v = (nv/2) + (\pi^2/4)(t^2 - a^2). \tag{2.29}$$

It remains to find the relations between n, v and a. We write

$$m = \min(a, t), \quad M = \max(a, t), \quad k = (m/M). \tag{2.30}$$

Then (2.6), (2.10), (2.12) and (2.13) give

$$n = 2 \int_0^m [(a^2 - x^2)/(t^2 - x^2)]^{1/2} dx, \tag{2.31}$$

$$v = -\pi \text{sign}(a - t) \int_m^M [(x^2 - a^2)/(t^2 - x^2)]^{1/2} dx. \tag{2.32}$$

Thus n and v can be expressed in terms of complete elliptic integrals (K, E, K', E') with modulus k [Gradshteyn and Ryzhik, 1980, pp. 276, 277, 905]. The results are, for the regime (2.15),

$$n = 2t[E - (1 - k^2)K], \quad v = \pi t[E' - k^2 K'], \tag{2.33}$$

and for the regime (2.16),

$$n = 2aE, \quad v = -\pi a[K' - E']. \tag{2.34}$$

These results provide an exact equation of state for the Coulomb fluid, with entropy and energy given by (2.25) and (2.28).

From (2.27), (2.33) and (2.34) it is easy to derive expressions for W, n and v in various limiting cases, using [Gradshteyn and Ryzhik, 1980, pp. 905, 906] for series expansions of the elliptic integrals. It is convenient to take n as the independent variable and express the other quantities in terms of n. There are now three limiting regimes. First, the case in which the interval I is almost empty of charge and the potential v is close to the limit (2.18). Then

$$n \ll 2t = s, \quad k \ll 1, \tag{2.35}$$

which gives

$$a^2 = (ns/\pi) - (n^2/2\pi^2) + O(n^3/s), \tag{2.36}$$

$$v = (\pi s/2) - (n/2)(L + 1) + O(n^2 L/s), \quad L = \log(4\pi s/n), \tag{2.37}$$

$$W = (\pi^2 s^2/16) - (\pi s n/2) + (n^2/4)(L + (3/2)) + O(n^3 L/s). \tag{2.38}$$

Second, the case in which the defect or excess of charge in I is a small fraction of the total charge. Then

$$|n - s| \ll s, \quad (1 - k^2) \ll 1, \tag{2.39}$$

which gives to leading order in $(|n - s|/s)$

$$a = (s/2) + ((n - s)/L'), \quad L' = \log(s/|n - s|), \tag{2.40}$$

$$v = \pi^2(s - n)/(2L'), \tag{2.41}$$

$$W = \pi^2(s - n)^2/(4L'). \tag{2.42}$$

These results (2.40)–(2.42) are valid both for $n < s$ and for $n > s$. The quantities (a, v, W) remain continuous and differentiable at the logarithmic singularity which occurs at $n = s$. Third, the case in which the charge in I is very large. Then

$$n \gg s, \quad k \ll 1, \tag{2.43}$$

which gives to leading order in (s/n)

$$a = (n/\pi) + (\pi s^2)/(16n), \tag{2.44}$$

$$v = -n(L'' - 1), \quad L'' = \log(8n/\pi s), \tag{2.45}$$

$$W = (1/2)n^2(L'' - (3/2)). \tag{2.46}$$

We carried the expansions further for the first case since the case of small n is most interesting for comparison with other calculations [Tracy and Widom, 1992a, 1992b].

3. Painlevé and the Fluid Combined

We shall use the exact Mehta identities of Sec. 1 to improve upon the approximate Coulomb fluid estimates of Sec. 2. Since the Coulomb fluid estimates are valid in the limit of large t, we expand the quantities Φ, A, B in series

$$\Phi = \Phi_2 + \Phi_0 + \cdots, \quad A = A_1 + A_0 + \cdots, \quad B = B_0 + \cdots, \tag{3.1}$$

where the terms with suffix k are of order t^k, and the leading terms Φ_2, A_1 are the Coulomb fluid estimates. According to (1.33), (1.36) and (1.37) with $\beta = 2$, the Coulomb fluid estimate for $\Delta(z, t)$ is

$$\log \Delta(z, t) = -2\Phi_2, \quad \Phi_2 = \min(W + vn), \tag{3.2}$$

and by (2.29)

$$\Phi_2 = (nv/2) + (\pi^2/4)(t^2 - a^2). \tag{3.3}$$

According to (2.31) and (2.32), Φ_2 is homogeneous of degree 2 in t and v together, so that

$$t(\partial\Phi_2/\partial t) + v(\partial\Phi_2/\partial v) = 2\Phi_2. \tag{3.4}$$

Since Φ_2 is defined by the minimum of (3.2) with t fixed,

$$(\partial\Phi_2/\partial v) = n. \tag{3.5}$$

Note that n and v are here defined by the Coulomb fluid model to be quantities of order 1 in t and are not expanded in series as in (3.1). In a more exact treatment, lower-order corrections to n and v would be required. The derivatives (1.4) were defined as derivatives with respect to t holding z fixed, which by (1.35) is the same as holding v fixed. Therefore (3.2)–(3.5) imply

$$A_1 = (\partial\Phi_2/\partial t) = ((2\Phi_2 - nv)/t) = (\pi^2/2t)(t^2 - a^2). \tag{3.6}$$

We now proceed to solve (1.14)–(1.18) neglecting terms of relative order t^{-1}. We write

$$S = (t/z)^{1/2}(P + iQ), \tag{3.7}$$

with P and Q real, so that (1.17), (1.14) and (1.15) become

$$\pi B = PQ, \tag{3.8}$$

$$dA/dt = (2/\pi^2)P^2Q^2, \tag{3.9}$$

$$d(tA)/dt = t(P^2 + Q^2). \tag{3.10}$$

Then the product

$$H = (1 - (2/\pi^2)P^2)(1 - (2/\pi^2)Q^2) = 1 - (2/\pi^2)(A/t) \tag{3.11}$$

is determined by (3.6) to leading order in t. The leading term in H is

$$H_0 = (a/t)^2 = \hat{k}^2, \tag{3.12}$$

where $\hat{k}$ means k in the regime (2.15) and k^{-1} in the regime (2.16). The leading terms in the expansions of P and Q are then functions of a new variable x,

$$1 - (2/\pi^2)P_0^2 = \hat{k}\exp(x), \quad 1 - (2/\pi^2)Q_0^2 = \hat{k}\exp(-x), \tag{3.13}$$

and (1.16) implies

$$dB_0/dt = P_0^2 - Q_0^2 = -\pi^2\hat{k}\sinh x. \tag{3.14}$$

When (3.7) and (3.13) are substituted into (1.18), the result is

$$dx/dt = (4/\pi)P_0Q_0 = 4B_0, \tag{3.15}$$

so that x satisfies the second-order equation

$$d^2x/dt^2 = -4\pi^2\hat{k}\sinh x. \tag{3.16}$$

If $\hat{k}$ is constant, (3.16) is a standard equation for a hyperbolic pendulum and the solutions are Jacobian elliptic functions,

$$B_0 = (f\lambda/2)\text{Sn}(y, \lambda), \tag{3.17}$$

$$x = 2\log[(\text{Dn}(y, \lambda) - \lambda\text{Cn}(y, \lambda))/\lambda']. \tag{3.18}$$

Here we use the notation Sn to mean sn with the argument scaled to make the period equal to 2 instead of $4K(\lambda)$, thus

$$\text{Sn}(y, \lambda) = \text{sn}(2K(\lambda)y, \lambda), \tag{3.19}$$

and similarly for Cn and Dn. If $\hat{k}$ is constant, (3.16) is satisfied by (3.18) provided that

$$2K(\lambda)y = ft, \quad (2\pi/f)^2\hat{k} = (\lambda')^2 = 1 - \lambda^2. \tag{3.20}$$

Although $\hat{k}$ is not strictly constant, it is a slowly varying function of t when t is large, and so the expressions (3.17), (3.18) satisfy (3.16) with error of order t^{-1}, provided that (3.20) is replaced by

$$2K(\lambda)y = \int f\,dt, \quad f = (d/dt)(2K(\lambda)y). \tag{3.21}$$

Note that (3.17) with (3.21) is not the adiabatic solution of (3.16) as $\hat{k}$ varies. The adiabatic solution of (3.16) would hold constant the action integral $\int (dx/dt)^2 dt$ integrated over a period of the motion. The adiabatic condition is here irrelevant since the equation (3.16) itself is only an approximation neglecting terms of order t^{-1}. We choose the condition (3.21) so that the period of the motion as a function of y is held constant as $\hat{k}$ varies. The reason for this choice is that we shall identify the variable y with the number of charges n of the Coulomb gas inside the interval $[t, -t]$, and we expect the properties of the Coulomb gas to be periodic in n. According to (1.4), (3.15) and (3.18), the odd Fredholm determinant Δ_R becomes

$$\Delta_R = C_0 \exp(-x/2) = C_0[(\text{Dn}(y, \lambda) + \lambda\text{Cn}(y, \lambda))/\lambda'], \tag{3.22}$$

with the coefficient C_0 independent of t. To fix the solution uniquely, we still have to determine three parameters, namely C_0, the lower limit of the indefinite integral (3.21), and the modulus λ of the elliptic functions in (3.17) and (3.22).

The solutions (3.17) and (3.22) show that the odd Fredholm determinant is an oscillating function of t with the period

$$T = (4/f)K(\lambda), \quad T(dy/dt) = 2. \tag{3.23}$$

The oscillations are a manifestation of the discrete nature of the Coulomb gas. They do not appear in the Coulomb fluid approximation. We may picture the Coulomb gas responding as t increases with a discrete bump each time a single charge moves across the gap from outside to inside the interval $[t, -t]$. In the odd Fredholm determinant the bumps will appear with alternating signs according to the parity of n. In the even Fredholm determinant Δ, we expect the bumps to appear with period $(T/2)$. The relation between period and charge-number is therefore

$$T(dn/dt) = 2. \tag{3.24}$$

Comparing (3.23) with (3.24), we see that

$$y = n + c, \tag{3.25}$$

with the integration-constant c still to be determined.

The relation (3.24) enables us to determine λ. Eliminating T and f from (3.20), (3.23) and (3.24), we find

$$K(\lambda)(dn/dt) = \pi(\hat{k}^{1/2}/\lambda'). \tag{3.26}$$

The derivative (dn/dt) is to be taken with v constant. In (2.33) and (2.34), n and v are given in the form

$$n = tp(k), \quad v = tq(k), \tag{3.27}$$

where p and q are known functions of k. We have then

$$(dn/dt) = [(pq' - qp')/q'], \tag{3.28}$$

where the primes mean derivatives with respect to k. After some manipulation of elliptic integrals, (3.28) becomes

$$(dn/dt) = (\pi/K'), \quad (dn/dt) = k^{-1}(\pi/K'), \tag{3.29}$$

in the two regimes (2.15) and (2.16). The condition (3.26) then reduces in both regimes to

$$K(\lambda) = (k^{1/2}/\lambda')K', \tag{3.30}$$

which is a standard elliptic integral identity [Gradshteyn and Ryzhik, 1980, p. 908] when λ has the value

$$\lambda = [(1 - k)/(1 + k)] = [|t - a|/(t + a)], \quad \lambda' = [2k^{1/2}/(1 + k)]. \tag{3.31}$$

With this value of λ, (3.20) and (3.21) become

$$f = 2\pi(1 + \epsilon\lambda)^{-1}, \quad K(\lambda)y = \pi \int (1 + \epsilon\lambda)^{-1} dt, \tag{3.32}$$

with
$$\epsilon = \text{sign}(v) = \text{sign}(t - a), \quad \epsilon\lambda = [(t - a)/(t + a)]. \qquad (3.33)$$

The two regimes (2.33) and (2.34) can now be combined into the expressions

$$n = 2t(1 + \epsilon\lambda)^{-1}(E'(\lambda) - \epsilon\lambda K'(\lambda)), \qquad (3.34)$$

$$v = 2\pi t(1 + \epsilon\lambda)^{-1}(E(\lambda) - (1 - \epsilon\lambda)K(\lambda)), \qquad (3.35)$$

$$\Phi_2 = \pi^2 t^2 \epsilon\lambda(1 + \epsilon\lambda)^{-2} + (nv/2). \qquad (3.36)$$

When $t \to \infty$ with v held fixed, (3.35) gives

$$\epsilon\lambda \approx (v/(t\pi^2)). \qquad (3.37)$$

The difference between $K(\lambda)$ and $(\pi/2)$ is of order t^{-2}, so that the function Sn becomes an ordinary sine as $t \to \infty$. To fix the constant c in (3.25) we compare the behavior of B_0 given by (3.17) and (3.37),

$$B_0 \approx |z|(2\pi t)^{-1}\text{Sn}(y, \lambda) \qquad (3.38)$$

with the leading term in the expansion of (1.4) in powers of z

$$B_0 \approx z(2\pi t)^{-1}\sin(2\pi t) \approx z(2\pi t)^{-1}\sin(\pi n). \qquad (3.39)$$

The comparison shows agreement when $c = 0$, except that the sign of B_0 is wrong when z and v are negative. It is convenient to take $c = 0$ in all cases, identifying y with n, and to insert the necessary change of sign into (3.17). The corrected B_0 is then

$$B_0 = \pi\epsilon\lambda(1 + \epsilon\lambda)^{-1}\text{Sn}(n, \lambda), \qquad (3.40)$$

and the corrected Δ_R is now, instead of (3.22),

$$\Delta_R = C_0[(\text{Dn}(n, \lambda) + \epsilon\lambda\text{Cn}(n, \lambda))/\lambda']. \qquad (3.41)$$

The expression in square brackets in (3.41) tends to unity as $t \to \infty$. We can calculate C_0 by finding the limit of Δ_R directly from (1.8). When $t \to \infty$, the kernel L defined by (1.7) becomes

$$L(\infty) = \delta(x - y)P_1. \qquad (3.42)$$

This implies for every positive integer j

$$\text{Tr}(R(L(\infty))^j) = \int_{-1}^{1} \delta(2x)dx = (1/2),$$

$$\text{Tr}(R\log(1 - zL(\infty))) = (1/2)\log(1 - z), \qquad (3.43)$$

and therefore by (1.8) and (1.35)

$$C_0 = (1 - z)^{1/2} = \exp(-v). \tag{3.44}$$

This completes the calculation of the odd quantities B and Δ_R given by (3.40) and (3.41). It remains to deal with the even quantities A and Δ.

From (1.14) and (3.40) we deduce

$$A = 2 \int (\pi\lambda/(1 + \epsilon\lambda))^2 (\text{Sn}(n, \lambda))^2 dt. \tag{3.45}$$

According to [Gradshteyn and Ryzhik, 1980, p. 630],

$$\int (\text{sn}(u, \lambda))^2 du = \lambda^{-2}(u - E(\text{am} u, \lambda)), \tag{3.46}$$

which implies that the mean-square value of $\text{sn}(u, \lambda)$ over its period $4K(\lambda)$ is

$$V(\lambda) = \lambda^{-2}((K(\lambda) - E(\lambda))/K(\lambda)). \tag{3.47}$$

When we replace the square of $\text{sn}(u, \lambda)$ in (3.45) by its mean value, the result is the leading contribution to A,

$$A_1 = 2 \int (\pi\lambda/(1 + \epsilon\lambda))^2 V(\lambda) dt. \tag{3.48}$$

Using (3.35), (3.36) and (3.47), we can evaluate the integral (3.48) exactly and obtain

$$A_1 = 2\pi^2 \epsilon\lambda t/(1 + \epsilon\lambda)^2, \tag{3.49}$$

in agreement with (3.6).

The remaining contribution to A is

$$A_0 = A - A_1 = 2 \int [\pi\lambda/(1 + \epsilon\lambda)]^2 [(\text{Sn}(n, \lambda))^2 - V(\lambda)] dt. \tag{3.50}$$

We now introduce the Jacobi theta-function defined by

$$\theta_4(z, q) = \sum_{m=-\infty}^{\infty} (-1)^m q^{m^2} \exp(2imz). \tag{3.51}$$

The connection between the theta-function and the function $\text{Sn}(y)$ is according to [Prudnikov *et al.*, 1990, p. 35, Eq. (18)]

$$(d^2/dy^2) \log \theta_4(\pi y, q) = -(2\lambda K(\lambda))^2 [(\text{Sn}(y, \lambda))^2 - V(\lambda)], \tag{3.52}$$

with

$$q = \exp(-\pi K'(\lambda)/K(\lambda)). \tag{3.53}$$

Inserting (3.52) into (3.50), and neglecting terms of order t^{-1} arising from differentiation of slowly varying quantities, we find

$$A_0 = -(1/2) \int (d^2/dt^2)(\log \theta_4(\pi y, q)) dt, \tag{3.54}$$

which can be immediately integrated to give

$$A_0 = -\pi((1 + \epsilon\lambda)K(\lambda))^{-1}(d/dn)(\log \theta_4(\pi n, q)). \tag{3.55}$$

When t is large and λ small according to (3.37), (3.55) becomes by (3.53)

$$A_0 \approx -4\pi q \sin(2\pi n) \approx -(\pi/4)\lambda^2 \sin(2\pi n), \tag{3.56}$$

showing the expected oscillation with twice the frequency of B_0 according to (3.39). Finally, we have to evaluate the oscillating component of Φ,

$$\Phi_0 = \int A_0 dt. \tag{3.57}$$

By (3.54) this reduces to the remarkably simple form

$$\Phi_0 = -(1/2) \log \theta_4(\pi n, q), \tag{3.58}$$

and according to (1.4) the even Fredholm determinant is

$$\Delta = C\theta_4(\pi n, q) \exp(-2\Phi_2), \tag{3.59}$$

with a coefficient C which may depend on v but is independent of t. From (3.59) we see the oscillations in Δ of relative magnitude

$$-(\lambda/2)^2 \cos(2\pi n) \tag{3.60}$$

when $t \to \infty$. Such oscillations have been seen before in related contexts, for example in the density matrix of a gas of impenetrable bosons [Vaidya and Tracy, 1979]. The result (3.59) completes our calculation of the Fredholm determinants. Unfortunately, we are not able to calculate C by the simple argument that led to (3.44) in the case of the odd determinant. The behavior of Δ for large t is much more delicate than the behavior of Δ_R. It is likely that C could be calculated by the method of inverse scattering discussed in chapter 12 of [Mehta, 1991]. We hope to come back to this question later.

The author is indebted to Madan Lal Mehta, Harold Widom and Craig Tracy for sending him their unpublished results and for their friendly help and encouragement.

4. Summary of Results

The Fredholm determinants (1.1) and their logarithmic derivatives (1.4) are evaluated in the limit when t is large and the potential

$$v = -(1/2)\log(1 - z) \tag{4.1}$$

is also large. In this limit, the determinants are given by (3.59) and (3.41), being in each case the product of an oscillating factor with a smooth factor. The oscillating factors are elliptic functions with modulus λ given by (3.35) and depending only on the ratio (v/t). The smooth factors are given by (3.36) and (3.44). The smooth factor for the even determinant is the exponential of the total energy of the Coulomb fluid model, an energy which is a combination of elliptic integrals with the same modulus λ. The smooth factor for the odd determinant depends only on the potential v.

References

E. L. Basor, C. A. Tracy and H. Widom [1992] "Asymptotics of level spacing distributions for random matrices", *Phys. Rev. Lett.* **69**, 5–8.

F. J. Dyson [1962a] *J. Math. Phys.* **3**, 140–156.

F. J. Dyson [1962b] *J. Math. Phys.* **3**, 157–165.

I. S. Gradshteyn and I. M. Ryzhik [1980] *Table of Integrals, Series and Products*, 4th edition, ed. A. Jeffrey (Academic Press, New York).

M. Jimbo, T. Miwa, Y. Môri and M. Sato [1980] *Physica D* **1**, 80–158.

M. L. Mehta [1991] *Random Matrices*, 2nd revised and enlarged edition (Academic Press, New York).

G. Mahoux and M. L. Mehta [1992] "Level spacing functions and non-linear differential equations", Preprint, Service de Physique Théorique, Saclay, France.

M. L. Mehta [1992] "A non-linear differential equation and a Fredholm determinant", *Jour. de Phys. I, France* **2**, 1721–1729.

P. Painlevé [1902] *Acta Mathematica* **25**, 1–85.

A. P. Prudnikov, Yu. A. Brychkov and O. I. Marichev [1990] *Integrals and Series, Vol. 3*, trans. G. G. Gould (Gordon and Breach, New York).

C. A. Tracy and H. Widom [1992a] "Introduction to random matrices", Preprint, Institute of Theoretical Dynamics, University of California, Davis.

C. A. Tracy and H. Widom [1992b] "Level-spacing distributions and the airy kernel", Preprint, Institute of Theoretical Dynamics, University of California, Davis.

H. G. Vaidya and C. A. Tracy [1979] *Phys. Rev. Lett.* **42**, 3–6.

C. N. Yang [1952] "The spontaneous magnetization of a two-dimensional Ising model", *Phys. Rev.* **85**, 808–816.

ELSEVIER

Nuclear Physics B 480 (1996) 37–54

The Oklo bound on the time variation of the fine-structure constant revisited

Thibault Damour [a,b,c], Freeman Dyson [b]

[a] *Institut des Hautes Etudes Scientifiques, F-91440 Bures-sur-Yvette, France*
[b] *School of Natural Sciences, Institute for Advanced Study, Olden Lane, Princeton, NJ 08540, USA*
[c] *DARC, Observatoire de Paris–CNRS, F-92195 Meudon, France*

Received 1 July 1996; accepted 5 September 1996

Abstract

It has been pointed out by Shlyakhter that data from the natural fission reactors which operated about two billion years ago at Oklo (Gabon) had the potential of providing an extremely tight bound on the variability of the fine-structure constant α. We revisit the derivation of such a bound by (i) reanalyzing a large selection of published rare-earth data from Oklo, (ii) critically taking into account the very large uncertainty of the temperature at which the reactors operated, and (iii) connecting in a new way (using isotope shift measurements) the Oklo-derived constraint on a possible shift of thermal neutron-capture resonances with a bound on the time variation of α. Our final (95% C.L.) results are: $-0.9 \times 10^{-7} < (\alpha^{\text{Oklo}} - \alpha^{\text{now}})/\alpha < 1.2 \times 10^{-7}$ and $-6.7 \times 10^{-17} \text{yr}^{-1} < \dot{\alpha}^{\text{averaged}}/\alpha < 5.0 \times 10^{-17} \text{yr}^{-1}$.

PACS: 06.20.Jr; 04.80.Cc; 28.41.-i; 12.10.-g
Keywords: Variation of fundamental constants; Fine-structure constant; Natural fission reactors

1. Introduction

Since Dirac [1] first suggested it as a possibility, the time variation of the fundamental constants has remained a subject of fascination which motivated numerous theoretical and experimental researches. For general discussions and references to the literature see, e.g. Refs. [2–5]. Superstring theories have renewed the motivation for a variation of the "constants" by suggesting that most of the dimensionless coupling constants of physics, such as the fine structure constant $\alpha = 1/137.0359895(61)$, are functions of the vacuum expectation values of some scalar fields (see, e.g. Ref. [6]). Recently, a mechanism for fixing the vacuum expectation values of such *massless* stringy scalar fields (dilaton or moduli) has been proposed [7]. This mechanism predicts that the time variation of the coupling constants, at the present cosmological epoch, should be much smaller than the Hubble time scale, but maybe not unmeasurably so. In this model, the time variations of all the coupling constants are correlated, and the ones of most observational significance are the fine structure constant α and the gravitational coupling constant G. In the present paper, we revisit the current best bounds on the variation of α.

One of the early ideas for setting a bound on the variation of α was to consider the fine-structure splittings in astronomical spectra [8]. With this method, Bahcall and Schmidt [9] concluded that α had varied by at most a fraction 3×10^{-3} of itself during the last 2×10^9 years. A recent update of this method has given the result $\Delta\alpha/\alpha = (0.2 \pm 0.7) \times 10^{-4}$ at redshifts $2.8 \leqslant z \leqslant 3.1$, i.e. the bound $|\dot{\alpha}/\alpha| < 1.6 \times 10^{-14} \, \mathrm{yr}^{-1}$ (2σ level) on the time derivative of α averaged over the last $\sim 10^{10} \, \mathrm{yr}$ [10]. See also Ref. [11] which obtains $-4.6 \times 10^{-14} \, \mathrm{yr}^{-1} < \dot{\alpha}/\alpha < 4.2 \times 10^{-14} \, \mathrm{yr}^{-1}$ from fine-structure splittings, and, denoting $x \equiv \alpha^2 g_p (m_e/m_p)$, $-2.2 \times 10^{-15} \, \mathrm{yr}^{-1} < \dot{x}/x < 4.2 \times 10^{-15} \, \mathrm{yr}^{-1}$ by comparing redshifts obtained from hyperfine (21 cm) and optical data.

One of us obtained the upper limit $|\dot{\alpha}/\alpha| < 5 \times 10^{-15} \, \mathrm{yr}^{-1}$ from an analysis of the abundance ratios of rhenium and osmium isotopes in iron meteorites and molybdenite ores [2]. The most recent direct laboratory test of the variation of α has obtained $|\dot{\alpha}/\alpha| < 3.7 \times 10^{-14} \, \mathrm{yr}^{-1}$ by comparing hyperfine transitions in Hydrogen and Mercury atoms [12]. For more references on the variation of constants see Refs. [2–5].

On the other hand, the much more stringent bound $|\dot{\alpha}/\alpha| < 10^{-17} \, \mathrm{yr}^{-1}$ has been claimed by Shlyakhter [13–15] to be derivable (at the three standard deviations level) from an analysis of data from the Oklo phenomenon. The Oklo phenomenon denotes a natural fission reactor (moderated by water) that operated about two billion years ago in the ore body of the Oklo uranium mine in Gabon, West Africa. This phenomenon was discovered by the French Commissariat à l'Energie Atomique (CEA) in 1972. The results of a thorough, multi-disciplinary investigation of this phenomenon have been presented in two conference proceedings [16,17]. See also Refs. [18,19] for summaries of the first phase of investigation.

In view of the importance of Shlyakhter's claim, of the lack of publication of a detailed analysis, [1] and of our dissatisfaction with some important aspects of the analysis presented in two preprints [14,15], we decided to revisit the Oklo bound on α. The main conclusions of our work are the following: (i) we confirm the basic claim of Shlyakhter that the Oklo data is an extremely sensitive probe of the time variation of α; (ii) after taking into account various sources of uncertainty (notably temperature effects) in the analysis of data, and connecting in a improved way the raw results of this analysis to a possible variation of α, we derive what we think is a secure (95% C.L.) bound on the change of α:

$$-0.9 \times 10^{-7} < \frac{\alpha^{\mathrm{Oklo}} - \alpha^{\mathrm{now}}}{\alpha} < 1.2 \times 10^{-7}. \tag{1}$$

---

[1] The very brief account published in Ref. [13] omits most of the analysis that is presented in the two preprints [14,15].

In terms of an averaged rate of variation, this reads

$$-6.7 \times 10^{-17} \text{ yr}^{-1} < \frac{\dot{\alpha}}{\alpha} < 5.0 \times 10^{-17} \text{ yr}^{-1} . \tag{2}$$

2. Extracting the neutron capture cross section of samarium 149 from Oklo data

The proof of the past existence of a spontaneous chain reaction in the Oklo ore consists essentially of (i) a substantial depletion of the uranium isotopic ratio[2] U^{235}/U^{238} with respect to the current standard value in terrestrial samples; and (ii) a correlated peculiar distribution of some rare-earth isotopes. The rare-earth isotopes are abundantly produced in the fission of U^{235} and the observed isotopic distribution is beautifully consistent with calculations of the effect of a strong neutron flux on the fission yields of U^{235} (see e.g. Refs. [20–25]). In particular, the strong neutron absorbers Sm^{149}, Eu^{151}, Gd^{155} and Gd^{157} are found in very small quantities in the central regions of the Oklo reactors (see, e.g., Fig. 2 of Ref. [23]). These isotopes were evidently burned up by the large neutron fluence produced by the fission process. Following Shlyakhter's suggestion [14,15], we concentrate on the determination of the neutron capture cross section of Sm^{149}: $Sm^{149}(n, \gamma) Sm^{150}$.

The evolution of the concentrations of the various samarium isotopes (sharing the common atomic number $Z = 62$) in the Oklo ore is especially simple to describe because of the absence of a stable chemical element with atomic number $Z = 61$. The most stable promethium nuclide is Pm_{61}^{145} with a half-life of 17.7 years. If there had existed, before the reaction started, some natural concentration of promethium it could, via neutron absorption and subsequent β^- decay, have generated some samarium. In absence of this, the final values of the Sm concentrations are determined by (i) their initial concentrations, before the reaction; (ii) the yields from the fissions; and (iii) the effect of neutron captures.

Following Refs. [16,17], one characterizes the neutron absorbing power of an isotope by the *effective* cross section

$$\hat{\sigma} \equiv \frac{\int \sigma(E) v n_E \, dE}{v_0 \int n_E \, dE} , \tag{3}$$

where v is the (relative) velocity of incident neutrons, $n_E \, dE$ the energy distribution of the neutrons, and v_0 the fiducial (thermal) velocity $v_0 = 2200$ m/s corresponding to a kinetic energy $E_0 = 0.0253$ eV. The advantage of the definition (3) is that, in the case of a "$1/v$ absorber", $\sigma(E) = C/v$, the effective cross section equals $\hat{\sigma} = C/v_0 = \sigma(E_0)$ *independently* of the neutron spectrum. [To a good approximation, this is the case for the thermal fission cross section of U^{235}]. On the other hand, in the case of nuclides exhibiting resonances in the thermal region (these are the strong absorbers, Sm^{149},

---

[2] For typographical convenience, we indicate atomic mass numbers as right, rather than left, superscripts.

Eu^{151}, Gd^{155}, Gd^{157}), the value of the effective cross section (3) is very sensitive to the neutron spectrum, especially to its thermal part. [3]

Associated to the introduction of the effective cross section (3), one defines an effective neutron flux $\hat{\phi} \equiv n v_0$ with $n = \int n_E\, dE$, and an effective infinitesimal fluence (integrated flux)

$$d\tau = \hat{\phi}\, dt = n v_0\, dt .\tag{4}$$

With this notation, the general equation describing the evolution of the total number N_A of nuclides of mass number A (for some fixed atomic number Z) in some sample reads

$$\frac{dN_A}{d\tau} = y_A N_5 \sigma_{f5} + \sigma_{A-1} N_{A-1} - \sigma_A N_A .\tag{5}$$

Here, y_A denotes the yield of the element A in the fission of U^{235}, N_5 and σ_{f5} are short-hand notations for N_{235} and the (effective) fission cross section of U^{235}, and the last two terms describe the effects of neutron captures within isotopes of some chemical element Z. For simplicity, we drop the carets over the cross sections. The evolution equation (5) neglects any contribution $\propto N_{A-1}(Z')$ coming from the β^- decay of the neighbouring chemical element $Z' = Z - 1$ after absorption of a neutron. As we said above, this approximation applies well to the samarium case. Eq. (5) neglects also the yields due to the fractionally small number of fissions of U^{238} and Pu^{239}. [See, e.g., Ref. [23] which estimates that, in a particular sample, 2.5% and 3% of the fissions were due to U^{238} and Pu^{239}, respectively.]

Samples in the cores of the various Oklo reactors were exposed to a total effective fluence $\tau = \int d\tau = \int n v_0\, dt$ of the order of 10^{21} neutrons/cm^2 = 1 inverse kilobarn. This means, roughly speaking, that processes with effective cross sections comparable or larger than 1 kb have led to a significant number of reactions, while processes with $\sigma \ll 1$ kb had a negligible effect. The former category includes the fission of U^{235} ($\sigma_{f5} \sim 0.6$ kb), the capture of neutrons by Nd^{143} ($\sigma_{143} \sim 0.3$ kb) and by the strong absorbers (such as Sm^{149}; $\sigma_{149} \gtrsim 70$ kb), while the latter category includes neutron captures by weak absorbers such as Sm^{144} and Sm^{148} with cross sections of only a few barns.

This allows one to neglect σ_{144} and σ_{148} (for $Z = 62$) in Eq. (5). Further simplification comes from the fact that the stable isotopes 144, 146 and 148 of neodymium prevent the formation of the long-lived [4] Sm^{144}, Sm^{146} and Sm^{148} as end points of (β^- decay) fission chains. The stable isotopes [5] obey the simple evolution equations

$$\frac{dN_{144}}{d\tau} = 0 ,\tag{6}$$

---

[3] The spectrum of moderated neutrons in a fission reactor consists of a Maxwell–Boltzmann thermal distribution up to energies of order a few times kT, followed by a tail $n_E\, dE \propto dE/(vE)$ due to neutrons still in the process of moderation.

[4] The half-life of Sm^{146} is 1.03×10^8 yr and therefore long with respect to the duration of the Oklo phenomenon.

[5] "Stable" means, in this context, a half-life much larger than the *age* of the Oklo phenomenon. For example, the half-life of Sm^{147} is 1.06×10^{11} yr$\gg 2 \times 10^9$ yr.

$$\frac{dN_{147}}{d\tau} = y_{147}N_5\sigma_{f5} - \sigma_{147}N_{147}, \tag{7}$$

$$\frac{dN_{148}}{d\tau} = \sigma_{147}N_{147}, \tag{8}$$

$$\frac{dN_{149}}{d\tau} = y_{149}N_5\sigma_{f5} - \sigma_{149}N_{149}, \tag{9}$$

$$\frac{dN_5}{d\tau} = -N_5\sigma_5^*. \tag{10}$$

To close the system, we have followed Refs. [22,23] in describing the burn up of U^{235} by means of a modified absorption cross section $\sigma_5^* = \sigma_5(1 - C)$, where σ_5 is the normal absorption cross section (fission plus capture) and C is a conversion factor representing the formation of U^{235} from the decay of Pu^{239} formed by neutron capture in U^{238}.

In the approximation where the (effective) cross sections, and the conversion factor C, are constant, the system (6)–(10) is easily solved and gives

$$N_5(\tau) = N_5(0)e^{-\sigma_5^*\tau}, \tag{11}$$

$$N_{144}(\tau) = N_{144}(0), \tag{12}$$

$$N_{147}(\tau) + N_{148}(\tau) = N_{147}(0) + N_{148}(0) + y_{147}\sigma_{f5}N_5(0)\frac{1 - e^{-\sigma_5^*\tau}}{\sigma_5^*}, \tag{13}$$

$$N_{149}(\tau) = N_{149}(0)e^{-\sigma_{149}\tau} + y_{149}\sigma_{f5}N_5(0)\frac{e^{-\sigma_5^*\tau} - e^{-\sigma_{149}\tau}}{\sigma_{149} - \sigma_5^*}. \tag{14}$$

Eq. (12) shows that the quantity of Sm^{144} measured in a sample now is equal to the quantity of natural Sm^{144} present in the sample before the nuclear reactions. Assuming that the natural samarium present in the sample at the beginning had the normal isotopic ratios (say $n_{144} = 3.1\%$, $n_{147} = 15.0\%$, $n_{148} = 11.3\%$, $n_{149} = 13.8\%$, etc. [26]), we can use (12) to correct (13) for the initial concentrations in Sm^{147} and Sm^{148}. The effect of $N_{149}(0)$ in Eq. (14) is totally negligible (as is the last term) because the exponent $\sigma_{149}\tau \gg 1$. We then derive the intermediate result

$$\frac{N_{147}(\tau) + N_{148}(\tau) - [(n_{147} + n_{148})/n_{144}]N_{144}(\tau)}{N_{149}(\tau)} = \frac{y_{147}}{y_{149}}\frac{e^{\sigma_5^*\tau} - 1}{\sigma_5^*}(\sigma_{149} - \sigma_5^*). \tag{15}$$

One can finally obtain an expression for σ_{149} in terms of "measured" quantities by connecting $\sigma_5^*\tau$ to the observed ratio between the numbers of U^{235} and U^{238} atoms in the Oklo sample. If, following Ref. [22], we define

$$w \equiv \frac{0.00725}{(N_5/N_8)_{\text{now}}^{\text{Oklo}}}, \tag{16}$$

where 0.00725 is the usual U^{235}/U^{238} ratio in natural uranium now, it is easy to verify that $w = e^{\sigma_5^* \tau}$. Finally, we get[6]

$$\sigma_{149} = \frac{1}{\tau} \left[\ln w + y \frac{\ln w}{w-1} \frac{N_{147} + N_{148} - nN_{144}}{N_{149}} \right], \tag{17}$$

where (using Refs. [27,26])

$$y = \frac{y_{149}}{y_{147}} = \frac{1.080384}{2.261681} \simeq 0.478, \tag{18}$$

$$n = \frac{n_{147} + n_{148}}{n_{144}} = \frac{15.0 + 11.3}{3.1} \simeq 8.48. \tag{19}$$

The quantities N_A in Eq. (17) denote the present values of the isotopic concentrations, or, equivalently, the present values of the isotopic ratios. The isotopic ratios of samarium have been measured in many Oklo samples [28,20,23,29,24]. Note that the quantity which is, at this stage, directly obtainable from observations is the dimensionless product $\sigma_{149}\tau = \int dt \int \sigma_{149}(E) v n_E \, dE$.

Similarly, by considering the fission yields of neodymium and the neutron-capture reaction $Nd^{143} \to Nd^{144}$ (using, e.g., the Nd^{142} content to subtract the contribution from the natural concentrations present before the reaction), several authors [20–22] have shown how to obtain the dimensionless product $\sigma_{143}\tau$ (where $\sigma_{143} \equiv \sigma_{(n,\gamma)}(Nd^{143})$) in terms of quantities observed in Oklo samples. Combining these two results, we see that the value two billion years ago of the ratio $\sigma_{149}\tau/\sigma_{143}\tau = \sigma_{149}/\sigma_{143}$ can be computed in terms of present Oklo data.

Although it would be conceptually clearer to deal only with the dimensionless ratio $\sigma_{149}/\sigma_{143}$, we shall follow previous usage in working with the dimensionful quantity σ_{149} obtained by inserting in Eq. (17) the value of the (effective) fluence τ deduced from neodymium data by previous authors. This procedure is justified by the fact that the effective cross-section σ_{143}, defined by Eq. (3), depends very little on the neutron spectrum because $\sigma_{143}(E)$ follows the $1/v$ law over most of the range of interest. Therefore the lack of knowledge of the temperature of the moderated neutrons is of no importance (contrary to the case of σ_{149}) and the effect of epithermal neutrons is also very small.[7] In other words, the extraction of $\sigma_{143}\tau$ from Oklo data is approximately done by assuming a fixed, fiducial value for σ_{143}, say $\sigma_{143} \simeq 325$ b, so that the use of Eq. (17) for computing a dimensionful σ_{149} is approximately equivalent to computing the dimensionless quantity $325\sigma_{149}/\sigma_{143}$.

The detailed isotopic analysis of Oklo data [16,17] has shown that, generally speaking, the ore composition has changed very little since the end of the nuclear reactions.

---

[6] Eq. (17) is equivalent to equations appearing in Refs. [22,14,15] apart from the facts that Shlyakhter's equations contain misprints (e.g. $(w-1)/\ln w$ instead of its inverse in Eq. (17)). The fractionally small first contribution on the right-hand side of (17) is neglected in the above references.

[7] In the analysis of Oklo data, it has been customary to parametrize the contribution of epithermal neutrons to the spectrum by a parameter called r. This parameter is found to be small, $r \sim 0.15$, and its effect on σ_{143} is only a few percent [30].

Table 1
Effective neutron cross sections of Sm^{149} computed for 15 Oklo samples using published data

Sample	Reference	$\hat{\sigma}_{149}$ (kb)
KN50-3548	[23]	93
SC36-1408/4	[24]	73
SC36-1410/3	[24]	73
SC36-1413/3	[24]	83
SC36-1418	[24]	64
SC39-1383	[28,29]	66
SC39-1385	[28,29]	69
SC39-1387	[28,29]	36
SC39-1389	[28,29]	64
SC39-1390	[28,29]	82
SC39-1391	[28,29]	82
SC39-1393	[28,29]	68
SC35bis-2126	[28,29]	57
SC35bis-2130	[28,29]	81
SC35bis-2134	[28,29]	71
SC52 1472	[31]	72

This is established by studying the correlation between the fluence τ and the uranium isotopic ratio N_5/N_8, and by showing that it can be explained by neutronics considerations (see, e.g. Ref. [29]). However, in some cases there is evidence for a partial reshuffling of chemical elements after the end of the reactions. We have examined these results and selected 16 samples as especially suitable for extracting a reliable value of σ_{149}. These samples are all core samples with high uranium content, large depletions of U^{235}, and high fluences, $\tau \gtrsim 0.7 \times 10^{21}$ n/cm^2. In all cases, the natural element correction in Eq. (17) is small, the observed samarium having been produced almost entirely by fission. The very small content of Sm^{149} (and, when data are available, of other strong absorbers such as Gd^{155} and Gd^{157}) is also a confirmation of the absence of chemical reshuffling after the reaction. The data we took come from Refs. [28,22,23,29,24], and Ref. [31].[8] The result of calculating σ_{149} from these data is exhibited in Table 1.

The large scatter of the values exhibited in Table 1 is compatible with the strong temperature dependence of σ_{149} (see below). The only exception is the 36 kb obtained for the sample SC39-1387. This value is a clear outlier which, most plausibly, has been contaminated in some way. Excluding this result, the other 15 results are all contained in the range

$$57 \text{ kb} \leqslant \hat{\sigma}_{149} \leqslant 93 \text{ kb}. \tag{20}$$

We think that it is conservative to consider the full range (20) as a "2σ" (or 95% C.L.) interval for $\hat{\sigma}_{149}$. [For clarity, we reestablish the caret meaning that we are dealing with an effective cross section, Eq. (3).] Actually, in view of what is known from Oklo, it is very plausible that the range (20) is to be attributed to a mixture of temperature

---

[8] The sample SC521472 taken from this last reference was exposed to a smaller fluence than the others. It was included because this sample has been used to estimate the temperature of the neutrons.

Table 2
Oklo data corresponding to the extreme cross-section results of Eq. (20). Notation as in Eqs. (16), (17)

Sample	N_5/N_8	N_{144}	N_{147}	N_{148}	N_{149}	τ (10^{21} n/cm^2)	$\hat{\sigma}_{149}$ (kb)
SC35bis-2126	0.00568	0.22	53.86	2.39	0.44	0.92	57
KN50-3548	0.00465	0.16	52.63	6.90	0.19	1.25	93

effects and a small amount of post-reaction chemical reshuffling. For our purpose, we will use the full range (20) to define a conservative bound on the variation of α. For completeness, and in view of the special use we make of the interval (20) we give in Table 2 the complete set of data allowing one (using (17)) to compute $\hat{\sigma}_{149}$ for the samples giving the extreme values (20).

Let us note that the values we obtain for $\hat{\sigma}_{149}$ are different from the result claimed by Shlyakhter [14,15], namely $\hat{\sigma}_{149} = (55 \pm 8)$ kb. As he did not mention the data he used, we could not trace the origin of this difference. We note that most of the values in Table 1 are compatible with thermal effects ($\hat{\sigma}_{149}$ increases from ~ 70 kb to ~ 99 kb when the temperature varies from 20°C to ~ 400°C, and then decreases for higher temperatures).[9]

3. Bounding a possible shift of the lowest resonance in the capture cross section of samarium 149

Following Shlyakhter's suggestion [13–15], we shall translate the range of "Oklo" values of $\hat{\sigma}_{149}$, Eq. (20), into a bound on the possible shift, between the time of the Oklo phenomenon and now, of the lowest resonance in the monoenergetic cross section $\sigma_{149}(E)$. The large values of the thermal capture cross sections of Sm^{149}, Gd^{155} and Gd^{157} are due to the existence of resonances in the thermal region. In presence of such a resonance, the monoenergetic capture cross section is well described, in the thermal region, by the Breit–Wigner formula

$$\sigma_{(n,\gamma)}(E) = \pi \frac{\hbar^2}{p^2} g \frac{\Gamma_n(E)\Gamma_\gamma}{(E - E_r)^2 + \frac{1}{4}\Gamma^2} . \qquad (21)$$

Here p is the momentum of the neutron, $E = p^2/2m_n$ its kinetic energy, $g = (2I' + 1)(2s+1)^{-1}(2I+1)^{-1}$ a statistical factor depending upon the spins of the compound nucleus I', of the incident neutron $s = \frac{1}{2}$, and of the target nucleus I, $\Gamma_n(E)$ is a neutron partial width, Γ_γ a radiative partial width, and Γ the total width. The neutron partial width $\Gamma_n(E)$ varies approximately as $E^{1/2}$

$$\Gamma_n(E) = \frac{2\gamma_n^2}{\hbar} p , \qquad (22)$$

---

[9] The fact that we did not find values between 93 and 99 kb is probably explained by some post-reaction remobilization. Anyway, the limit we shall derive on α depends only on the lower bound on $\hat{\sigma}_{149}$.

where γ_n^2 is a "reduced partial width" (see, e.g., Ref. [32]). With sufficient approximation, the total width is given by

$$\Gamma \simeq \Gamma_\gamma + \Gamma_n(E_r). \tag{23}$$

As we shall see explicitly below, the position of the resonance E_r (with respect to the
threshold defined by zero-kinetic-energy incident neutrons) is extremely sensitive to the
value of the fine-structure constant. By contrast, the other quantities entering the Breit–
Wigner formula, γ_n^2, Γ_γ, have only a mild (polynomial) dependence on α. Therefore
the sensitivity to α of the effective cross sections $\hat{\sigma}$, Eq. (3), measured with Oklo data
is totally dominated, for strong absorbers, by the dependence of $\sigma(E)$ upon the position
of the lowest lying resonance E_r. As we said above, we should, more rigorously, work
with dimensionless ratios such as $\hat{\sigma}_{149}/\hat{\sigma}_{143}$. However, the mild absorber Nd^{143} has no
resonances in the thermal region. Therefore the α-sensitivity of the ratio $\hat{\sigma}_{149}/\hat{\sigma}_{143}$ is
completely dominated by the α-sensitivity of $\hat{\sigma}_{149}$ inherited from the dependence on
$E_r^{149}(\alpha)$.

The main problem is to use the range (20) of values of $\hat{\sigma}_{149}$ to put a limit on a
possible shift of the lowest lying resonance in Sm^{149},

$$\Delta \equiv E_r^{149(\mathrm{Oklo})} - E_r^{149(\mathrm{now})}. \tag{24}$$

Previous attempts [13–15,4] at relating Oklo-deduced ranges of values of $\hat{\sigma}_{149}$ to Δ are
unsatisfactory because they did not properly take into account the very large uncertainty
in $\hat{\sigma}_{149}$ due to poor knowledge of the neutron temperature in the Oklo reactors. The
original analysis of Shlyakhter assumed a temperature $T \simeq 20°C$ (which is much too
low), and the analysis of [4] took $T \simeq 1000$ K, i.e. $T \simeq 725°C$ (which is possible,
but on the high side) and assumed that one could work linearly in the fractional shift
Δ/E_r. We think that one should neither fix the neutron temperature T (which could have
varied over a wide range), nor work linearly in Δ/E_r (which could have been larger
than unity).

Several studies, using independent data, have tried to constrain the value of the
temperature in the Oklo reactors [17]. Mineral phase assemblages observed within a
few meters of the Oklo reactor zones 2 and 5 indicate a minimum temperature in
these regions of about 400°C, while relict textures in the reactor zone rock suggest
that temperatures $T \simeq 650–700°C$ may have been reached within the reactors [33]. A
study of fluid inclusions and petrography of the sandstones suggest pressures $p \simeq 800$–
1000 bar [10] and temperatures ranging between 180°C and at least 600°C. On the other
hand, the temperature of the water-moderated neutrons during the fission reactions has
been evaluated by a study of the Lu^{176}/Lu^{175} and Gd^{156}/Gd^{155} isotope ratios in several
samples. The values obtained range between 250±40°C and 450±20°C depending upon
the sample and the isotope ratio considered [31]. It is to be noted that the concentration
of strong absorbers such as Sm^{149} or Gd^{155} (which are burned very efficiently) is

---

[10] It is thought that the Oklo phenomenon took place while the uranium deposits were buried ~ 4 km deep
[34].

determined by the values of the effective cross sections $\hat{\sigma}_{149}$ or $\hat{\sigma}_{155}$ *at the end of the fission phenomenon*. Therefore, we cannot exclude that the temperature to be used in evaluating $\hat{\sigma}_{149}$ or $\hat{\sigma}_{155}$ be on the low side of the allowed range.

Summarizing, we consider that the temperature to be used to determine $\hat{\sigma}_{149}$ or $\hat{\sigma}_{155}$ could be in the full range

$$180°C \leqslant T \lesssim 700°C. \tag{25}$$

Inserting a Maxwell–Boltzmann spectrum [11]

$$\frac{n_E}{n} dE = \frac{2\pi}{(\pi kT)^{3/2}} e^{-E/kT} E^{1/2} dE, \tag{26}$$

in the definition (3), with $\sigma(E)$ of the Breit–Wigner form (21), we find that the dependence of the effective cross section of a strong absorber on resonance shift Δ, Eq. (24), and temperature is given by

$$\hat{\sigma}(\Delta, T) = \frac{2\pi}{(\pi kT)^{3/2}} \sigma_0 (1 + y_0^2) \int_0^\infty \frac{e^{-E/kT} E^{1/2} dE}{1 + y^2(E, \Delta)}. \tag{27}$$

Here σ_0 denotes the "thermal" radiative cross section (as observed now), i.e. the monoenergetic $\sigma_{(n,\gamma)}(E_0)$ evaluated at $E_0 \equiv 0.0253$ eV, y_0 denotes $2(E_0 - E_r^{\text{now}})/\Gamma$, and $y(E, \Delta) \equiv 2(E - E_r^{\text{now}} - \Delta)/\Gamma$. The $E^{1/2}$ in the integrand comes from combining several different factors: a factor $E^{-1/2}$ coming from $\Gamma_n(E)/p^2$ $(1/v$ law), a factor $E^{1/2}$ coming from the factor v in Eq. (3), and the factor $E^{1/2}$ in the Maxwell spectrum (26).

In the case of Sm^{149} the numerical values needed to evaluate (27) are (from Ref. [35]): $E_r^{\text{now}} = 0.0973$ eV, $\sigma_0 = 40.14$ kb, and $\Gamma \simeq 0.061$ eV. [The latter being estimated from $\Gamma \simeq \Gamma_\gamma + \Gamma_n(E_r)$ with $\Gamma_\gamma = 60.5 \times 10^{-3}$ eV, $2g\Gamma_n(E_r) = 0.6 \times 10^{-3}$ eV, with $2g = 9/8$ corresponding to $I' = 4$ and $I = 7/2$.] The dependence of $\hat{\sigma}_{149}$ upon the resonance shift Δ is shown in Fig. 1 for several temperatures spanning the range (25). In this figure we have indicated the conservative range of values of $\hat{\sigma}_{149}$, Eq. (20).

The limits on Δ shown in Fig. 1 are -0.12 eV$< \Delta < 0.08$ eV. The lower limit depends on the minimal allowed temperature. Given the temperature estimates quoted above, we consider $180°C$ as a firm minimal temperature and therefore -0.12 eV as a firm lower bound. As the upper limit 0.08 eV depends on the maximum allowed temperature which is more uncertain, we have also explored temperatures higher than $700°C$. We found that when $\Delta = 0.09$ eV $\hat{\sigma}$ never exceeds 57 kb even if T is allowed to take values much larger than $700°C$. [$\hat{\sigma}(0.09, T)$ reaches a maximum < 57 kb somewhere around $T \sim 1000°C$.] Therefore to be conservative, we take $\Delta < 0.09$ eV as firm upper limit. We conclude that the Oklo samarium data constrain a possible resonance shift to be in the range

$$-0.12 \text{ eV} < \Delta < 0.09 \text{ eV}. \tag{28}$$

---

[11] We do not consider the effect of epithermal neutrons, which introduce only a rather small fractional correction (spectrum index of order $r \sim 0.15$ [16,17]). This correction is negligible compared to the wide range we consider.

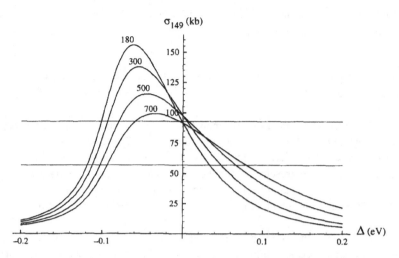

Fig. 1. Variation of the effective neutron capture cross section of Sm^{149}, σ_{149}, as a function of a possible shift $\Delta = E_r^{Oklo} - E_r^{now}$ in the lowest resonance energy, for several values of the neutron temperature T. σ_{149}, Δ and T (labelling the curves) are measured in kbarn, eV and degree Celsius, respectively. The two horizontal lines represent a conservative range of values of σ_{149} compatible with Oklo data.

For comparison, let us mention that Refs. [13–15] estimate a 2σ range $|\Delta| < 0.02$ eV from the samarium data alone, and a 3σ range $|\Delta| < 0.05$ eV from combining samarium and europium data.

We tried to make use of Oklo gadolinium data to restrict further the samarium-derived range (28). A priori, one could think of making use of both Gd^{155} and Gd^{157} which are strong absorbers of neutrons. In fact, Gd^{157} is such a strong absorber that its final concentration $N_{157} \propto y_{157}/\hat{\sigma}_{157}$ (generalizing Eq. (14)) is too small to be measured reliably. The case of Gd^{155} is more favorable, its effective cross section being comparable to that of Sm^{149}. However, its fission yield is much smaller ($y_{155} = 0.032\%$ instead of $y_{149} = 1.08\%$). The absolute concentration of all isotopes of gadolinium is about ten times smaller than that of samarium (see, e.g., Ref. [23]). This implies that gadolinium data are much more prone to various contaminations (natural element contamination due to a post-reaction remobilization, and uncertainties in the isotopic analysis measurements). To make a meaningful analysis of gadolinium data, one should probably restrict oneself to samples that were exposed to rather mild fluences. Such samples are SC361901 and SC521472 which have been studied in detail in [31]. The effective cross sections $\hat{\sigma}(Gd^{155})$ obtained in the latter reference are $\hat{\sigma}_{155} = (42.0 \pm 0.5)$ kb in SC 361901, and $\hat{\sigma}_{155} = (32.5 \pm 0.5)$ kb in SC 521472. These values are compatible with the present values of $\hat{\sigma}_{155}$ if the temperatures in these samples were $T_{361901} \simeq 380°C$ and $T_{521472} \simeq 450°C$. [Actually, these temperatures disagree with the lutetium-derived ones: $T_{361901}^{Lu} \simeq 250°C$ and $T_{521472}^{Lu} \simeq 280°C$. This difference is probably to be explained by a moderate amount of contamination of natural lutetium after the reaction [31].] However, we could not use these data to derive more stringent limits on

α because the Gd^{155} resonance turns out to be less than half as sensitive as samarium to changes in α (see next section).

4. Translating possible resonance shifts into a bound on the variation of the fine-structure constant

Let us finally translate the allowed range (28) into a bound on a possible difference between the value of α during the Oklo phenomenon and its value now. The treatments given in previous analyses are unsatisfactory. The original analysis of Shlyakhter [13–15] rested on a coarse representation of the nucleus as a square potential well, together with dubious assumptions about nuclear compressibility, while the analysis of Ref. [4] used an ill-motivated finite-temperature description of the excited state of the compound nucleus.

The observed neutron-resonance energy $E_r^{\text{now}} = 0.0973$ eV, for the radiative capture of neutrons by Sm_{62}^{149}, corresponds to the existence of a particular excited quantum state of Sm_{62}^{150}. More precisely, if we write the total mass-energy of the relevant excited state of Sm_{62}^{150} as [12] $E_{150}^* = 62m_p + 88m_n + E_1$ (with $E_1 < 0$), and the total mass-energy of the ground state of Sm_{62}^{149} as $E_{149} = 62m_p + 87m_n + E_2$ (with $E_2 < 0$), we have

$$E_r = E_{150}^* - E_{149} - m_n = E_1 - E_2 . \tag{29}$$

Both E_1 and E_2 are eigenvalues of the Hamiltonian

$$H = H_n + H_c , \tag{30}$$

where H_c is the Coulomb energy

$$H_c = e^2 \sum R_{ij}^{-1} , \tag{31}$$

summed over the pairs of protons in the nucleus, and H_n is, to a good accuracy, independent of e^2. We neglect small effects such as the magnetic-moment interactions or the QED corrections to the masses, such as m_n and m_p, entering the nuclear Hamiltonian H_n.

Now let e^2 vary while H_n remains fixed. Then for any eigenstate of H with eigenvalue E,

$$e^2 \, dE/de^2 = \langle H_c \rangle , \tag{32}$$

and therefore

$$e^2 dE_r/de^2 = \langle H_c \rangle_1 - \langle H_c \rangle_2 . \tag{33}$$

The Coulomb energies on the right of (33) are not directly measurable. The quantities that can be directly measured by optical spectroscopy [36] are the mean-square radii

---

[12] We set $c = 1$.

$\langle r^2 \rangle$ of the charge distributions of the protons in the various isotopes of samarium. Let us recall that "isotope shifts" in heavy atoms are related to the effect of the finite extension of the nucleus on electron energies. A first-order perturbation analysis of the latter effect (see, e.g., Ref. [37]) yields $\Delta E = (2\pi/3)\psi_e^2(0)Ze^2\langle r^2 \rangle$ where $\psi_e(r)$ is an (s-state) electron wave function, and where $\langle r^2 \rangle = Z^{-1}\int \rho r^2\, dv$ with ρ denoting the proton charge distribution in the nucleus.

To connect the expectation values in (33) with the mean-square radii, we use the semi-classical approximation

$$\langle H_c \rangle_i = \frac{1}{2}e^2 \int V_i \rho_i\, dv, \quad i = 1, 2, \tag{34}$$

where ρ_i is the density of protons in the nuclear state i, normalized to

$$\int \rho_i\, dv = Z = 62, \tag{35}$$

for samarium, and V_i is the electrostatic potential generated by ρ_i. From (33) and (34),

$$\frac{dE_r}{de^2} = \frac{1}{2}\int (V_1\rho_1 - V_2\rho_2)\, dv$$

$$= -\left(\frac{1}{2}\right)\int \delta V\, \delta\rho\, dv + \int V_1\, \delta\rho\, dv < \int V_1\, \delta\rho\, dv, \tag{36}$$

with

$$\delta\rho = \rho_1 - \rho_2, \qquad \delta V = V_1 - V_2. \tag{37}$$

The term that is dropped in (36) is negative because it is minus an electrostatic self-energy. The integrand on the right side of (36) is a small difference $\delta\rho$ multiplied by the smooth potential V_1. With an error of second order in small quantities, we may approximate V_1 by the classical potential of a uniformly charged sphere with radius R_1,

$$V_1(r) = Z\left[\frac{3R_1^2 - r^2}{2R_1^3}\right]. \tag{38}$$

Then (36) becomes

$$\frac{dE_r}{de^2} < -\left[\frac{Z^2}{2R_1^3}\right]\delta_{12}(r^2), \tag{39}$$

where

$$\delta_{12}(r^2) = \frac{1}{Z}\int r^2\, \delta\rho\, dv \tag{40}$$

is the difference in mean-square charge radius between the states 1 and 2. Let the label 3 denote the ground-state of Sm^{150}. Then

$$\delta_{12}(r^2) = \delta_{13}(r^2) + \delta_{32}(r^2). \tag{41}$$

The difference $\delta_{13}(r^2)$ cannot be calculated because we do not know the shape of the excited state of Sm^{150}. But it seems safe to assume that the proton charge distribution

will not be more tightly concentrated in the excited state than in the ground state. That
is to say,

$$\delta_{13}(r^2) \geqslant 0. \tag{42}$$

An inequality stronger than (42) could be deduced from more dubious assumptions
about nuclear compressibility, but a stronger inequality is not needed. From (39) and
(42) we have

$$\frac{dE_r}{de^2} < - \left[\frac{Z^2}{2R_1^3} \right] \delta_{32}(r^2), \tag{43}$$

and this is sufficient for our purposes. The experimental isotope-shift measurements
reported by [38] give directly $\delta_{34}(r^2) = 0.303 \pm 0.016$ fm^2 and $\delta_{24}(r^2) = 0.092 \pm 0.005$ fm^2, where the label 4 denotes the ground-state of Sm^{148}. Taking the difference
gives

$$\delta_{32}(r^2) = 0.211 \pm 0.017 \text{ fm}^2 \quad (3\sigma \text{ error}). \tag{44}$$

For the radius of the Sm^{150} nucleus to insert in (15), we use Eqs. (50) and (51) on
page 568 of Ref. [36], which give

$$R_1 = 8.11 \text{ fm}. \tag{45}$$

From (43)–(45), we find

$$\alpha \frac{dE_r}{d\alpha} < -(1.09 \pm 0.09) \text{ MeV}. \tag{46}$$

The estimate (46), obtained here directly from measurements of the small charge-
radius difference (44) between Sm^{150} and Sm^{149}, agrees with the result obtained by
differencing the phenomenological Bethe–Weizsäcker formula (droplet model). The latter formula estimates the nuclear Coulomb energy as $\langle H_c \rangle = 0.717\, Z(Z-1)A^{-1/3}$ MeV
[39]. Taking the difference between Sm$_{62}^{150}$ and Sm$_{62}^{149}$ (and arguing as above that excited
states are less charge concentrated) yields the inequality $\alpha\, dE_r/d\alpha < -1.14$ MeV, which
is compatible with the result (46). By contrast, the Bethe–Weizsäcker formula overesti-
mates by about a factor two the α-sensitivity of the resonance energy $E_r^{\text{now}} = 0.0268$ eV,
for the radiative capture of neutrons by Gd$_{64}^{155}$. This follows (using the same method as
above) from the fact that isotope-shift measurements reported in [40] yield

$$\langle r^2 \rangle_{156} - \langle r^2 \rangle_{155} = 0.097 \pm 0.005 \text{ fm}^2, \tag{47}$$

which is less than half the samarium difference (44). As a consequence, the α-sensitivity
parameter $|\alpha\, dE_r/d\alpha|$ of the Gd^{155} resonance is less than half that of the Sm^{149} reso-
nance.

We are now in position to convert the bound (28) obtained above from our analysis
of Oklo data into a bound on the variation of α. To be conservative we use the worst
3σ limit on the α-sensitivity of E_r obtainable from (46), namely

$$\left| \alpha \frac{dE_r}{d\alpha} \right| > (1.09\text{--}0.09) \text{ MeV} = 1.0 \text{ MeV}. \tag{48}$$

Combining (48) with the bound (28) on the shift $\Delta = E_r(\alpha^{\text{Oklo}}) - E_r(\alpha^{\text{now}}) = -|\alpha \, dE_r/d\alpha|(\alpha^{\text{Oklo}} - \alpha^{\text{now}})/\alpha$ yields our final result

$$-0.9 \times 10^{-7} < \frac{\alpha^{\text{Oklo}} - \alpha^{\text{now}}}{\alpha} < 1.2 \times 10^{-7}, \tag{49}$$

which we consider as a 95% C.L. limit.

Though we have been very conservative in our analysis, our result (49) confirms the main claim of Refs. [13–15], namely that Oklo rare-earth data are extremely sensitive probes of a possible variation of α: the surprisingly good $\sim 10^{-7}$ bound comes mainly from the 10^7 amplification factor between the MeV level in $\alpha \, dE_r/d\alpha$ (which is physically clearly understood from the Bethe–Weizsäcker formula) and the 0.1 eV level of the value of the samarium resonance (with respect to the threshold).

Let us note also the consistency of the approximations we made: a change $\delta\alpha/\alpha \sim 10^{-7}$ has a totally negligible effect in all the quantities (such as γ_n or Γ) depending at most polynomially on α, and its effect on the dimensionless ratio $\hat{\sigma}_{149}/\hat{\sigma}_{143}$ is dominated by the Breit–Wigner denominator of $\hat{\sigma}_{149}$, i.e. by the change $\partial E_r^{149}/\partial\alpha \, \delta\alpha$. [The contribution coming from the shift of neutron capture resonances on Nd^{143} is relatively negligible because, from the approximate $Z(Z-1)A^{-4/3}$ dependence expected from the Bethe–Weizsäcker formula, $\delta E_r^{143} \lesssim \delta E_r^{149} \sim 0.1$ eV which is small compared to the near-threshold resonances in Nd^{143}, specifically the one below the threshold at $E_r^{143} \sim -6$ eV.]

In deriving the bound (49), we have implicitly assumed that, during the Oklo phenomenon, α took some fixed value α^{Oklo} (possibly different from α^{now}). The situation would be more complicated if, at the time, $\alpha(t)$ were oscillating on a time scale smaller than the duration of Oklo. As Sm^{149} data depend essentially on the value of $\hat{\sigma}_{149}$ at the end of the fission reaction, we expect that the bound (49) restricts the amplitude of the variation of α in many extended scenarios comprising α-oscillations.

Dividing (49) by the age of the Oklo phenomenon, we can convert it into a bound on the time derivative of α averaged over the time span separating us from the end of the Oklo phenomenon. [In scenarios where α varies on the Hubble time scale, this averaged time derivative is nearly equal to the present time derivative.] This conversion introduces a further uncertainty, because the age of Oklo is not determined with precision. The geochronological studies suggest an age around 1.8×10^9 yr [16], while several studies based on nuclear decay time scales gave $\sim 10\%$ higher values: for instance, 1.98×10^9 yr [24], 1.93×10^9 yr [41], and 2.05×10^9 yr [42]. To remain conservative in our bounds, we shall use the lower, geochronological value. Dividing Eq. (49) by 1.8×10^9 yr, we get the following conservative (95% C.L.) limit on the time derivative of α averaged over the time since the Oklo reactor was running:

$$-6.7 \times 10^{-17} \text{ yr}^{-1} < \frac{\dot{\alpha}}{\alpha} < 5.0 \times 10^{-17} \text{ yr}^{-1}. \tag{50}$$

This is weaker than Shlyakhter's estimates (which ranged between $\pm 5 \times 10^{-18}$ yr^{-1} [14] and $\pm 10^{-17}$ yr^{-1} [13,15]) but rests on a firmer experimental basis. On the other hand, this is between two and three orders of magnitude stronger than the other constraints on the variability of α (see the Introduction). Thanks to recent advances in atomic clock technology, it is conceivable (and desirable) that direct laboratory tests might soon compete with the Oklo bound (50).

We have focused in this paper on the time variation of the fine-structure constant because one can estimate with some confidence the effect of a change of α on resonance energies. It is more difficult to estimate the effect of a change in the Fermi coupling constant G_F, or, better, in the dimensionless quantity $\beta = G_F m_p^2 c/\hbar^3 \simeq 1.03 \times 10^{-5}$. The estimates of Ref. [43] for the (Weinberg–Salam) weak-interaction contribution to nuclear ground state energies yield $E_{weak}^{150} - E_{weak}^{149} \simeq 5.6\,\mathrm{eV}$. If one assumes that this gives an approximate estimate of the difference involving the relevant excited state of Sm^{150}, and that there is no cancellation between the effects of changes in α and β, one finds from Eq. (28) the approximate bound

$$\frac{|\beta^{Oklo} - \beta^{now}|}{\beta} < 0.02, \tag{51}$$

$$\left|\frac{\dot{\beta}}{\beta}\right| < 10^{-11}\ \mathrm{yr}^{-1}. \tag{52}$$

This bound is more stringent than the limit $|\dot{\beta}/\beta| < 10^{-10}\ \mathrm{yr}^{-1}$ obtained from the constancy of the K^{40} decay rate [2], and is comparable to the limit derived from Big Bang nucleosynthesis: $|\beta^{BBN} - \beta^{now}|/\beta < 0.06$ [44].

Deducing from Oklo data a limit on the time variation of the "strength of the nuclear interaction" poses a greater challenge. First, one must notice that, as remarked in Section 2, only dimensionless ratios of nuclear quantities, such as $\sigma_{149}/\sigma_{143}$, can be extracted from Oklo data. Within the QCD framework, one generally expects any such dimensionless ratio to become a (truly constant) pure number in the "chiral" limit of massless quarks.

Time variation of such a dimensionless ratio is then linked (in QCD) with possible changes in the subleading terms proportional to the mass ratios m_q/m_p, where m_q denotes the masses of the light quarks. However, the chiral limit of nuclear binding energies is tricky because of non-analyticity effects in m_q. The present chiral perturbation technology does not allow one to estimate the dependence of nuclear quantities such as E_r^{149} or $\sigma_{149}/\sigma_{143}$ on m_q/m_p. One, however, anticipates that Oklo data might provide a very stringent test (probably at better than the 10^{-7} level) on the time variation of m_q/m_p. To separate unambiguously the effects of variations in α and m_q/m_p, it would be necessary to extract from Oklo data several independent measured quantities (e.g. by analyzing in detail the effects of resonance shifts in Gd^{155} and Gd^{157}).

Acknowledgements

We thank A. Michaudon and B. Pichon for informative communications about nuclear data and M. Ganguli and E. Hansen for helping us to obtain the information relevant to the Oklo phenomenon. The work of T.D. at the Institute for Advanced Study was supported by the Monell Foundation.

References

[1] P.A.M. Dirac, Nature 139 (1937) 323; Proc. Roy. Soc. A 165 (1938) 199.

[2] F.J. Dyson, The fundamental constants and their time variation, in Aspects of Quantum Theory, ed. A. Salam and E.P. Wigner (Cambridge University Press, Cambridge, 1972), pp. 213–236.

[3] F.J. Dyson, Variation of constants, in Current Trends in the Theory of Fields, ed. J.E. Lannutti and P.K. Williams (American Institute of Physics, New York, 1978) pp. 163–167.

[4] P. Sisterna and H. Vucetich, Phys. Rev. D 41 (1990) 1034.

[5] D.A. Varshalovich and A.Y. Potekhin, Space Science Reviews 74 (1995) 259.

[6] M.B. Green, J.H. Schwarz and E. Witten, Superstring theory (Cambridge University Press, Cambridge, 1987).

[7] T. Damour and A.M. Polyakov, Nucl. Phys. B 423 (1994) 532; Gen. Rel. Grav. 26 (1994) 1171.

[8] M.P. Savedoff, Nature 178 (1956) 688.

[9] J.N. Bahcall and M. Schmidt, Phys. Rev. Lett. 19 (1967) 1294.

[10] D.A. Varshalovich, V.E. Panchuk and A.V. Ivanchick, Astronomy Letters 22 (1996) 6.

[11] L.L. Cowie and A. Songaila, Astrophys. J. 453 (1995) 596.

[12] J.D. Prestage, R.L. Tjoelker and L. Maleki, Phys. Rev. Lett. 74 (1995) 3511.

[13] A.I. Shlyakhter, Nature 264 (1976) 340.

[14] A.I. Shlyakhter, Direct test for the constancy of fundamental nuclear constants using the Oklo natural reactor, Preprint n°260, Leningrad Nuclear Physics Institute, Leningrad, September 1976.

[15] A.I. Shlyakhter, Direct test of the time-independence of fundamental nuclear constants using the Oklo natural reactor, ATOMKI Report A/1 (1983).

[16] Le Phénomène d'Oklo (The Oklo Phenomenon) Proceedings of a symposium on the Oklo phenomenon, Libreville, Gabon, June 1975 (International Atomic Energy Agency, Vienna, 1975).

[17] Les Réacteurs de Fission Naturels (Natural Fission Reactors), Proceedings of a meeting on natural fission reactors, Paris, France, December 1977 (International Atomic Energy Agency, Vienna, 1978).

[18] M. Maurette, Ann. Rev. Nuc. Sci 26 (1976) 319.

[19] Y.V. Petrov, Sov. Phys. Usp. 20 (1977) 937.

[20] J.C. Ruffenach et al., in Ref. [16], p. 371.

[21] H. Bassière et al., in Ref. [16], p. 385.

[22] M. Neuilly and R. Naudet, in Ref. [16], p. 541

[23] J.C. Ruffenach et al., Earth and Planetary Science Letters 30 (1976) 94.

[24] J.C. Ruffenach, in Ref. [17], p. 441.

[25] J. Cesario, D. Poupard and R. Naudet, in Ref. [17], p. 473.

[26] F.W. Walker, J.R. Parrington and F. Feiner, Nuclides and Isotopes, Fourteenth Edition (General Electric Company, San Jose, 1989).

[27] B.F. Rider, Compilation of Fission Product Yields, NEDO-12154-3(C) (Vallecitos Nuclear Center, Pleasanton, California, 1981).

[28] J.F. Dozol and M. Neuilly, in Ref. [16], p. 357.

[29] M. Neuilly, J.F. Dozol and R. Naudet, in Ref. [17], p. 433.

[30] M. Lucas et al., in Ref. [17], p. 407.

[31] P. Holliger, C. Devillers and G. Retali, in Ref. [17], p. 553.

[32] A.M. Weinberg and E.P. Wigner, The Physical Theory of Neutron Chain Reactors (The University of Chicago Press, Chicago, 1958).

[33] R.J. Vidale, in Ref. [17], p. 235.

[34] B. Poty, in Ref. [17], p. 636.

[35] S.F. Mughabghab, Neutron Cross Sections, Vol. 1, Part B (Academic Press, New York, 1984).

[36] E.W. Otten, Nuclear Radii and Moments of Unstable Nuclei, in Treatise on Heavy-Ion Science, Vol. 8, ed. D. Allan Bromley, (Plenum, New York, 1989), pp. 517–638.

[37] L.D. Landau and E.M. Lifshitz, Quantum Mechanics, Non-relativistic Theory (Pergamon Press, Oxford, 1965).

[38] H. Brand, B. Seibert and A. Steudel, Z. Phys. A 296 (1980) 281.

[39] P. Möller and J.R. Nix, Atomic Data and Nuclear Data Tables 39 (1988) 213.

[40] S.K. Borisov et al., Soviet Physics JETP, 66 (1987) 882.

[41] C. Devillers and J. Menes, in Ref. [17], p. 495.

[42] A.J. Gancarz, in Ref. [17], p. 513.

[43] M.P. Haugan and C.M. Will, Phys. Rev. Lett. 37 (1976) 1.

[44] R.A. Malaney and G. Mathews, Phys. Rep. 229 (1993) 147;
     H. Reeves, Rev. Mod. Phys. 66 (1994) 193.

THE SIXTH FERMAT NUMBER AND PALINDROMIC CONTINUED FRACTIONS*

An elementary argument is presented, verifying the known factorization of the sixth Fermat number $2^{64} + 1$. The verification is made more elegant by using an old theorem of Serret concerning palindromic continued fractions.

It is easy to explain why the fifth Fermat number $2^{32} + 1$ is divisible by 641 [Hardy and Wright, 1938]. The prime 641 has the two representations

$$641 = 5.2^7 + 1 = 5^4 + 2^4. \tag{1}$$

The congruence

$$2^{32} = 2^4 . 2^{28} \equiv -5^4 . 2^{28} = -(5.2^7)^4 \equiv -1 \quad (\text{mod } 641) \tag{2}$$

follows immediately from (1).

The purpose of this note is to explain in a similarly elementary way why the sixth Fermat number $2^{64} + 1$ is divisible by 274177. The divisor has the representation

$$q = 274177 = 1 + 2^8 f, \quad f = (2^6 - 1)(2^4 + 1), \tag{3}$$

and it is easily verified that

$$2^{24} - 1 = fg, \quad g = (2^6 + 1)(2^8 - 2^4 + 1). \tag{4}$$

We look for a factorization of the form

$$2^{64} + 1 = (x^2 + y^2)(z^2 + w^2), \quad 2^{32} - i = (x + iy)(z - iw), \tag{5}$$

so that we require integers x, y, z, w satisfying

$$xz + yw = 2^{32}, \quad xw - yz = 1, \quad x^2 + y^2 = q. \tag{6}$$

---

*Accepted for publication in *L'Enseignement Mathématique*, July 2000.

When $z = gx, w = gy$, the right side of (5), $(x + iy)g(x - iy)$, becomes gq; and $gq = 2^{32} + a$, very close to 2^{32}, with the difference $a = g - 2^8 = 15409$ by (3) and (4). This implies that the ratios (z/x) and (w/y) must both be very close to g. We therefore try to set

$$z = gx - s, \quad w = gy - t, \tag{7}$$

with integers s, t expected to be small. Then (5) will hold, provided that

$$x^2 + y^2 = q, \quad xs + yt = a, \quad ys - xt = 1. \tag{8}$$

From (8) it follows that

$$a^2 + 1 = qu, \quad u = s^2 + t^2. \tag{9}$$

Since the integers q and a are known and comparatively small, it is not difficult to find an integer solution of (8) and (9) by trial and error, namely $x = 516, y = 89, s = 29, t = 5, u = 866$. The solution of (8) and (9), with z and w given by (7), provides a solution of (5). This completes the verification of the factorization of the sixth Fermat number. The factorization (4) of $2^{24} - 1$ leads directly to the factorization (5) of $2^{64} + 1$.

Instead of using trial and error to solve (8) and (9), we can obtain a solution more elegantly by using an old theorem of Serret [Serret, 1848; Perron, 1954] on continued fractions. In the statement of the theorem, q and a are any pair of coprime integers with $0 < a < q$. The ratio q/a may be expressed as a continued fraction with partial quotients $(a_j, j = 1, \ldots, n)$ in precisely two ways, one with n even and one with n odd. In one way, the last two quotients are $[j, 1]$, where j may be any integer. In the other way these two quotients are replaced by the single quotient $[j + 1]$, all other quotients remaining the same. We say that the continued fraction q/a with n quotients is a palindrome if

$$a_j = a_{n+1-j}, \quad j = 1, \ldots, n. \tag{10}$$

SERRET'S Theorem. Suppose that the continued fraction q/a with n partial quotients is given. It will be a palindrome if and only if an integer u exists with

$$qu = a^2 + (-1)^n. \tag{11}$$

Proof of Theorem. Let A be any finite ordered set of positive integers, and let $S(A)$ be the numerator of the continued fraction whose partial quotients are the members of A. Euler discovered the explicit formula for $S(A)$, [Euler, 1764; Perron, 1954],

$$S(A) = \sum_B P(B), \tag{12}$$

where the sum extends over certain ordered subsets B of A, and $P(B)$ is the product of the members of B. The subsets included in the sum (12) are A itself, and those B for which the complement $A - B$ is a union of non-overlapping pairs, each pair consisting of two consecutive members of A. The empty set E will appear in the sum (12) if and only if the number of members of A is even. Thus, for example, $S(E) = 1, S(a) = a, S(a,b) = 1 + ab, S(a,b,c) = a + c + abc, S(a,b,c,d) = 1 + ab + ad + cd + abcd$. For a lucid and accessible explanation of Euler's formula, see [Davenport, 1982, chapter 4, pp. 82–84].

Let now A be the set $(a_j, \quad j = 1, \ldots, n)$ of partial quotients of one of the two continued fractions q/a. We write

$$N(k, m) = S(a_k, \ldots, a_m), \tag{13}$$

so that

$$q = N(1, n), \quad a = N(2, n). \tag{14}$$

Euler's formula (12) implies the symmetry relation

$$S(a_1, \ldots, a_n) = S(a_n, \ldots, a_1), \tag{15}$$

and the recurrence relations

$$N(1, n) = N(1, k)N(k + 1, n) + N(1, k - 1)N(k + 2, n), \tag{16}$$

$$N(1, n)N(2, n - 1) - N(1, n - 1)N(2, n) = (-1)^n. \tag{17}$$

Suppose that (11) holds. Define

$$b = N(1, n - 1), \quad v = N(2, n - 1). \tag{18}$$

The symmetry (15) implies

$$q/b = S(a_n, \ldots, a_1)/S(a_{n-1}, \ldots, a_1), \tag{19}$$

so that one of the two continued fractions for q/b has the same partial quotients as the continued fraction for q/a, in reversed order. Then (11), (17) and (18) imply

$$qv = ab + (-1)^n, \quad q(u - v) = a(a - b). \tag{20}$$

Now $q > a > 0$ by hypothesis, and $q > b > 0$ by (19), so that $q > |a - b|$. If $a \neq b$, the fraction $(u - v)/(a - b)$ is equal to a/q, which is impossible because a and q are coprime and $|a - b| < q$. Therefore $a = b, u = v$, and the continued fractions for q/a and q/b are identical, so that (10) holds and the continued fraction is a palindrome. Conversely, if (10) holds, (13) and (18) imply $a = b$, and then (17) implies (11) with $u = v$ defined by (18). *End of Proof.*

Applying the theorem to q and a given by (3) and (9), we deduce that the continued fraction q/a with n even must be a palindrome. The partial quotients are easily calculated, with the result

$$q = S(17, 1, 3, 1, 5, 5, 1, 3, 1, 17), \quad a = S(1, 3, 1, 5, 5, 1, 3, 1, 17). \quad (21)$$

We now apply the recurrence relation (16) to (21) with $k = n/2 = 5$. As a result of the symmetry (15) and the palindrome property of q, (16) gives a solution of (8) with

$$x = S(17, 1, 3, 1, 5) = 516, \quad y = S(17, 1, 3, 1) = 89,$$
$$s = S(1, 3, 1, 5) = 29, \quad t = S(1, 3, 1) = 5. \quad (22)$$

In this way Serret's theorem leads to a solution of (8), following a deductive route rather than trial and error.

The same method leads to the factorization of the fifth Fermat number, starting from

$$q = 1 + 2^7 f, \quad 2(2^8 - 1) = fg, \quad f = 2^2 + 1, \quad g = 2(2^2 - 1)(2^4 + 1). \quad (23)$$

The palindromic continued fraction q/a has numerator $q = S(4, 6, 6, 4)$, and the factorization of $2^{32} + 1$ results with $x = S(4, 6) = 25$, $y = S(4) = 4$, $s = S(6) = 6$, $t = S(E) = 1$, $x^2 + y^2 = 641$. But in this case the argument using (1) and (2) gives the same result more quickly. It would be more interesting if we could understand in a similar way the factorizations of the seventh and higher Fermat numbers. For a summary of the known factorizations, see [Brent, 1999]. The seventh Fermat number has two prime factors with 17 and 22 decimal digits.

According to (21), the partial quotients of q/a in the case of the sixth Fermat number are all odd, while in the case of the fifth Fermat number the partial quotients are all even. To decide whether this unexpected behavior of the partial quotients is a numerical accident or a general rule, we need to find more examples.

The factorization of the sixth Fermat number was originally published in [Landry, 1880], without any explanation of how it was found. For a conjectured reconstruction of Landry's method see [Williams, 1993]. Landry worked for several months to find the factorization. The argument presented in this note only verifies the factorization after the factors are known.

Acknowledgment

The author is grateful to a referee for informing him of the existence of Serret's theorem and providing the references to Perron, Serret and Davenport.

References

R. P. Brent [1999] "Factorization of the tenth Fermat number", *Math. Comp.* **68**, 429–451.

H. Davenport [1982] *The Higher Arithmetic*, 5th edition (Cambridge University Press, Cambridge).

L. Euler [1764] "Specimen algorithmi singularis, Commentatio 281", *Novi Commentarii Academiae scientiarum Imperialis Petropolitanae* **9**, (1762–1763), 53–69; Reprinted in L. Euler, Opera omnia, series I, Vol. 15, ed. G. Faber [Leipzig and Berlin, B. G. Teubner, 1927], 31–49. In the Opera Omnia edition, our equations (12) and (15) appear on page 36, (17) on page 41, (16) on page 47.

G. H. Hardy and E. M. Wright [1938] *An Introduction to the Theory of Numbers* (Oxford University Press, Oxford). For the fifth and sixth Fermat numbers, see pp. 14, 15.

F. Landry [1880] "Sur la décomposition du nombre $2^{64} + 1$", *C. R. Acad. Sci. Paris* **91**, 138.

O. Perron [1954] *Die Lehre von den Kettenbrüchen, Volume 1*, third edition [Stuttgart, B. G. Teubner], pp. 6–7, 28–29.

J. A. Serret [1848] "Sur un théorème relatif aux nombres entiers", *J. de Mathématiques Pures et Appliquées* **13**, 12–14.

H. C. Williams [1993] "How was F_6 factored"? *Math. Comp.* **61**, 463–474.

CHAPTER 5.4

THOUGHT-EXPERIMENTS — IN HONOR OF JOHN WHEELER*

1. Beyond the Black Hole

In 1979, we held a symposium at the Institute for Advanced Study to celebrate the hundredth birthday of Albert Einstein. Unfortunately Einstein could not be there, but John Wheeler made up for Einstein's absence. Wheeler gave a marvelous talk with the title "Beyond the Black Hole", sketching with poetic prose and Wheelerian pictures his grand design for the future of science. Wheeler's philosophy of science is much more truly relativistic than Einstein's. Wheeler would make all physical law dependent on the participation of observers. He has us creating physical laws by our existence. This is a radical departure from the objective reality in which Einstein believed so firmly. Einstein thought of nature and nature's laws as transcendent, standing altogether above and beyond us, infinitely higher than human machinery and human will.

One of the questions that has always puzzled me is this. Why was Einstein so little interested in black holes? To physicists of my age and younger, black holes are the most exciting consequence of general relativity. With this judgment the man-in-the-street and the television commentators and journalists agree. How could Einstein have been so indifferent to the promise of his brightest brainchild? I suspect that the reason may have been that Einstein had some inkling of the road along which John Wheeler was traveling, a road profoundly alien to Einstein's philosophical preconceptions. Black holes make the laws of nature contingent on the mechanical accident of stellar collapse. John Wheeler embraces black holes because they show most sharply the contingent and provisory character of physical law. Perhaps Einstein rejected them for the same reason.

---

*Contribution to Wheeler Symposium held at Princeton, March 15–18, 2002.

Let me quote a few sentences from Wheeler's Varenna lectures, published in 1979 with the title "Frontiers of Time" [Wheeler, 1979]. His talk at the Einstein symposium was a condensed version of these lectures. Here is Wheeler: "Law without law. It is difficult to see what else than that can be the plan of physics. It is preposterous to think of the laws of physics as installed by a Swiss watchmaker to endure from everlasting to everlasting when we know that the universe began with a big bang. The laws must have come into being. Therefore they could not have been always a hundred per cent accurate. That means that they are derivative, not primary . . . Events beyond law. Events so numerous and so uncoordinated that, flaunting their freedom from formula, they yet fabricate firm form . . . The universe is a self-excited circuit. As it expands, cools and develops, it gives rise to observer-participancy. Observer-participancy in turn gives what we call tangible reality to the universe . . . Of all strange features of the universe, none are stranger than these: time is transcended, laws are mutable, and observer-participancy matters".

Wheeler unified two streams of thought which had before been separate. On the one hand, in the domain of cosmology, the anthropic principle of Bob Dicke and Brandon Carter constrains the structure of the universe. On the other hand, in the domain of quantum physics, atomic systems cannot be described independently of the experimental apparatus by which they are observed. The Einstein–Rosen–Podolsky paradox showed once and for all that it is not possible in quantum mechanics to give an objective meaning to the state of a particle, independent of the state of other particles with which it may be entangled. Wheeler has made an interpolation over the enormous gap between the domains of cosmology and atomic physics. He conjectures that the role of the observer is crucial to the laws of physics, not only at the two extremes where it has hitherto been noticeable, but over the whole range. He conjectures that the requirement of observability will ultimately be sufficient to determine the laws completely. He may be right, but it will take a little while for particle physicists, astronomers and string-theorists to fill in the details in his grand picture of the cosmos.

There are two kinds of science, known to historians as Baconian and Cartesian. Baconian science is interested in details, Cartesian science is interested in ideas. Bacon said, "All depends on keeping the eye steadily fixed on the facts of nature, and so receiving their images as they are. For God forbid that we should give out a dream of our own imagination for a pattern of the world". Descartes said, "I showed what the laws of nature were, and without basing my arguments on any principle other than the infinite perfections of God, I tried to demonstrate all those laws about which we could have any doubt, and to show that they are such that, even if God created many worlds, there could not be any in which they failed to be observed". Modern science leapt ahead in the seventeenth century as a result of fruitful competition between Baconian and Cartesian viewpoints. The relation between Baconian science and Cartesian science is complementary, where I use the word complementary as Niels Bohr used it. Both viewpoints are true, and both are necessary, but they cannot be seen simultaneously. We need Baconian scientists to explore the universe and find out what is there to be explained. We need Cartesian scientists to explain and unify what we have found. Wheeler, as you can tell from the passage that I quoted, is a Cartesian. I am a Baconian. I admire the majestic Cartesian style of his thinking, but I cannot share it. I cannot think the way he thinks. I cannot debate with Wheeler the big questions that he is raising, whether science is based on logic or on circumstances, whether the laws of nature are necessary or contingent. In this chapter I do not try to answer the big questions. I write about details, about particles traveling through detectors, about clocks in boxes, about black holes evaporating, about the concrete objects that are the subject-matter of Baconian science. Only intermittently, in honor of John Wheeler, I interrupt the discussion of details with a few Cartesian remarks about ideas.

The subject of this chapter is a set of four thought-experiments that are intended to set limits to the scope of quantum mechanics. Each of the experiments explores a situation where the hypothesis, that quantum mechanics can describe everything that happens, leads to an absurdity. The conclusion that I draw from these examples is that quantum mechanics cannot be a complete description of nature. This conclusion is, of course, controversial. I do not expect everyone, or even a majority, to agree with me. The purpose of writing about a controversial subject is not to compel agreement but to provoke discussion. Being myself a Baconian, I am more interested in the details of the thought-experiments than in the philosophical inferences that may be drawn from them. The details are as solid as the classical apparatus with which the experiments are done. The philosophy, like quantum mechanics, is always a little fuzzy.

I have observed in teaching quantum mechanics, and also in learning it, that students go through an experience that divides itself into three distinct stages. The students begin by learning the tricks of the trade. They learn how to make calculations in quantum mechanics and get the right answers, how to calculate the scattering of neutrons by protons or the spectrum of a rigidly rotating molecule. To learn the mathematics of the subject and to learn how to use it takes about six months. This is the first stage in learning quantum mechanics, and it is comparatively painless. The second stage comes when the students begin to worry because they do not understand what they have been doing. They worry because they have no clear physical picture in their heads. They get confused in trying to arrive at a physical explanation for each of the mathematical tricks they have been taught. They work very hard and get discouraged because they do not seem to be able to think clearly. This second stage often lasts six months or longer. It is strenuous and unpleasant. Then, unexpectedly, the third stage begins. The students suddenly say to themselves,

"I understand quantum mechanics", or rather they say, "I understand now that there isn't anything there to be understood". The difficulties which seemed so formidable have vanished. What has happened is that they have learned to think directly and unconsciously in quantum-mechanical language.

The duration and severity of the second stage are decreasing as the years go by. Each new generation of students learns quantum mechanics more easily than their teachers learned it. There is less resistance to be broken down before the students feel at home with quantum ideas. Ultimately the second stage will disappear entirely. Quantum mechanics will be accepted by students from the beginning as a simple and natural way of thinking, because we shall all have grown used to it. I believe the process of getting used to quantum mechanics will become quicker and easier, if the students are aware that quantum mechanics has limited scope. Much of the difficulty of the second stage resulted from misguided attempts to find quantum-mechanical descriptions of situations to which quantum mechanics does not apply.

Unfortunately, while the students have been growing wiser, some of the older physicists of my generation have been growing more foolish. Some of us have regressed mentally to the second stage, the stage which should only be a disease of adolescence. We tend then to get stuck in the second stage. If you are a real adolescent, you may spend six months floundering in the second stage, but then you grow up fast and move on the third stage. If you are an old-timer returning to your adolescence, you do not grow up any more. Meanwhile, we may hope that the students of today are moving ahead to the fourth stage, which is the new world of ideas explored by John Wheeler. In the fourth stage, you are at home in the quantum world, and you are also at home in the brave new world of black holes and mutable laws that Wheeler has imagined.

2. Complementarity and Decoherence

Roughly speaking, there are two schools of thought about the meaning of quantum mechanics. I call them broad and strict. I use the words in the same way they are used in American constitutional law, where the broad interpretation says the constitution means whatever you want it to mean, while the strict interpretation says the constitution means exactly what it says and no more. The broad school says that quantum mechanics applies to all physical processes equally, while the strict school says that quantum mechanics covers only a small part of physics, namely the part dealing with events on a local or microscopic scale. Speaking again roughly, one may say that the extreme exponent of the broad view is Stephen Hawking, who is trying to create a theory of quantum cosmology with a single wave-function for the whole universe. If a wave-function for the whole universe makes sense, then any restriction on the scope of quantum mechanics must be nonsense. The historic exponent of the strict view of quantum mechanics was Niels Bohr, who maintained that quantum mechanics can only describe processes occurring within a larger framework that must be defined classically. According to Bohr, a wave-function can only exist for a piece of the world that is isolated in space and time from the rest of the world. In Bohr's view, the notion of a wave-function for the whole universe is an extreme form of nonsense.

As often happens in the history of religions or philosophies, the disciples of the founder established a code of orthodox doctrine that is more dogmatic and elaborate than the founder intended. Bohr's pragmatic view of quantum mechanics was elaborated by his disciples into a rigid doctrine, the so-called "Copenhagen Interpretation". When I use the word strict to describe Bohr's view, I have in mind the orthodox Copenhagen dogma rather than Bohr himself. If you read what Bohr himself wrote, you find that he is much more tentative and broad-minded than his disciples. Bohr's approach to science is based on the principle of complementarity, which says that nature is too subtle to be described adequately by any single viewpoint. To obtain a true description of nature you have to use several alternative viewpoints that cannot be seen simultaneously. The different viewpoints are complementary in the sense that they are all needed to tell a complete story, but they are mutually exclusive in the sense that you can only see them one at a time. In Bohr's view quantum mechanics and classical physics are complementary aspects of nature. You cannot describe what is going on in the world without using both. Quantum language deals with probabilities while classical language deals with facts. Our knowledge of the world consists of an inseparable mixture of probabilities and facts. So our description of the world must be an inseparable mixture of quantum and classical pictures.

Against this dualistic philosophy of Bohr, putting strict limits to the scope of quantum descriptions, the quantum cosmologists take a hard line. They say the quantum picture must include everything and explain everything. In particular, the classical picture must be built out of the quantum picture by a process which they call decoherence. Decoherence is the large-scale elimination of the wave-interference effects that are seen in quantum systems but not in classical systems. For the benefit of any literary scholars who may be among my readers, decoherence is to science as deconstruction is to literature, a fashionable buzz-word that is used by different people to mean different things. I quote a few sentences from a classic article by Bryce DeWitt [DeWitt 1992], explaining decoherence, from the point of view of the quantum cosmologists, with unusual clarity.

"In the old Copenhagen days one seldom worried about decoherence. The classical realm existed *a priori* and was needed as a basis for making sense of quantum mechanics. With the emergence of quantum cosmology, it became important to understand how the classical realm emerges from quantum mechanics itself...The formalism is able to generate its own interpretation". After some simple mathematics describing a particular quantum system that first decoheres and thereafter exhibits classical behavior, DeWitt goes on, "The above results have the following implications for decoherence in quantum cosmology: (1) Although complexity (metastability, chaos, thermal baths, wave packets) can only help in driving massive bodies to localized states, it is massiveness, not complexity, that is the key to decoherence. (2) Given the fact that the elementary particles of matter tend, upon cooling, to form stable bound states consisting of massive agglomerations, decoherence at the classical level is a natural phenomenon of the quantum cosmos. (3) Given the fact that the interaction described here is a simple scattering interaction and not at all specially designed like a... measurement interaction, the universe is likely to display decoherence in almost all states that it may find itself in. (4) An arrow of time has no basic role to play in decoherence".

I have tried to give you a fair and balanced statement of the two points of view, Bohr on one side and DeWitt on the other. Personally, I find both of them entirely reasonable. As usual, when people are engaged in philosophical argument, what they do is more reasonable than what they say. DeWitt rejects with scorn what he calls "the old Copenhagen days", but his mathematical analysis of decoherence does not contradict the analysis of quantum processes made sixty years earlier by Bohr. From Bohr's point of view, decoherence is just another example of complementarity, showing in detail how quantum and classical descriptions give complementary pictures of events in the early universe. From DeWitt's point of view, complementarity is just a complicated way of talking about decoherence, giving a spurious importance to the classical description which is only an approximation to the true quantum universe. My Princeton colleague Stephen Adler [Adler, 2001] has written a paper with the title, "Why decoherence has not solved the measurement problem", which I recommend as a clear statement of what decoherence can and cannot do.

The first of my four thought-experiments is an old one, invented long ago by Schrödinger and not by me. To sharpen the issues between Bohr and DeWitt, I look again at the experiment known to experts as Schrödinger's Cat. Schrödinger's Cat is imprisoned in a cage with a bottle of hydrogen cyanide, arranged with a quantum-mechanical device connected to a hammer so that a single atom decides whether the hammer falls and breaks the bottle. The atom is prepared in a coherent state with equal probabilities for its spin to be up or down. If the spin is up, the hammer falls, and if the spin is down, the hammer stays still. The cat is then in a coherent state, with equal probabilities of being dead or alive. Two questions then arise. What does it mean to be in a coherent state of life and death? What does the cat think about the experiment?

From the point of view of Bohr the answers are simple. When you open the cage and examine the cat, or when the cat inside the cage examines itself to see whether it is alive, the experiment is over and the result can only be stated in classical terms. The coherent state lasts only as long as the examination of the cat is in the future. The cat cannot be aware of the coherent state, because as soon as the cat is aware the state is a matter of fact and not a matter of probability. From the point of view of DeWitt, the cat itself, just because it is a massive object with complicated interactions, achieves its own decoherence. It destroys the paradoxical coherence automatically, as a consequence of Schrödinger's equation.

I like to remain neutral in this philosophical debate. If I were forced to make a choice, I would choose to follow Bohr rather than DeWitt, because I find the idea of complementarity more illuminating than the idea of decoherence. Complementarity is a principle that has wide applications extending beyond physics into biology, anthropology and ethics, wherever problems exist that can be understood in depth only by going outside the limits of a single viewpoint or a single culture. Decoherence, so far as I know, has not yet been adopted by anthropologists as a slogan, although one might consider the loss of traditional family and tribal loyalties, when people migrate from farms and villages into city slums, to be the cultural equivalent of decoherence.

3. Two More Thought-Experiments

That is enough about philosophy. I now move on to technical issues which are to me more interesting than philosophy. The next two thought-experiments could be carried out with real apparatus if anybody found them worth the money and time that they would require. The results of the experiments are clear and simple. They seem to show that in some sense Bohr is right, that limits exist to the scope of quantum descriptions of events. This does not mean that DeWitt is necessarily wrong. According to Bohr, there are two kinds of truth, ordinary truth and deep truth. You can tell the difference between the two kinds of truth by looking at their opposites. The opposite of an ordinary truth is a falsehood, but the opposite of a deep truth is another deep truth. The essence of Bohr's idea of complementarity is that you need deep truths to describe nature correctly. So my thought-experiments do not prove that DeWitt is wrong, only that he is not telling us an ordinary truth. His picture of decoherence may be correct as far as it goes, but it can only be a deep truth, giving a partial view of the way nature works.

The second thought-experiment consists of two small Čerenkov counters separated by a distance D with empty space in between. An electron is fired through the first counter at time T_1 and hits the second counter at time T_2. The positions of the counters and the times of arrival of the electron are measured.

First I give you a simple qualitative argument and then a more careful quantitative argument. The qualitative argument goes like this. Suppose the positions and times are known precisely. Then the velocity of the electron between the counters is also known precisely. If we assume that the mass of the electron is known precisely, the momentum is also known precisely. This contradicts the uncertainty principle, which says that the position and momentum of an electron cannot both be known precisely. The contradiction means that it is not legitimate to use a quantum description of the motion of the electron between the two counters.

To make the conclusion firmer, I now give you a quantitative argument, which takes account of the inevitable inaccuracy of the measurements. This argument is an exercise in elementary quantum mechanics, and the experiment is just an old-fashioned two-slit experiment with the slits arranged in series instead of in parallel. The time-interval between the two measurements does not need to be measured accurately. We assume only that the time-interval is known to be greater than T. Suppose that there are two parallel slits of width L, one placed at the exit from the first counter and the other at the entrance to the second counter. When an electron is counted in both counters, we know that it has passed through both slits. Let x be the coordinate of the electron perpendicular to the plane containing the slits. The positions of the slits are measured with sufficient accuracy, so that the uncertainty of x as the electron passes through either slit is less than L. Let p be the momentum of the electron conjugate to x. We consider the mean-square dispersions

$$D(t) = \langle |\Delta x|^2 \rangle, \quad K = \langle |\Delta p|^2 \rangle, \tag{1}$$

as a function of time t as the electron travels between the slits. Since the electron is traveling freely, K is independent of time. According to the virial theorem, which holds for a free electron in non-relativistic quantum mechanics,

$$(d^2 D/dt^2) = (2/m^2)K, \tag{2}$$

where m is the electron mass. The right side of (2) is independent of time, so that

$$D(t) = D(t_0) + K(t - t_0)^2/m^2, \tag{3}$$

where t_0 is the time when D is smallest. But the Heisenberg uncertainty principle says

$$D(t_0)K \geq (1/4)\hbar^2, \tag{4}$$

which together with (3) gives

$$D(t) \geq (\hbar/m)|t - t_0|. \tag{5}$$

The value of D at either counter is less than L^2, and the values of $|t - t_0|$ at the two counters add to at least T. Therefore (5) implies

$$(2L^2/T) \geq (\hbar/m), \tag{6}$$

which must be valid if the electron is described by a wave-function satisfying the Schrödinger equation. But it is easy to choose L and T so that (6) is violated. Then the uncertainty principle (4) will also be violated, and a wave-function describing the passage of the electron through the two slits cannot exist.

Let us put in some numbers to show that the violation of the uncertainty principle could be achieved with apparatus small enough to sit on a tabletop. The numbers are not absurd. The right side of (6) is about 1 square centimeter per second. Without stretching the state of the art, we may take the width L of the slits to be 1 micron or 10^{-4} centimeters. Then (6) will be violated for any time-interval T longer than 20 nanoseconds. For an electron with energy one kilovolt, a travel time of 20 nanoseconds corresponds to a travel distance of 20 centimeters between the two counters. We can easily imagine doing the experiment on a tabletop with a travel distance longer than this. To make the purpose of the thought-experiment clear, I hasten to add that it does not prove quantum-mechanics wrong. It only proves that quantum-mechanics is wrongly applied to this particular situation.

Before discussing the meaning of this second thought-experiment, I go on to the third experiment. The third experiment uses the Einstein Box (see Fig. 1), a device invented by Einstein for the purpose of violating the uncertainty principle. Einstein wanted to use his box to prove that quantum mechanics was inconsistent, since he did not believe that quantum mechanics was true. Einstein confronted Bohr with this box at a public meeting in 1930 [Bohr 1949]. Bohr won the argument with a dramatic counter-attack, pointing out that Einstein had forgotten to take into account his own theory of general relativity when he discussed the behavior of the box. When Bohr included the gravitational effects that follow from general relativity, it turned out that the uncertainty principle was not violated after all. So Einstein was defeated and the box became a victory trophy for Bohr.

The idea of the Einstein Box was that you hang it from a spring-balance and measure its mass by measuring its weight in a known gravitational field. You measure its weight by measuring the momentum p transferred to the balance by the spring in a given time T. The uncertainty in the mass is then

$$\Delta m = \Delta p / gT, \tag{7}$$

where g is the gravitational field. The box has a window with a shutter that can be opened and closed from the inside, and a clock that measures the times when the shutter is opened and closed. It is important that the clock sits inside the box, so that

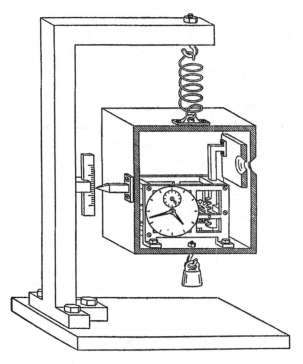

Fig. 1.   The Einstein Box. Taken from [Bohr, 1949].

the weighing is not disturbed by time-signals coming into the box from the outside. At the time when the shutter is open, a photon leaves the box and carries away with it a mass proportional to its energy E. The weighing of the box before and after the emission determines the energy of the photon with an uncertainty

$$\Delta E = c^2 \Delta m = c^2 \Delta p / gT. \tag{8}$$

Einstein thought he could violate the uncertainty principle between energy and time,

$$\Delta E . \Delta t \geq (1/2)\hbar, \tag{9}$$

because he thought he could set the internal clock to make the uncertainty Δt in the time of emission of the photon as small as he pleased. Bohr defeated this scheme by pointing out that the rate of the clock would be affected by the position of the box in the gravitational potential according to general relativity. If the uncertainty in the position of the box is Δx, then the resulting uncertainty in the clock-time during the weighing is

$$\Delta t = Tg\Delta x / c^2. \tag{10}$$

Putting together (8) and (10), we see that the uncertainty relation (9) between time and energy for the photon follows immediately from the ordinary uncertainty relation between position and momentum for the box. Point, set and match to Bohr.

My third thought-experiment is nothing more than a repetition of the Einstein Box experiment with one measurement added. You arrange a photon detector with an accurate clock outside the box, and measure the time at which the photon arrives at the detector. The uncertainty in the arrival-time is then independent of the movement of the box. The uncertainty in the emission-time is determined by the uncertainty in the travel-time of the photon. The travel-time is uncertain, according to general relativity again, because the route of the photon in the gravitational potential is uncertain. However, we can arrange an optical system with f-number f that will focus the photon onto a fixed point at a distance

$$l = f \Delta x \qquad (11)$$

from the window of the box, no matter where the box happens to be at the moment of emission. The travel-time of the photon from the window to the focus will be (l/c), with an uncertainty introduced by the gravitational potential as before. The travel of the photon from the focus to the detector is along a known path and introduces no additional uncertainty. The travel-time uncertainty is then

$$\Delta t = (l/c)(g\Delta x/c^2) = (fg/c^3)(\Delta x)^2, \qquad (12)$$

which with (8) gives

$$\Delta E \cdot \Delta t = (f/cT)(\Delta x)^2 \cdot \Delta p. \qquad (13)$$

We can now choose the ratio $(\Delta x/cT)$ to be as small as we like, so that the uncertainty-relation (9) will be violated for the photon even when the usual relation between position and momentum is valid for the box. In this way we achieve the violation that Einstein intended when he introduced his box. In this third experiment, just like the second, the violation is easily achieved with apparatus of desktop size. The optical system that focuses the photon can be a simple telescope with a length of a few centimeters. After seventy years, Einstein is finally vindicated.

Bohr would not have been disturbed for a moment by these two thought-experiments. Both of them only violate the uncertainty principle by violating the rules that Bohr laid down for a legitimate use of quantum mechanics. Bohr's rules say that a quantum-mechanical description of an object can only be used to predict the probabilities that the object will behave in specified ways when it has been prepared in a specified quantum-state. The quantum-mechanical description cannot be used to say what actually happened after the experiment is finished. The two thought-experiments merely confirm that this restriction of the use of quantum-mechanics is necessary. If, in the second experiment, it were possible to define a wave-function for the electron traveling between the two counters at the observed times, this wave function could be proved to satisfy the uncertainty principle by the usual mathematical argument. But we saw that the uncertainty principle was violated, and therefore no such wave-function can exist. Similarly, in the third thought-experiment, there can be no wave-function describing the travel of the photon from the box to the detector. A wave-function can only say that a photon has a certain probability of arriving, not that a photon has arrived. Although Bohr would say that the two experiments only confirm the correctness of his strict interpretation of quantum-mechanics, Einstein might also claim that they justify his distrust. They prove in a simple and convincing fashion the contention of Einstein that quantum-mechanics is not a complete description of nature. Perhaps Einstein would be happy to learn that his box is still alive and well after seventy years, and still making trouble for believers in quantum-mechanics.

Let me now summarize the results of these two thought-experiments. They lead to two conclusions. First, statements about the past cannot in general be made in quantum-mechanical language. For example, we can describe a uranium nucleus by a wave-function including an outgoing alpha-particle wave which determines the probability that the nucleus will decay tomorrow. But we cannot describe by means of a wave-function the statement, "This nucleus decayed yesterday at 9 a.m. Greenwich time". As a general rule, knowledge about the past can only be expressed in classical terms. Lawrence Bragg, a shrewd observer of the birth of quantum mechanics, summed up the situation in a few words, "Everything in the future is a wave, everything in the past is a particle". Since a large fraction of science, including most of geology and astronomy as well as the whole of paleontology, is knowledge of the past, quantum-mechanics must always remain a small part of science. The second conclusion is that the "role of the observer" in quantum-mechanics is solely to make the distinction between past and future. Since quantum-mechanical statements can be made only about the future, all such statements require a precise definition of the boundary separating the future from the past. Every quantum-mechanical statement is relative, in the sense that it describes possible futures predicted from a particular past-future boundary. Only in a classical description can the universe be viewed as an absolute spacetime continuum without distinction between past and future. All quantum-mechanical descriptions are partial. They refer only to particular regions of spacetime, separated from other regions within which the description is classical.

These conclusions of mine contradict both the extreme DeWitt view and the extreme Copenhagen view of quantum-mechanics. I contradict DeWitt when he says it makes sense to speak about a wave-function for the universe as a whole. I contradict the orthodox Copenhagen view, that the role of the observer in quantum mechanics is to cause an abrupt "reduction of the wave-packet" so that the state of the system appears to jump discontinuously at the instant when it is observed. This picture of the observer interrupting the course of natural events is unnecessary and misleading. What really happens is that the quantum-mechanical description of an event ceases to be meaningful as the observer changes the point of reference from before the event to after it. We do not need a human observer to make quantum mechanics work. All we need is a point of reference, to separate past from future, to separate what has happened from what may happen, to separate facts from probabilities.

4. Black Holes and Quantum Causality

Twenty-six years ago, Stephen Hawking published a remarkable paper [Hawking 1976], with the title, "Breakdown of predictability in gravitational collapse". He had then recently discovered the phenomenon of Hawking radiation, which led him to the prediction that black holes should slowly evaporate and finally vanish into a puff of gamma-rays. Some years earlier, gamma-ray bursts had been detected by orbiting satellites, but we then had no clue concerning how and where the bursts originated. We only knew that they came from some violent process occurring at a great distance from the earth. When Hawking made his prediction, we hoped at first that the bursts might be the final display of fireworks giving direct evidence of the evaporation of small black holes by the Hawking process. To agree with the observed brightness of the bursts, the evaporating black holes would have to be at distances of the order of one light-year, far beyond the planets but still loosely attached to the gravitational field of the sun. Such a population of small black holes in our neighborhood would have been a wonderful laboratory for studying black hole physics. Unfortunately, the Hawking process would give a final burst of gamma-rays of much higher energy than the observed bursts. Most of the bursts are certainly not produced by the Hawking process. We now know that the bursts are events of stupendous violence occurring in remote galaxies at cosmological distances. The nature of the sources is one of the outstanding problems of astronomy.

I return now to Hawking's 1976 paper. Hawking asked three important questions about the process of black hole evaporation. None of his questions is yet definitively answered. First, when the horizon around the black hole disappears at the end of the evaporation process, does a naked spacetime singularity exist at the point of disappearance, and is the singularity exposed to view from the outside? Second, when a real star composed of ordinary matter collapses into a black hole that later evaporates, what happens to the law of conservation of baryons? Third, is the process of black-hole evaporation consistent with quantum-mechanical causality? Each of these three questions has given rise to important progress in our understanding of the universe. The first question led to improved understanding of the geometrical structure of horizons. The second question led to a gradual abandonment of the belief that baryon-conservation could be an exact law of nature. The third question placed a new limit on the validity of quantum mechanics. Since the validity of quantum mechanics is the subject of this chapter, I am mainly concerned with the third question, but I will have something to say about all three.

Following the good example of Bohr and Einstein, Hawking clarified his questions by means of a thought-experiment. This is the last of the four experiments that I am discussing. The experiment is a simple one, although it requires a rather large laboratory and a long-lived experimenter. The experimenter prepares a massive object in a pure quantum-state at zero temperature, and keeps it isolated from all contact with the rest of the universe. The Schrödinger equation, being a linear equation, predicts that the object will remain in a pure state so long as it is not disturbed from the outside. The object is assumed to be so massive that its internal pressure is insufficient to keep it from collapsing into a black hole. After the collapse, the black hole slowly evaporates into thermal radiation as predicted by Hawking's theory. Then the thermal radiation should also be in a pure quantum-state. But this is a contradiction in terms. Thermal radiation is in a state of maximum entropy, as far removed as possible from a pure state. Therefore, Hawking concludes, the Schrödinger equation cannot be correct. At some stage in the process of collapse and evaporation, the Schrödinger equation must fail. The failure of the Schrödinger equation is what Hawking means by the phrase, "violation of quantum causality". This was a highly unwelcome conclusion for Hawking, who believed that quantum mechanics should be universally valid. With his habitual honesty and open-mindedness, he did not conceal his discomfort but published his disagreable conclusion for all of us to ponder.

What do we learn from Hawking's thought-experiment? It throws light on all three of his questions. First, the question of naked singularities. There exists a tentative model of the evaporation of a black hole which avoids the appearance of a naked singularity. The model consists of a sequence of black hole configurations (see Fig. 2), described by John Wheeler thirty-seven years ago [Harrison *et al.*, 1965]. Wheeler introduced these configurations as a model of rapid gravitational collapse, but they work equally well as a model of slow evaporation. The configurations are described by an exterior Schwarzschild metric with mass M joined onto an interior uniform-curvature metric with mass-density ρ. The uniform-curvature part of the spacetime is a polar cap, a piece of a hypersphere containing all points within an angular distance x from one pole. The relations connecting M, ρ and x are

$$M = M_0(2/3)(\sin x)^3(x - \sin x \cos x)^{-1}, \tag{14}$$

$$\rho = \rho_0(x - \sin x \cos x)^2, \tag{15}$$

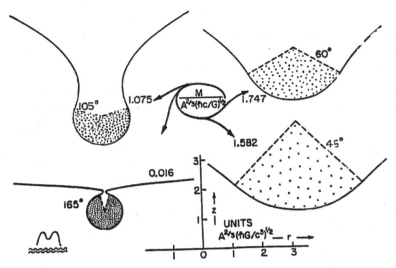

Fig. 2. The Wheeler model of gravitational collapse, borrowed here as a model of black hole evaporation. Taken from [Harrison *et al.*, 1965].

with the initial mass M_0 and the density-coefficient ρ_0 constant. The model of the evaporation process has the angle x slowly increasing from 0 to π as evaporation proceeds. The interior metric describes a distribution of cold matter with zero pressure and with local conservation of mass. The total amount of matter in the interior does not vary with x. But M, the gravitational mass of the object as seen from the outside, decreases steadily to zero as x approaches π. When x is close to π, the Schwarzschild radius of the exterior metric becomes small and there is only a narrow neck connecting the interior spacetime with the world outside. At the instant when $x = \pi$, the neck is pinched off and evaporation of the black hole is complete. We are then left with two disconnected pieces of spacetime, a flat piece containing no black hole and only outgoing waves of radiation from the evaporation process, and a completely closed hypersphere containing the original matter of mass M_0 in splendid isolation. There is no naked singularity remaining. A discontinuity in the spacetime curvature only occurs momentarily at the instant of separation, when the interior metric floats away into nothingness. The discontinuity is actually hidden from any possibility of observation from the outside, because the radiation emitted immediately before the end of the evaporation forms an opaque fireball. The size and energy content of the fireball can be roughly calculated. The energy content turns out to be of the order of a few megatons of TNT, comparable to a large hydrogen bomb. The fireball thus provides a highly effective screen, enforcing the rule of "cosmic censorship" which forbids the observation of naked singularities.

Second, we come to the question of baryon conservation. In Wheeler's simple model of the evaporation process, the baryon-number is everywhere locally conserved. The baryon-number density is proportional to the mass-density ρ of the interior metric. At the end of the evaporation, the baryons contained in the black hole have disappeared from the part of spacetime that is connected with the outside universe. Whether they still "exist" is a question of words and not a question of physics. The question would only become physically meaningful if the detached hypersphere should for some reason reattach itself to our universe at some other point of spacetime. If this happened, we would observe it as a "white hole", and we would be able to verify that the baryons contained in it had been conserved. In the absence of white holes, the model says that baryons are conserved locally but may disappear globally.

The evaporation model is purely schematic and not based upon dynamical calculations. The Wheeler configurations are treated as if they were static or slowly varying. In reality, these configurations are subject to dynamical collapse with a rapid timescale. In order to justify the model as a quantitatively correct description of Hawking's thought-experiment, the evolution of the evaporating black hole should be calculated with a consistent dynamical treatment of the spacetime geometry and of the matter. Since I am not an expert in the solution of the Einstein equations for black hole dynamics, I have not tried to make the model dynamically consistent.

Lastly I come to Hawking's third question, the question that is directly relevant to this chapter. What does his thought-experiment tell us about quantum causality? I claim that it gives strong support to my main thesis, that quantum-mechanical description can be consistently applied only to the future and not to the past. More precisely, I am saying that quantum-mechanical description is limited to parts of the universe that are confined within the future of a conceivable observer. Hawking's experiment illustrates this principle, because no global separation between past and future is possible in any region of spacetime containing a black hole. The observer who defines the separation between past and future has to make a choice, either to plunge into the black hole or to stay outside it. It is impossible to include the interior of a black hole in the past or the future of an observer outside. This impossibility implies as a corollary that no complete description of a black hole together with its outside environment can be made in quantum-mechanical language. So we confirm Hawking's conclusion that black holes are not subject to quantum causality as he defines it. The Schrödinger equation is not violated locally. The equation remains valid wherever it can be meaningfully stated. It breaks down in Hawking's experiment only because the notion of quantum-mechanical coherence between events occurring inside and outside a black hole has no physical meaning. The processes of formation and evaporation of a black hole provide us with a good model for DeWitt's concept of decoherence. That is my answer to Hawking's third question. His discovery of the laws of radiation from black holes does not imply a breakdown of quantum theory. The Hawking process is, on the contrary, entirely consistent with the known limitations of the classical and quantum-mechanical descriptions of nature.

5. Concluding Remarks

I hope this chapter has left you as confused as it leaves me. Once, when Bohr was accused of confusing people with his convoluted sentences, he replied that one should not speak more clearly than one can think. This wise remark applies particularly to speaking about quantum theory. I am usually reluctant to engage in discussions about the meaning of quantum theory, because I find that the experts in this area have a tendency to speak with dogmatic certainty, each of them convinced that one particular solution to the problem has a unique claim to be the final truth. I have the impression that they are less wise than Bohr. They tend to speak more clearly than they think. Each of them presents to us one particular version of quantum theory as the definitive description of the way nature works. Their efforts do not convince me, because I am a working physicist. As a physicist, I am much more impressed by our ignorance than by our knowledge. During the last hundred years we have made tremendous progress in our understanding of nature, but there is no reason to fear that our progress is coming close to an end. During the twenty-first century we shall probably meet with as many rude surprises as we have met in the twentieth. The discovery of Hawking radiation was the most recent big surprise in theoretical physics, but there is no reason to expect that it will be the last. The structure of theoretical physics as a whole, and of quantum theory in particular, looks to me like a makeshift agglomeration of bits and pieces, not like a finished design. If the structure of science is still provisional, still growing and changing as the years go by, it makes no sense to impose on the structure a spurious philosophical coherence. That is why I am skeptical of all attempts to squeeze quantum theory into a clean and tidy philosophical doctrine. I prefer to leave you with the feeling that we still have a lot to learn.

One question that I have not discussed in this chapter is the existence of gravitons. Theoretical physicists have almost unanimously assumed that the gravitational field must be a quantum field, with associated particles called gravitons. The statement that gravitons exist can only have meaning if one can devise a thought-experiment to demonstrate their existence. I have searched in vain for such a thought-experiment. For example, if one tries to imagine an experiment to detect the emission of a single graviton in a high-energy particle collision, one needs a detector of such astronomical dimensions that it cannot be prevented from collapsing into a black hole. Detection of single gravitons appears always to be frustrated by the extraordinary weakness of the gravitational interaction. Feasible detectors can detect gravitational waves only when the source is massive and the waves are classical. If it turned out to be true that no conceivable thought-experiment could detect effects of quantum gravity, then quantum gravity would have no physical meaning. In that case, the gravitational field would be a purely classical field, and efforts to quantize it would be illusory. Particles with the properties of gravitons would not exist. This conclusion is a hypothesis to be tested, not a statement of fact. To decide whether it is true, a careful and complete analysis of possible thought-experiments is needed. I leave this task to the physicists of the twenty-first century.

References

S. Adler [2001] "Why decoherence has not solved the measurement problem: a response to P. W. Anderson", Institute for Advanced Study preprint, quant-ph/0112095v1.

N. Bohr [1949] "Discussion with Einstein on epistemological problems in atomic physics", in *Albert Einstein, Philosopher-Scientist*, ed. P. A. Schilpp (Library of Living Philosophers, Evanston, Illinois).

B. DeWitt [1992] "Decoherence without complexity and without an arrow of time", talk given at the Workshop on Time Asymmetry, Mazagón, Huelva, Spain, September 1991 (University of Texas Center for Relativity preprint, Austin, Texas).

K. Harrison, K. Thorne, M. Wakano and J. Wheeler [1965] *Gravitation Theory and Gravitational Collapse* (University of Chicago Press), pp. 75–79.

S. Hawking [1976] "Breakdown of predictability in gravitational collapse", *Phys. Rev. D* **14**, 2460–2473.

J. A. Wheeler [1979] *Frontiers of Time* (North-Holland, Amsterdam, for the Società Italiana di Fisica, Bologna), pp. 11, 13, 22, 25, 44.

CHAPTER 5.5

OPACITY BOUNDS[*]

A bound on the Rosseland mean opacity is derived, assuming that Thomson scattering can be ignored and that radiation can be described by a non-relativistic dipole approximation. The same method yields an inequality between the Planck mean opacity and the Rosseland mean. Finally, unitarity gives another bound on opacities in the very high temperature regime.

1. Introduction

In the course of some work done in connection with the Orion project many years ago (see [Dyson, 2002]), we computed the opacities and equations of state of several light elements at low densities. We noted that quantum mechanics leads to a bound on these opacities and that this bound is a useful test of approximation methods. Our study [Bernstein and Dyson, 1959] was never published in the open literature. Recently our attention was again called to this work, and we wondered if, in the many years since we found it, this bound had been rediscovered. The only mention of it that we found in the published literature is [Armstrong, 1962], who states our bound and gives our unpublished study as a reference. The purpose of Armstrong's paper is to prove a different bound, relating the Rosseland mean opacity to the Planck mean opacity. We here present a proof of our old bound, and explain how this is related to Armstrong's proof of his bound. Before turning to the formal arguments, we give a simple dimensional analysis to show why such bounds are to be expected.

Suppose that radiation with frequency bandwidth B is passing through matter composed of atoms with atomic number Z and atomic weight A. The opacity, which has the dimensions of square centimeters per gram, is $(Z\sigma/AM)$, where M is the mass of a hydrogen atom and σ is an average cross-section for absorption of radiation by an electron. The product σB is the total oscillator strength of one electron and

---

[*]Stevens Institute of Technology, Hoboken, NJ, 07030.

is bounded by a universal constant F. For thermal radiation of temperature T, the bandwidth is given approximately by (kT/h), where k and h are Boltzmann's and Planck's constants. Thus the opacity is bounded by $(ZhF/AMkT)$. The quantity F has dimensions of square centimeters per second and must be composed of the constants e, m, h and c, where e and m are the charge and mass of an electron. In the dipole approximation, F is proportional to e^2, so it must be equal to (fe^2/mc) where f is a pure number. The opacity is then bounded by $(fZe^2h/AMmckT)$. Whether such a bound has any utility depends on the magnitude of f, which we can only discover with a formal derivation to which we now turn. It turns out, as we see in (18) below, that $f = 0.288$, a satisfactorily small number. The physical reason why f is smaller than unity is that the effective bandwidth of thermal radiation is actually many times (kT/h). The high-energy photons in the tail of the Planck distribution make an important contribution to the penetration of radiation through a thick absorber.

2. The Bound on the Rosseland Mean

We begin with several definitions and notational conventions. The absorption per unit length, for a photon of frequency v, is $k(v) = n\sigma(v)$, where $\sigma(v)$ is the cross-section for the absorption of a photon by a single electron, and n is the electron density. We can define the mass absorption coefficient $\kappa(v)$ by the relation $\kappa(v) = (k(v)/\rho) = (Z/AM)\sigma(v)$, where, as above, (Z/AM) is the number of electrons per gram of absorber and ρ is its mass density. We will now see how the Rosseland mean enters the radiation transport problem. The transport equation for the specific intensity $I(v)$ (measured in ergs per square centimeter per second per steradian per Hertz) is [Chandrasekhar, 1958]

$$dI(v)/ds = \kappa(v)\rho(B(v) - I(v)), \tag{1}$$

where

$$B(v) = (2hv^3/c^2)(1 - \exp(-hv/kT))^{-1} \tag{2}$$

is the intensity of radiation in thermal equilibrium at the local temperature T. In the interior of a thick absorber, $I(v)$ and $B(v)$ do not differ greatly, and (1) becomes

$$I(v) = B(v) - (1/\kappa(v)\rho)\nabla B(v). \tag{3}$$

To obtain the flux $F(v)_s$ in the s direction, we multiply (3) by a direction cosine and integrate over all directions. Since $B(v)$ is isotropic, we find

$$F(v)_s = -(4\pi/3)(1/\kappa(v)\rho)dB(v)/ds. \tag{4}$$

The integral of (4) over frequency is the total flux

$$F_s = -(4\pi/3\rho) \int_0^\infty (1/\kappa(v))(dB(v)/dT)(dT/ds)dv = -(1/\kappa_r\rho)(16S/3)T^3 dT/ds,$$
$$(5)$$

where $S = (2\pi^3 k^4/15c^2h^3)$. In (5) we have introduced the Rosseland mean opacity κ_r, or the Rosseland mean mass absorption coefficient, defined by

$$(1/\kappa_r) = \int_0^\infty (1/\kappa(v))(dB(v)/dT)dv \Big/ \int_0^\infty (dB(v)/dT)dv, \qquad (6)$$

which is a harmonic mean of $\kappa(v)$ as we shall see later. There are four sources of absorption contributing to $\kappa(v)$, namely bound-free, bound-bound, free-free (bremsstrahlung), and the scattering of photons from free electrons which in the non-relativistic regime is given by the frequency-independent Thomson cross-section. The latter becomes important, indeed dominant, in the high temperature, low density regime. In the regime we are interested in, Rydberg temperatures, it can be neglected and in the sequel we shall do so. Thus when we mention an absorption coefficient we shall mean $\kappa = \kappa_{bf} + \kappa_{bb} + \kappa_{ff}$.

Implicit in the previous discussion has been a treatment of the effects of stimulated emission. We shall now make this treatment explicit. The net absorption of photons with frequency v is equal to the gross absorption minus the stimulated emission. We define $\kappa(v)$ to be the net absorption coefficient, and use the notation $\kappa^*(v)$ for the gross absorption coefficient. These definitions apply when v is positive. We can then define $\kappa^*(-v)$ to be the stimulated emission coefficient for photons of frequency v, so that

$$\kappa(v) = \kappa^*(v) - \kappa^*(-v). \qquad (7)$$

But the principle of detailed balance implies that in thermal equilibrium at temperature T

$$\kappa^*(-v) = \exp(-hv/kT)\kappa^*(v), \qquad (8)$$

and therefore

$$\kappa(v) = (1 - \exp(-hv/kT))\kappa^*(v). \qquad (9)$$

In integrals representing the total net absorption of photons, we can either let the photon frequency v run from zero to infinity and use the net absorption coefficient $\kappa(v)$, or we can let v run from minus to plus infinity and use the gross coefficient $\kappa^*(v)$. In the work above we have chosen to use $\kappa(v)$.

To evaluate the denominator in (6) we let $u = (h\nu/kT)$ and obtain

$$\int_0^\infty u^4 \exp u (\exp u - 1)^{-2} du = (4\pi^4/15) = 25.976. \tag{10}$$

Thus

$$(1/\kappa_r) = \int_0^\infty (W(u)/\kappa(u)) du, \quad W(u) = (15/4\pi^4) u^4 \exp u (\exp u - 1)^{-2}. \tag{11}$$

In the dipole approximation, [Rybicki 1979], $\kappa^*(u) = (\pi e^2/mc)(Z/AM) D^*(\nu)$ is the gross absorption coefficient, where

$$D^*(\nu) = (2/m\nu) \sum_f \delta(E_f - E_i - h\nu) |\langle p.\epsilon \rangle_{if}|^2 \tag{12}$$

is the normalized probability of absorption of a photon with frequency ν and polarization-vector ϵ by an electron moving from initial state i to final state f. The same formula (12) also describes induced emission when ν is negative. We now prove that $D^*(\nu)$ satisfies the sum-rule

$$\int_{-\infty}^\infty D^*(\nu) d\nu = 1. \tag{13}$$

We make use of the relations $p = m\dot{r} = (1/i\hbar)[r, H]$ and $[r_j, p_k] = i\hbar \delta_{jk}$. Thus

$$|\langle p.\epsilon \rangle_{if}|^2 = (m/2i\hbar)(E_f - E_i)\epsilon_j \epsilon_k [\langle r_j \rangle_{if} \langle p_k \rangle_{fi} - \langle p_k \rangle_{if} \langle r_j \rangle_{fi}]. \tag{14}$$

If we sum over the final states f, the bracket reduces to the canonical commutator $[r_j, p_k]$ and the polarization factor becomes $(\epsilon \cdot \epsilon) = 1$. Thus (13) is proved. Note that if we had multiplied the terms in (14) by the stimulated emission factor $(1 - \exp((E_i - E_f)/kT))$, we would have a sum rule with the integral running from zero to infinity, $\int_0^\infty D(\nu) d\nu = \int_{-\infty}^\infty D^*(\nu) d\nu = 1$, with $D(\nu)$ describing net rather than gross absorption.

When we sum over all electrons, the sum-rule (13) which holds for a single electron becomes a sum-rule for the mass absorption coefficient,

$$\int_0^\infty \kappa(\nu) d\nu = (\pi e^2 Z/mcAM). \tag{15}$$

The Rosseland mean opacity is given by (11). We can now use the Schwarz inequality

$$\left(\int abdu \right)^2 \le \left(\int a^2 du \right) \left(\int b^2 du \right), \tag{16}$$

with equality when a and b are proportional. Thus

$$\left(\int_0^\infty W(u)(\kappa(u))^{-1} du \right) \left(\int_0^\infty \kappa(u) du \right) \geq \left(\int_0^\infty (W(u))^{1/2} du \right)^2, \tag{17}$$

with equality when $\kappa(u)$ is proportional to $(W(u))^{1/2}$. Putting together (11), (16) and (17), we have

$$\kappa_r \leq (\pi e^2 hZ/mcAMkTs^2), \tag{18}$$

with

$$s = \int_0^\infty (W(u))^{1/2} du = 3.302. \tag{19}$$

If we now put in all the numerical factors and introduce the Rydberg energy defined by $\mathrm{Ry} = (me^4/2\hbar^2) = 13.6\mathrm{eV}$, we find the bound

$$\kappa_r \leq (Z/A)(\mathrm{Ry}/kT)(4.43)10^5 (\mathrm{cm}^2/\mathrm{g}). \tag{20}$$

This bound is a useful check on approximation methods for computing opacity.

3. Other Bounds

The Schwarz inequality (16) is a continuous version of the inequality

$$A \geq H, \tag{21}$$

relating the arithmetic and harmonic means,

$$A = n^{-1} \sum x_i, \quad H = n \left(\sum x_i^{-1} \right)^{-1}, \tag{22}$$

of n positive numbers x_i. To convert (21) into (17) we replace the sums by integrals and take

$$x_i = \kappa(u)(W(u))^{-1/2}. \tag{23}$$

The inequality (17) puts a bound on the Rosseland opacity by virtue of the sum-rule (15). Another bound, found by [Armstrong, 1962], relates the Rosseland opacity κ_r to the Planck opacity κ_p. The Rosseland opacity measures the penetration of radiation when the absorbing medium is optically thick, when the mean free path of the photons is much smaller than the scale over which the temperature, pressure and density change. The flux does not then depart much from isotropy and the radiation field is locally close to a Planck distribution. We can then ignore the

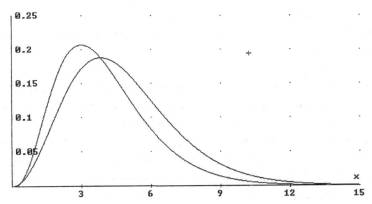

Fig. 1. This is a plot of the weight functions for the Rosseland and Planck means. The Planck weight peaks at about three while the Rosseland weight peaks at about four. Note the similarity in shape. Since the effective band width is much larger than kT the opacity is small.

difference between (1) and (3). The Planck opacity measures absorption when the medium is optically thin. It is defined by

$$\kappa_p = \int_0^\infty W_p(u)\kappa(u)du, \quad W_p(u) = (15/\pi^4)u^3 \exp(-u). \tag{24}$$

Thus the Planck opacity is an arithmetic mean of the absorption function $\kappa(u)$, while the Rosseland opacity is a harmonic mean according to (11). If the weight functions $W(u)$ and $W_p(u)$ were the same, the inequality (21) would imply $\kappa_p \geq \kappa_r$. The weight functions are not the same, but they are not very different, as we see from Fig. 1 where they are compared. The Planck weight function has its maximum at three while the Rosseland weight function has its maximum at about four, but the shapes are remarkably close. The Schwarz inequality (16) gives

$$\kappa_p/\kappa_r = \left(\int_0^\infty W_p(u)\kappa(u)du\right)\left(\int_0^\infty W(u)(\kappa(u))^{-1}du\right)$$

$$\geq \left(\int_0^\infty (W_p(u)W(u))^{1/2}\right)^2 = 0.950. \tag{25}$$

Thus we have proved Armstrong's inequality,

$$\kappa_r \leq 1.053\kappa_p. \tag{26}$$

Let us see what this means in two limiting cases. If absorption is due to Thomson scattering, $\kappa(u)$ is independent of u, κ_r and κ_p are equal, and (26) holds with only a little to spare. On the other hand, if absorption is due to photo-ionization (bound-free absorption), then $\kappa(u)$ is roughly proportional to u^{-3}, and κ_p is about four hundred times larger than κ_r. In an optically thin medium, the absorption is dominated by the abundant low-energy photons with large cross-sections, while in an optically thick medium the penetration is dominated by the rare high-energy photons with small cross-sections.

The most striking fact demonstrated by Fig. 1 is that the weight-functions $W(u)$ and $W_p(u)$ are both very low and broad, with maxima less than a quarter. That is why the integral s defined by (19) is large compared with unity, and the bound (18) for the Rosseland opacity, which has s^2 in the denominator, is small. Translating this remark from mathematics into physics, we may say that the reason opacities are so small is that the effective bandwidth of thermal radiation is large compared with kT. According to Fig. 1, photons with energies up to $10kT$ are making significant contributions to opacity.

Finally we note that at very high energies opacities are bounded by unitarity. The unitarity bound on photon absorption cross-sections is roughly the square of the photon wave-length. For thermal radiation with temperature T, the bound is roughly $(hc/kT)^2$, so both Rosseland and Planck opacities are bounded by

$$\kappa \leq (Z/AM)(hc/kT)^2. \tag{27}$$

For very large kT, substantially larger than the electron rest energy mc^2, this bound becomes stronger than (20). But at such high temperatures the electrons are relativistic and the opacity will be dominated by an electron-positron plasma.

References

B. H. Armstrong [1962] "A maximum opacity theorem", *Astrophys. J.* **136**, 309–310.

J. Bernstein and F. J. Dyson [1959] "The continuous opacity and equations of state of light elements at low densities", General Atomic Report GA-848, unpublished.

S. Chandrasekhar [1958] *The Introduction to the Study of Stellar Structure* (Dover Publications, New York), Chapter 5, pp. 183–215.

G. B. Dyson [2002] *Project Orion* (Henry Holt, New York).

G. B. Rybicki and A. P. Lightman [1979] *Radiative Processes in Astrophysics* (John Wiley and Sons, New York), Chapter 10, pp. 267–293.

CHAPTER 5.6

ITERATED PRISONER'S DILEMMA CONTAINS STRATEGIES THAT DOMINATE ANY EVOLUTIONARY OPPONENT

William H. Press* and Freeman J. Dyson[†]

*Department of Computer Science and School of Biological Sciences,
University of Texas at Austin, Austin, TX 78712

[†] School of Natural Sciences,
Institute for Advanced Study, Princeton, NJ 08540

*Contributed by William H. Press, April 19, 2012 (sent for review March 14, 2012)

The two-player Iterated Prisoner's Dilemma game is a model for both sentient and evolutionary behaviors, especially including the emergence of cooperation. It is generally assumed that there exists no simple ultimatum strategy whereby one player can enforce a unilateral claim to an unfair share of rewards. Here, we show that such strategies unexpectedly do exist. In particular, a player X who is writing of these strategies can (i) deterministically set her opponent Y's score, independently of his strategy or response, or (ii) enforce an extortionate linear relation between her and his scores. Against such a player, an evolutionary player's best response is to accede to the extortion. Only a player with a theory of mind about his opponent can do better, in which case Iterated Prisoner's Dilemma is an Ultimatum Game.

evolution of cooperation | game theory | tit for tat

Iterated 2×2 games, with Iterated Prisoner's Dilemma (IPD) as the notable example, have long been touchstone models for elucidating both sentient human behaviors, such as cartel pricing, and Darwinian phenomena, such as the evolution

---

Author contributions: W.H.P. and F.J.D. designed research, performed research, contributed new reagents/analytic tools, analyzed data, and wrote the paper.

The authors declare no conflict of interest.

Freely available online through the PNAS open access option.

*To whom correspondence should be addressed. E-mail: wpress@cs.utexas.edu.

of cooperation [1–6]. Well-known popular treatments [7–9] have further established IPD as foundational lore in fields as diverse as political science and evolutionary biology. It would be surprising if any significant mathematical feature of IPD has remained undescribed, but that appears to be the case, as we show in this paper.

Figure 1A shows the setup for a single play of Prisoner's Dilemma (PD). If X and Y cooperate (c), then each earns a reward R. If one defects (d), the defector gets an even larger payment T, and the naive cooperator gets S, usually zero. However, if both defect, then both get a meager payment P. To be interesting, the game must satisfy two inequalities: $T > R > P > S$ guarantees that the Nash equilibrium of the game is mutual defection, whereas $2R > T + S$ makes mutual cooperation the globally best outcome. The "conventional values" $(T, R, P, S) = (5, 3, 1, 0)$ occur most often in the literature. We derive most results in the general case, and indicate when there is a specialization to the conventional values.

Figure 1B shows an iterated IPD game consisting of multiple, successive plays by the same opponents. Opponents may now condition their play on their opponent's strategy insofar as each can deduce it from the previous play. However, we give each

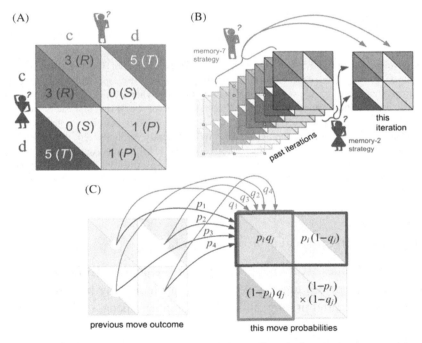

Fig. 1. (A) Single play of PD, Players X (blue) and Y (red) each choose to cooperate (c) or defect (d) with respective payoffs R, T, S, or P as shown (along with the most common numerical values), (B) IPD, where the same two players play arbitrarily many times; each has a strategy based on a finite memory of the previous plays. (C) Case of two memory-one players. Each player's strategy is a vector of four probabilities (of cooperation), conditioned on the four outcomes of the previous move.

player only a finite memory of previous play [10]. One might have thought that a player with longer memory always has the advantage over a more forgetful player. In the game of bridge, for example, a player who remembers all of the cards played has the advantage over a player who remembers only the last trick; however, that is not the case when the same game (same allowed moves and same payoff matrices) is indefinitely repeated. In fact, it is easy to prove (*Appendix A*) that, for any strategy of the longer-memory player Y, shorter-memory X's score is exactly the same as if Y had played a certain shorter-memory strategy (roughly, the marginalization of Y's long-memory strategy: its average over states remembered by Y but not by X), disregarding any history in excess of that shared with X. This fact is important. We derive strategies for X assuming that both players have memory of only a single previous move, and the above theorem shows that this involves no loss of generality. Longer memory will not give Y any advantage.

Figure 1C, then, shows the most general memory-one game. The four outcomes of the previous move are labeled $1, \ldots, 4$ for the respective outcomes at $xy \in (cc, cd, dc, dd)$, where c and d denote cooperation and defection. X's strategy is $\mathbf{p} = (p_1, p_2, p_3, p_4)$, her probabilities for cooperating under each of the previous outcomes. Y's strategy is analogously $\mathbf{q} = (q_1, q_2, q_3, q_4)$ for outcomes seen from his perspective, that is, in the order of $yx \in (cc, cd, dc, dd)$. The outcome of this play is determined by a product of probabilities, as shown in Fig. 1.

Methods and Results

Zero-Determinant Strategies. As is well understood [10], it is not necessary to simulate the play of strategies $\mathbf{p}$ against $\mathbf{q}$ move by move. Rather, $\mathbf{p}$ and $\mathbf{q}$ imply a Markov matrix whose stationary vector $\mathbf{v}$, combined with the respective payoff matrices, yields an expected outcome for each player. (We discuss the possibility of nonstationary play later in the paper.) With rows and columns of the matrix in X's order, the Markov transition matrix $\mathbf{M}(\mathbf{p}, \mathbf{q})$ from one move to the next is shown in Fig. 2A.

Because $\mathbf{M}$ has a unit eigenvalue, the matrix $\mathbf{M}' \equiv \mathbf{M} - \mathbf{I}$ is singular, with thus zero determinant. The stationary vector $\mathbf{v}$ of the Markov matrix, or any vector proportional to it, satisfies

$$\mathbf{v}^T \mathbf{M} = \mathbf{v}^T, \quad \text{or} \quad \mathbf{v}^T M' = \mathbf{0}. \tag{1}$$

Cramer's rule, applied to the matrix $\mathbf{M}'$, is

$$\text{Adj}(\mathbf{M}') \mathbf{M}' \equiv \det(\mathbf{M}') \mathbf{I} = 0, \tag{2}$$

where $\text{Adj}(\mathbf{M}')$ is the adjugate matrix (also known as the classical adjoint or, as in high-school algebra, the "matrix of minors"). Equation (2) implies that every row of

(A)

$$\begin{bmatrix} p_1q_1 & p_1(1-q_1) & (1-p_1)q_1 & (1-p_1)(1-q_1) \\ p_2q_3 & p_2(1-q_3) & (1-p_2)q_3 & (1-p_2)(1-q_3) \\ p_3q_2 & p_3(1-q_2) & (1-p_3)q_2 & (1-p_3)(1-q_2) \\ p_4q_4 & p_4(1-q_4) & (1-p_4)q_4 & (1-p_4)(1-q_4) \end{bmatrix}$$

(B) $\qquad \mathbf{v} \cdot \mathbf{f} \equiv D(\mathrm{p},\mathrm{q},\mathbf{f})$

$$= \det \begin{bmatrix} -1+p_1q_1 & -1+p_1 & -1+q_1 & f_1 \\ p_2q_3 & -1+p_2 & q_3 & f_2 \\ p_3q_2 & p_3 & -1+q_2 & f_3 \\ p_4q_4 & p_4 & q_4 & f_4 \end{bmatrix}$$

$$\equiv \tilde{\mathbf{p}} \qquad \equiv \tilde{\mathbf{q}}$$

Fig. 2. (A) Markov matrix for the memory-one game shown in Fig. 1C. (B) The dot product of any vector $\mathbf{f}$ with the Markov matrix stationary vector $\mathbf{v}$ can be calculated as a determinant in which, notably, a column depends only on one player's strategy.

Adj($\mathbf{M}'$) is proportional to $\mathbf{v}$. Choosing the fourth row, we see that the components of $\mathbf{v}$ are (up to a sign) the determinants of the 3×3 matrices formed from the first three columns of $\mathbf{M}'$, leaving out each one of the four rows in turn. These determinants are unchanged if we add the first column of $\mathbf{M}'$ into the second and third columns.

The result of these manipulations is a formula for the dot product of an arbitrary four-vector $\mathbf{f}$ with the stationary vector $\mathbf{v}$ of the Markov matrix, $\mathbf{v} \cdot \mathbf{f} \equiv D(\mathbf{p}, \mathbf{q}, \mathbf{f})$, where D is the 4×4 determinant shown explicitly in Fig. 2B. This result follows from expanding the determinant by minors on its fourth column and noting that the 3×3 determinants multiplying each f_i are just the ones described above. What is noteworthy about this formula for $\mathbf{v} \cdot \mathbf{f}$ is that it is a determinant whose second column,

$$\tilde{\mathbf{p}} \equiv (-1+p_1, -1+p_2, p_3, p_4), \tag{3}$$

is solely under the control of X; whose third column,

$$\tilde{\mathbf{q}} \equiv (-1+q_1, q_3, -1+q_2, q_4), \tag{4}$$

is solely under the control of Y; and whose fourth column is simply $\mathbf{f}$.

X's payoff matrix is $\mathbf{S}_X = (R, S, T, P)$, whereas Y's is $\mathbf{S}_Y = (R, T, S, P)$. In the stationary state, their respective scores are then

$$\begin{aligned} S_X &= \frac{\mathbf{v} \cdot \mathbf{S}_X}{\mathbf{v} \cdot \mathbf{1}} = \frac{D(\mathbf{p}, \mathbf{q}, \mathbf{S}_X)}{D(\mathbf{p}, \mathbf{q}, \mathbf{1})} \\ S_Y &= \frac{\mathbf{v} \cdot \mathbf{S}_Y}{\mathbf{v} \cdot \mathbf{1}} = \frac{D(\mathbf{p}, \mathbf{q}, \mathbf{S}_Y)}{D(\mathbf{p}, \mathbf{q}, \mathbf{1})}, \end{aligned} \tag{5}$$

332   *Birds and Frogs*

where **1** is the vector with all components 1. The denominators are needed because **v** has not previously been normalized to have its components sum to 1 (as required for a stationary probability vector).

Because the scores S in Eq. (5) depend linearly on their corresponding payoff matrices **S**, the same is true for any linear combination of scores, giving

$$\alpha s_X + \beta s_Y + \gamma = \frac{D(\mathbf{p}, \mathbf{q}, \alpha \mathbf{S}_X + \beta \mathbf{S}_Y + \gamma \mathbf{1})}{D(\mathbf{p}, \mathbf{q}, \mathbf{1})}. \tag{6}$$

It is Eq. (6) that now allows much mischief, because both X and Y have the possibility of choosing unilateral strategies that will make the determinant in the numerator vanish. That is, if X chooses a strategy that satisfies $\tilde{\mathbf{p}} = \alpha \mathbf{S}_X + \beta \mathbf{S}_Y + \gamma \mathbf{1}$, or if Y chooses a strategy with $\tilde{\mathbf{q}} = \alpha \mathbf{S}_X + \beta \mathbf{S}_Y + \gamma \mathbf{1}$, then the determinant vanishes and a linear relation between the two scores,

$$\alpha s_X + \beta s_Y + \gamma = 0 \tag{7}$$

will be enforced. We call these zero-determinant (ZD) strategies. We are not aware of any previous recognition of these strategies in the literature; they exist algebraically not only in IPD but in all iterated 2×2 games. However, not all ZD strategies are feasible, with probabilities **p** all in range $[0, 1]$. Whether they are feasible in any particular instance depends on the particulars of the application, as we now see.

X Unilaterally Sets Y's Score. One specialization of ZD strategies allows X to unilaterally set Y's score. From the above, X need only play a fixed strategy satisfying $\tilde{\mathbf{p}} = \beta \mathbf{S}_Y + \gamma \mathbf{1}$ (i.e., set $\alpha = 0$ in Eq. (7)), four equations that we can solve for p_2 and p_3 in terms of p_1 and p_4, that is, eliminating the nuisance parameters β and γ. The result, for general R, S, T, P (not necessarily a PD game), is

$$p_2 = \frac{p_1(T - P) - (1 + p_4)(T - R)}{R - P}$$

$$p_3 = \frac{(1 - p_1)(P - S) + p_4(R - S)}{R - P}. \tag{8}$$

With this substitution, Y's score (Eq. (5)) becomes

$$s_Y = \frac{(1 - p_1)P + p_4 R}{(1 - p_1) + p_4}. \tag{9}$$

All PD games satisfy $T > R > P > S$. By inspection, Eq. (8) then has feasible solutions whenever p_1 is close to (but $\leq$)1 and p_4 is close to (but $\geq$)0. In that case, p_2 is close to (but $\leq$)1 and p_3 is close to (but $\geq$) zero. Now also by inspection of Eq. (9), a weighted average of P and R with weights $(1 - p_1)$ and p_4, we see that all scores $P \leq s_Y \leq R$ (and no others) can be forced by X. That is, X can set Y's score to any value in the range from the mutual noncooperation score to the mutual cooperation score.

What is surprising is not that Y can, with X's connivance, achieve scores in this range, but that X can force any particular score by a fixed strategy **p**, independent of Y's strategy **q**. In other words, there is no need for X to react to Y, except on a timescale of her own choosing. A consequence is that X can simulate or 'spoof" any desired fitness landscape for Y that she wants, thereby guiding his evolutionary path. For example, X might condition Y's score on some arbitrary property of his last 1,000 moves, and thus present him with a simulated fitness landscape that rewards that arbitrary property. (We discuss the issue of timescales further, below.)

X Tries to Set Her Own Score. What if X tries to set her own score? The analogous calculation with $\tilde{\mathbf{p}} = \alpha \mathbf{S}_x + \gamma \mathbf{1}$ yields

$$p_2 = \frac{(1 + p_4)(R - S) - p_1(P - S)}{R - P} \geq 1$$

$$p_3 = \frac{-(1 - p_1)(T - P) - p_4(T - R)}{R - P} \leq 0.$$
(10)

This strategy has only one feasible point, the singular strategy $\mathbf{p} = (1, 1, 0, 0)$, "always cooperate or never cooperate." Thus, X cannot unilaterally set her own score in IPD.

X Demands and Gets an Extortionate Share. Next, what if X attempts to enforce an extortionate share of payoffs larger than the mutual noncooperation value P? She can do this by choosing

$$\tilde{\mathbf{p}} = \phi[(\mathbf{S}_X - P\mathbf{1}) - \chi(\mathbf{S}_Y - P\mathbf{1})],$$
(11)

where $\chi \geq 1$ is the extortion factor. Solving these four equations for the p's gives

$$p_1 = 1 - \phi(\chi - 1)\frac{R - P}{P - S}$$

$$p_2 = 1 - \phi\left(1 + \chi\frac{T - P}{P - S}\right)$$
(12)

$$p_3 = \phi\left(\chi + \frac{T - P}{P - S}\right)$$

$$p_4 = 0$$

Evidently, feasible strategies exist for any χ and sufficiently small ϕ. It is easy to check that the allowed range of ϕ is

$$0 < \phi \leq \frac{(P - S)}{(P - S) + \chi(T - P)}.$$
(13)

Under the extortionate strategy, X's score depends on Y's strategy $\mathbf{q}$, and both are maximized when Y fully cooperates, with $\mathbf{q} = (1, 1, 1, 1)$. If Y decides (or evolves) to maximize his score by cooperating fully, then X's score under this strategy is

$$s_X = \frac{P(T - R) + \chi[R(T - S) - P(T - R)]}{(T - R) + \chi(R - S)}. \tag{14}$$

The coefficients in the numerator and denominator are all positive as a consequence of $T > R > P > S$. The case $\phi = 0$ is formally allowed, but produces only the singular strategy $(1, 1, 0, 0)$ mentioned above.

The above discussion can be made more concrete by specializing to the conventional IPD values $(5, 3, 1, 0)$; then, Eq. (12) becomes

$$\mathbf{p} = [1 - 2\phi(\chi - 1), 1 - \phi(4\chi + 1), \phi(\chi + 4), 0], \tag{15}$$

a solution that is both feasible and extortionate for $0 < \phi \leq (4\chi + 1)^{-1}$. X's and Y's best respective scores are

$$s_X = \frac{2 + 13\chi}{2 + 3\chi}, \quad s_Y = \frac{12 + 3\chi}{2 + 3\chi}. \tag{16}$$

With $\chi > 1$, X's score is always greater than the mutual cooperation value of 3, and Y's is always less. X's limiting score as $\chi \to \infty$ is 13/3. However, in that limit, Y's score is always 1, so there is no incentive for him to cooperate. X's greed is thus limited by the necessity of providing some incentive to Y. The value of ϕ is irrelevant, except that singular cases (where strategies result in infinitely long "duels") are more likely at its extreme values. By way of concreteness, the strategy for X that enforces an extortion factor 3 and sets ϕ at its midpoint value is $\mathbf{p} = (\frac{11}{13}, \frac{1}{2}, \frac{7}{26}, 0)$, with best scores about $s_X = 3.73$ and $s_Y = 1.91$.

In the special case $\chi = 1$, implying fairness, and $\phi = 1/5$ (one of its limit values), Eq. (15) reduces to the strategy $(1, 0, 1, 0)$, which is the well-known tit-for-tat (TFT) strategy [7]. Knowing only TFT among ZD strategies, one might have thought that strategies where X links her score deterministically to Y must always be symmetric, hence fair, with X and Y rewarded equally. The existence of the general ZD strategy shows this not to be the case.

Extortionate Strategy Against an Evolutionary Player. We can say, loosely, that Y is an evolutionary player if he adjusts his strategy $\mathbf{q}$ according to some optimization scheme designed to maximize his score s_Y, but does not otherwise explicitly consider X's score or her own strategy. In the alternative case that Y imputes to X an independent strategy, and the ability to alter it in response to his actions, we can say that Y has a theory of mind about X [11–13].

Against X's fixed extortionate ZD strategy, a particularly simple evolutionary strategy for Y, close to if not exactly Darwinian, is for him to make successive small adjustments in $\mathbf{q}$ and thus climb the gradient in S_Y. [We note that true Darwinian evolution of a trait with multiple loci is, in a population, not strictly "evolutionary" in our loose sense [14]].

Because Y may start out with a fully noncooperative strategy $\mathbf{q}_0 = (0, 0, 0, 0)$, it is in X's interest that her extortionate strategy yield a positive gradient for Y's cooperation at this value of $\mathbf{q}$. That gradient is readily calculated as

$$\frac{\partial s_Y}{\partial \mathbf{q}}\bigg|_{\mathbf{q}=\mathbf{q}_0} = \left(0, 0, 0, \frac{(T-S)(S+T-2P)}{(P-S)+\chi(T-P)}\right). \tag{17}$$

The fourth component is positive for the conventional values $(T, R, P, S) = (5, 3, 1, 0)$, but we see that it can become negative as P approaches R, because we have $2R > S + T$. With the conventional values, however, evolution away from the origin yields positive gradients for the other three components.

We have not proved analytically that there exist in all cases evolutionary paths for Y that lead to the maximum possible scores (Eq. (16)) and that have positive directional derivatives everywhere along them. However, this assertion seems likely from numerical evidence, at least for the conventional values. Figure 3 shows a typical numerical experiment in which X plays an extortionate strategy (here, $\chi = 5$, with maximum scores $s_X = 3.94$ and $s_Y = 1.59$), and Y takes small steps that locally increase his score. Y has no unique gradient direction because the mapping from the score gradient (a covariant vector) to the step direction (a contravariant vector) involves an arbitrary metric, signifying how easily Y can evolve in each direction.

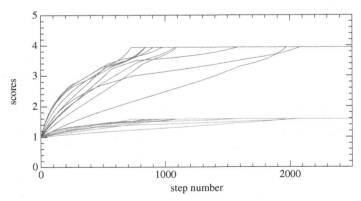

Fig. 3. Evolution of X's score (blue) and Y's score (red) in 10 instances. X plays a fixed extortionate strategy with extortion factor $\chi = 5$. Y evolves by making small steps in a gradient direction that increases his score. The 10 instances show different choices for the weights that Y assigns to different components of the gradient, i.e., how easily he can evolve along each. In all cases, X achieves her maximum possible (extortionate) score.

Figure 3 shows 10 arbitrary choices for this metric. In no case does Y's evolution get hung up at a local maximum. That is, all of the evolutions shown (and all of a much larger number tried) reach the value of Eq. (14).

Discussion

We have several times alluded to issues of timescale. The ZD strategies are derived mathematically under the assumption that the players' expected scores are generated by a Markov stationary state defined by their respective strategies $\mathbf{p}$ and $\mathbf{q}$. However, we have also suggested situations in which X may vary her ZD strategy so as to spoof Y with a fictitious fitness landscape. The question also arises whether Y can somehow vary his strategy on timescales faster than that for Markov equilibrium to be established. Perhaps by playing "inside the equilibration timescale" he can evade the linear constraint on scores (Eq. (7)) imposed by X.

Interestingly, it is easy to prove that this latter situation cannot occur (*Appendix* B). If X plays a constant ZD strategy, then any strategy of Y's, rapidly varying or not, turns out to be equivalent (from X's perspective) to a fixed strategy against which X's imposition of a constraint is effective.

In the former situation, where it is X whose strategies are changing (e.g., among ZD strategies that set Y's score), things are not as crisp. Because X must be basing her decisions on Y's behavior, which only becomes evident with averaging over time, the possibility of a race condition between X's and Y's responses is present with or without Markov equilibration. This reason is sufficient for X to vary her strategy only slowly. If X chooses components of $\mathbf{p}$ in Eqs. (9) and (8) that are bounded away from the extreme values 0 and 1, then the Markov equilibration time will not be long and thus not a consideration. In short, a deliberate X still has the upper hand.

The extortionate ZD strategies have the peculiar property of sharply distinguishing between "sentient" players, who have a theory of mind about their opponents, and "evolutionary" players, who may be arbitrarily good at exploring a fitness landscape (either locally or globally), but who have no theory of mind. The distinction does not depend on the details of any particular theory of mind, but only on Y's ability to impute to X an ability to alter her strategy.

If X alone is witting of ZD strategies, then IPD reduces to one of two cases, depending on whether Y has a theory of mind. If Y has a theory of mind, then IPD is simply an ultimatum game [15, 16], where X proposes an unfair division and Y can either accept or reject the proposal. If he does not (or if, equivalently, X has fixed her strategy and then gone to lunch), then the game is dilemma-free for Y. He can maximize his own score only by giving X even more; there is no benefit to him in defecting.

If X and Y are both witting of ZD, then they may choose to negotiate to each set the other's score to the maximum co-operative value. Unlike naive PD, there is no advantage in defection, because neither can affect his or her own score and each can punish any irrational defection by the other. Nor is this equivalent to the classical TFT strategy [7], which produces indeterminate scores if played by both players.

To summarize, player X, witting of ZD strategies, sees IPD as a very different game from how it is conventionally viewed. She chooses an extortion factor χ, say 3, and commences play. Now, if she thinks that Y has no theory of mind about her [13] (e.g., he is an evolutionary player), then she should go to lunch leaving her fixed strategy mindlessly in place. Y's evolution will bestow a disproportionate reward on her. However, if she imputes to Y a theory of mind about herself, then she should remain engaged and watch for evidence of Y's refusing the ultimatum (e.g., lack of evolution favorable to both). If she finds such evidence, then her options are those of the ultimatum game [16]. For example, she may reduce the value of χ, perhaps to its "fair" value of 1.

Now consider Y's perspective, if he has a theory of mind about X. His only alternative to accepting positive, but meager, rewards is to refuse them, hurting both himself and X. He does this in the hope that X will eventually reduce her extortion factor. However, if she has gone to lunch, then his resistance is futile.

It is worth contemplating that, though an evolutionary player Y is so easily beaten within the confines of the IPD game, it is exactly evolution, on the hugely larger canvas of DNA-based life, that ultimately has produced X, the player with the mind.

Appendix A: Shortest-Memory Player Sets the Rules of the Game. In iterated play of a fixed game, one might have thought that a player Y with longer memory of past outcomes has the advantage over a more forgetful player X. For example, one might have thought that player Y could devise an intricate strategy that uses X's last 1,000 plays as input data in a decision algorithm, and that can then beat X's strategy, conditioned on only the last one iteration. However, that is not the case when the same game (same allowed moves and same payoff matrices) is indefinitely repeated. In fact, for any strategy of the longer-memory player Y, X's score is exactly the same as if Y had played a certain shorter-memory strategy (roughly, the marginalization of Y's long-memory strategy), disregarding any history in excess of that shared with X.

Let X and Y be random variables with values x and y that are the players' respective moves on a given iteration. Because their scores depend only on (x, y) separately at each time, a sufficient statistic is the expectation of the joint probability of (X, Y) over past histories H (of course in their proportion seen). Let $H = [H_0, H_1]$, where H_0 is the recent history shared by both X and Y, and H_1 is the older history seen only by Y. Then a straightforward calculation is,

$$
\begin{aligned}
\langle P(x,y\,|\,H_0,H_1)\rangle_{H_0,H_1} &= \sum_{H_0,H_1} P(x,y\,|\,H_0,H_1)P(H_0,H_1) \\
&= \sum_{H_0,H_1} P(x\,|\,H_0)P(y\,|\,H_0,H_1)P(H_0,H_1) \\
&= \sum_{H_0} P(x\,|\,H_0)\left[\sum_{H_1} P(y\,|\,H_0,H_1)P(H_1\,|\,H_0)P(H_0)\right] \\
&= \sum_{H_0} P(x\,|\,H_0)\left[\sum_{H_1} P(y,H_1\,|\,H_0)\right]P(H_0) \\
&= \sum_{H_0} P(x\,|\,H_0)P(y\,|\,H_0)P(H_0) \\
&= \langle[P(x,y\,|\,H_0)\rangle_{H_0}
\end{aligned}
\tag{18}
$$

Here, the first line makes explicit the expectation, and the second line expresses conditional independence.

Thus, the result is a game conditioned only on H_0, where Y plays the marginalized strategy

$$
P(y\,|\,H_0) \equiv \sum_{H_1} P(y,H_1\,|\,H_0).
\tag{19}
$$

Because this strategy depends on H_0 only, it is a short-memory strategy that produces exactly the same game results as Y's original long-memory strategy.

Note that if Y actually wants to compute the short-memory strategy equivalent to his long-memory strategy, he has to play or simulate the game long enough to compute the above expectations over the histories that would have occurred for his long-memory strategy. Then, knowing these expectations, he can, if he wants, switch to the equivalent short-memory strategy.

To understand this result intuitively, we can view the game from the forgetful player X's perspective: If X thinks that Y's memory is the same as her own, she imputes to Y a vector of probabilities the same length as his own. Because the score for the play at time t depends only on expectations over the players' conditionally independent moves at that time, Y's use of a longer history, from X's perspective, is merely a peculiar kind of random number generator whose use does not affect either player. So Y's switching between a long- and short-memory strategy is completely undetectable (and irrelevant) to X.

The importance of this result is that the player with the shortest memory in effect sets the rules of the game. A player with a good memory-one strategy can force the game to be played, effectively, as memory-one. She cannot be undone by another player's longer-memory strategy.

Appendix B: ZD Strategies Succeed Without Markov Equilibrium. We here prove that Y cannot evade X's ZD strategy by changing his own strategy on a short timescale — even arbitrarily on every move of the game. The point is that Y cannot usefully "keep the game out of Markov equilibrium" or play "inside the Markov equilibration time scale."

For arbitrary $\mathbf{p}$ and $\mathbf{q}$, the Markov matrix is as shown in Fig. 2A. We suppose that X is playing a ZD strategy with some fixed $\mathbf{p}_z$ and write $\mathbf{M}(\mathbf{q}) \equiv \mathbf{M}(\mathbf{p}_z, \mathbf{q})$. The key point is that each row of $\mathbf{M}(\mathbf{q})$ is linear in exactly one component of $\mathbf{q}$. Thus, the average of any number of different $\mathbf{M}(\mathbf{q}_i)$'s satisfies

$$\langle \mathbf{M}(\mathbf{q}_i) \rangle_i = \mathbf{M}(\langle \mathbf{q}_i \rangle_i). \tag{20}$$

Now consider the result of N consecutive plays, $i = 1, 2, \ldots, N$, where N is a large number. The game goes through N states α_i, with $\alpha \in \{cc, cd, dc, dd\}$. Comparing times i and $i+1$, the game goes from state α_i, to state α_{i+1} by a draw from the four probabilities $M_{\alpha_i, \alpha_{i+1}}(q_{i\alpha_i})$, $\alpha_{i+1} \in \{cc, cd, dc, dd\}$ where $q_{i\alpha_i}$ is the α_ith component of $\mathbf{q}_i$, (at time i). So the expected number of times that the game is in state β is

$$\langle N_\beta \rangle = \sum_{i=1}^{N} M_{\alpha_i \beta}(q_{i\alpha_i})$$

$$= \sum_\alpha \sum_{i \mid \alpha} M_{\alpha\beta}(q_{i\alpha})$$

$$= \sum_\alpha N_\alpha \langle M_{\alpha\beta}(q_{i\alpha}) \rangle_{i \mid \alpha}$$

$$= \sum_\alpha N_\alpha M_{\alpha\beta}(\langle q_{i\alpha} \rangle_{i \mid \alpha}) \tag{21}$$

Here the notation $i \mid \alpha$ is to be read as "for values of i such that $\alpha_i = \alpha$."

Now taking the (ensemble) expectation of the right-hand side and defining probabilities

$$P_\alpha = \frac{1}{N} \langle N_\alpha \rangle, \tag{22}$$

Eq. (21) becomes

$$P_\beta = \sum_\alpha P_\alpha M_{\alpha\beta}(\langle q_{i\alpha} \rangle_{i \mid \alpha}). \tag{23}$$

This result shows that a distribution of states identical to those actually observed would be the stationary distribution of Y's playing the fixed strategy $q_\alpha = \langle q_{i\alpha} \rangle_{i \mid \alpha}$. Because X's ZD strategy is independent of any fixed strategy of Y's, we have shown that, for large N, X's strategy is not spoiled by Y's move-to-move strategy changes.

That the proofs in *Appendix A* and *Appendix B* have a similar flavor is not coincidental; both exemplify situations where Y devises a supposedly intricate strategy that an oblivious X automatically marginalizes over.

Acknowledgments

We thank Michael Brenner, Joshua Plotkin, Drew Fudenberg, Jeff Hussmann, and Richard Rapp for helpful comments and discussion.

References

1. Axelrod R, Hamilton WD (1981) The evolution of cooperation. *Science* 211:1390–1396.
2. Roberts K (1985) Cartel behavior and adverse selection. *J Industr Econ* 33:401–413.
3. Axelrod R, Dion D (1988) The further evolution of cooperation. *Science* 242: 1385–1390.
4. Nowak M, Sigmund K (1993) A strategy of win-stay, lose-shift that outperforms tit-for-tat in the Prisoner's Dilemma game. *Nature* 364:56–58.
5. Nowak MA (2006) Five rules for the evolution of cooperation. *Science* 314:1560–1563.
6. Kendall G, Yao X, Chong SY (2007) *The Iterated Prisoners' Dilemma 20 Years On* (World Scientific, Singapore).
7. Axelrod R (1984) *The Evolution of Cooperation* (Basic Books, New York).
8. Dawkins R (1988) *The Selfish Gene* (Oxford Univ Press, New York).
9. Poundstone W (1992) *Prisoner's Dilemma* (Doubleday, New York).
10. Hauert Ch, Schuster HG (1997) Effects of increasing the number of players and memory steps in the Iterated Prisoner's Dilemma, a numerical approach. *Proc Biol Sci* 264:513–519.
11. Premack DG, Woodruff G (1978) Does the chimpanzee have a theory of mind? *Behav Brain Sci* 1:515–526.
12. Saxe R, Baron-Cohen S, eds (2007) *Theory of Mind: A Special issue of Social Neuroscience* (Psychology Press, London).
13. Lurz RW (2011) *Mindreading Animals: The Debate over What Animals Know about Other Minds* (MIT Press, Cambridge, MA).
14. Ewens WJ (1989) An interpretation and proof of the fundamental theorem of natural selection. *Theor Popul Biol* 35:167–180.
15. Güth W, Schmittberger R, Schwarze B (1982) An experimental analysis of ultimatum bargaining. *J Econ Behav Organ* 3:367–388.
16. Nowak MA, Page KM, Sigmund K (2000) Fairness versus reason in the ultimatum game. *Science* 289:1773–1775.

PARTITIONS AND THE GRAND CANONICAL ENSEMBLE*

Two disconnected remarks about partitions. First, a pedagogical remark connecting pure mathematics with statistical physics. The grand canonical ensemble of statistical mechanics is applied to the counting of partitions. This picture borrowed from physics gives a simple approximation to the exact calculation of the partition function by Hardy and Ramanujan. Second, an exact formula is guessed for the function $N_S(m, n)$ defined in a recent paper by Andrews, Garvan and Liang. The formula was subsequently proved by Garvan. We hope that it may lead to a better understanding of the beautiful new congruence properties of partitions discovered by Andrews.

1. A Physical Model for Partitions

I am delighted to offer this small contribution to Ramanujan's one hundred and twenty fifth birthday. I follow his good example, jumping from one subject to another without any logical connection. First, I use a physical model to describe the partitions of a large integer n. The model is the Grand Canonical Ensemble, used to describe physical systems with a variable number of particles. Each part in the partition is regarded as a particle in the model. For each integer k, there is a separate species of particle of energy k. The particles are emitted and absorbed like the photons in a cavity filled with thermal radiation. The states of the entire system are labeled by the numbers $[m_1, m_2, m_3, \ldots]$, where m_k is the number of particles of energy k. So each state describes a partition with sum

$$n = \sum_k km_k. \tag{1}$$

---

*For *The Ramanujan Journal*, December 2012 issue.

In the physical model, n is the energy of the state. The model is in thermal equilibrium with temperature T, so the probability of the state $[m_k]$ is

$$f(m_k) = Z^{-1} \prod_k \exp(-bkm_k), \tag{2}$$

with

$$Z = \sum_m \prod_k \exp(-bkm_k) = \prod_k (1 - x^k)^{-1} = \sum_n p(n)x^n, \tag{3}$$

$$x = \exp(-b), \quad b = T^{-1}. \tag{4}$$

The name "partition function" is given by physicists to the sum Z and by mathematicians to the coefficients $p(n)$. For mathematicians, $p(n)$ is the number of partitions of n. For physicists, the mean value of the energy of the ensemble is

$$E = -Z^{-1}dZ/db = \sum_k k(\exp(bk) - 1)^{-1}, \tag{5}$$

the free energy is

$$F = -b^{-1} \log Z = b^{-1} \sum_k \log(1 - \exp(-bk)), \tag{6}$$

and the entropy is

$$S = b(E - F). \tag{7}$$

In the grand canonical ensemble, the mean number of parts of size k is

$$a(k) = (\exp(bk) - 1)^{-1}, \tag{8}$$

and the mean number of parts greater than k is

$$g(k) = \sum_{k+1}^{\infty} a(j) = \sum_1^{\infty} \exp(-bjk)(\exp(bj) - 1)^{-1}. \tag{9}$$

If b is small and k is large, (9) reduces to

$$g(k) = -b^{-1} \log(1 - \exp(-bk)), \tag{10}$$

which can be written in the symmetrical form

$$\exp(-bg(k)) + \exp(-bk) = 1. \tag{11}$$

When b is small, the number of parts becomes large and the shape of the average partition converges to the smooth curve

$$\exp(-x) + \exp(-y) = 1, \tag{12}$$

where (x/b) is the number of parts larger than (y/b). The symmetry of (12) is the well-known symmetry between a partition and its conjugate. According to (8), the mean energy E is equal to the mean sum of all the parts,

$$E = \sum_k ka(k) = \sum_k \sum_j k \exp(-jkb) = \sum_j (1/4)(\sinh^{-2}(jb/2)), \tag{13}$$

The sum (13) is separated into two parts, the first exactly summable, the second having a non-singular summand,

$$E = \sum_j (jb)^{-2} + \sum_j [(1/4)(\sinh^{-2}(jb/2) - (jb)^{-2}]. \tag{14}$$

The second sum can be approximated by an integral, so (14) becomes

$$E = (\pi^2/6)b^{-2} - (1/2)b^{-1} + (1/24), \tag{15}$$

with an error of the order of b. The coefficient $(\pi^2/6)$ in (15) is equal to the area under the curve (12). According to (5) and (6),

$$E = d(bF)/db, \tag{16}$$

so that the free energy is obtained by integrating (15),

$$F = -(\pi^2/6)b^{-2} - (1/2)b^{-1}\log(b/c), \tag{17}$$

where c is a constant of integration and terms of order unity are neglected. A rather tricky calculation of the asymptotic behavior of F for small b, using (6), gives the result

$$c = 2\pi. \tag{18}$$

The entropy is given by (7),

$$S = (\pi^2/3)b^{-1} + (1/2)(\log(b/2\pi) - 1), \tag{19}$$

neglecting terms of order b. The thermodynamics of the partitions of a large number are described by (15), (17)–(19), with temperature given by (4).

The statistical distribution of the energy n in the grand canonical ensemble can be exactly described by the cumulant generating function

$$C(y) = \sum c_k(y^k/k!) = \log\langle\exp(ny)\rangle,$$ (20)

where y is a formal expansion parameter. The cumulants are the mean c_1, the variance c_2, the skewness c_3, the kurtosis c_4, and so on. The definition (20) together with (1)–(3) gives

$$C(y) + \log Z(b) = \log Z(b - y),$$ (21)

and then by (6) and (17),

$$C(y) = (\pi^2/6)[y/(b(b - y))] - (1/2)\log[b/(b - y)],$$ (22)

$$c_k = [(\pi^2/6)b^{-k-1} - (2k)^{-1}b^{-k}]k!,$$ (23)

neglecting terms of higher order in b. According to (20), the scale of the distribution-function of the random variable n is set by the square-root of the variance c_2, so the scale is proportional to $b^{-3/2}$. When scaled to the size of the distribution, the skewness is of order $b^{1/2}$, the kurtosis is of order b, and all the higher cumulants c_k are of order $b^{(k/2)-1}$. So when b is small, the distribution of n becomes Gaussian with deviations that vanish as b tends to zero.

It remains to use the thermodynamic model to estimate the size of the partition-function $p(n)$ appearing in (3). Let N be a fixed large integer. The ensemble of states with the sum (1) exactly equal to N is the microcanonical ensemble $M(N)$, containing exactly $p(N)$ states. The grand canonical ensemble with inverse temperature b is obtained by adding together all the microcanonical ensembles $M(n)$ with the statistical weights x^n according to (3) and (4). We choose the temperature b^{-1} so that the quantity

$$\log(f(x)x^{-N}) = (b(N - F))$$ (24)

is a minimum as b varies with N fixed. The condition for (24) to be a minimum is according to (16)

$$N = F + b(dF/db) = E.$$ (25)

So the grand and microcanonical ensembles have the same mean energy N, and the value of (24) at the minimum is the entropy S of the grand ensemble according to (7). Thus (6) and (17) give

$$f(x) = \exp(\pi^2/6b)\sqrt{(b/2\pi)}.$$ (26)

The terms of the series

$$f(x)x^{-N} = \sum p(n)x^{n-N}, \tag{27}$$

have a maximum value $p(N)$ at $n = N$, and decrease on each side with a distribution that is Gaussian with variance given by (23),

$$V = (\pi^2/3)b^{-3}, \tag{28}$$

to leading order in b. Therefore

$$f(x) = p(N)\exp(-bN)\sqrt{(2\pi V)}, \tag{29}$$

with fractional error tending to zero with b. Putting together (29) with (15), (17) and (26), we find the asymptotic formula

$$p(N) = (4N\sqrt{3})^{-1} \exp(\pi\sqrt{(2N/3)}), \tag{30}$$

in agreement with [Hardy and Ramanujan, 1918, Eq. (1.41), p. 280]. The asymptotic formula (30) was the starting-point for the historic analysis by which Hardy and Ramanujan obtained an exact formula for $p(N)$. Physical arguments can only take us to the starting-point. To go further, Hardy and Ramanujan studied $f(x)$ as a function of the complex variable x analytic in the unit circle, and introduced the magic of modular forms to determine its behavior precisely all around the circle.

2. An Exercise in Experimental Mathematics

Recently, George Andrews, Frank Garvan and Jie Liang wrote a paper [Andrews et al., 2011], with the title, "Combinatorial interpretations of congruences for the SPT-function". The SPT-function spt(n) is the total number of smallest parts in all the partitions of an integer n. They proved that

$$\text{spt}(n) = \sum_m N_S(m, n), \tag{31}$$

where the integer m is called the spt-crank of a partition of n and takes values from $n - 1$ to $1 - n$. The $N_S(m, n)$ are defined as sums over partitions of contributions equal to plus one or minus one, and it came as a surprise to Andrews et al. when they were able to prove that all the $N_S(m, n)$ are positive. In their paper, they display a table of numerical values of $N_S(m, n)$ for n up to 20 and m up to 10. Since $N_S(m, n)$ is an even function of m, only non-negative values of m are displayed.

As an exercise in experimental mathematics, I looked for a simple pattern in the numerical values of the $N_S(m, n)$, and guessed a simple formula which makes it easy to calculate them. The formula also explains why they are all positive. The formula was proved correct by Garvan [2012]. The clue to guessing the formula was the behavior of the numbers on the right-hand side of the table where $m \geq n/2$. In this region the values are all equal to the standard partition function $p(n - m - 1)$. When I subtracted $p(n-m-1)$ from all the numbers in the table, the result was again simple in the region where $m \geq n/3$. In that region the residuals were all negative and equal to minus $(p(n - 2m - 3) + p(n - 2m - 5))$. Adding these terms to the residuals, I found the result to be again simple in the region $m \geq n/4$. In that region the second residuals were positive and equal to $(p(n-3m-6)+p(n-3m-9)+p(n-3m-12))$. The pattern was now clear. My guess for the general formula was

$$N_S(m, n) = \sum_{k=1}^{\infty} (-1)^{k-1} \sum_{j=0}^{k-1} p(n - k(m+j) - (k(k+1)/2)), \qquad (32)$$

which fitted all the data that I had available. For each n, the sum terminates after a small number of terms.

The formula (32) explains in a simple way why the $N_S(m, n)$ are all positive. It is convenient to use the generating function

$$A_m(x) = \sum_{n=1}^{\infty} N_S(m, n)x^n. \qquad (33)$$

Then (32) becomes

$$A_m(x) = Z(x) \sum_{k=1}^{\infty} (-1)^{k-1} \sum_{j=0}^{k-1} x^{(k(k+1)/2)+k(j+m)}, \qquad (34)$$

where $Z(x)$ is the generating function (3) of ordinary partitions. Now we group the terms of the sum (34) together in groups of four,

$$(k, j+1), \quad (k+1, j), \quad (k+1, j+1), \quad (k+2, j), \qquad (35)$$

with k and j both odd. These groups of four cover the double sum (34) exactly, except for the terms (k, k) which appear instead of the correct $(k, k-1)$. So the sum (34) becomes

$$A_m(x) = Z(x) \sum_{k=1}^{\infty} \left[\sum_{j=1}^{k} x^{(k(k+1)/2)+k(j+m)}(1 - x^{j+m})(1 - x^{j+m+k+1}) \right.$$

$$\left. + x^{(k(3k-1)/2)+mk}(1 - x^k) \right], \qquad (36)$$

with the sums taken over k and j both odd. The minus signs of (32) have now disappeared into factors like $(1 - x^{j+m})$ which merely cancel the corresponding factors in $Z(x)$. From (36) it follows immediately that $N_S(m, n)$ is positive. In fact $N_S(m, n)$ is a sum of terms counting the number of partitions of $(n - (k(k+1)/2) - k(m+j))$ into parts with $(j+m)$ and $(j+m+k+1)$ missing, plus partitions of $(n - (k(3k-1)/2) - mk)$ with k missing.

Another interesting consequence of (34) is the identity

$$A_m(x) - A_{m+1}(x) = Z(x) \sum_{k=1}^{\infty} (-1)^{k-1} \left(x^{(k(k+1)/2)+mk} - x^{(k(3k+1)/2)+mk} \right). \quad (37)$$

This shows the difference $(A_m - A_{m+1})$ to be closely related to the generating functions of ranks and cranks according to Eqs. (7.7) and (7.17) of [Garvan, 1988, pp. 68 and 70]. It looks as if there are more treasures to be found here if we dig a little deeper. I am grateful to Andrews, Garvan and Liang for sharing their results with me before publication and for encouraging me to publish this little addendum.

References

G. E. Andrews, F. G. Garvan and J. Liang, "Combinatorial interpretations of congruences for the SPT-function", to be published.

F. G. Garvan, , "New combinatorial interpretations of Ramanujan's partition congruences mod 5, 7 and 11", *Trans. Am. Math. Soc.* **305**, 47–77.

F. G. Garvan, Companion paper in this issue of the *Ramanujan Journal*.

G. H. Hardy and S. Ramanujan [1918] "Asymptotic formulae in combinatory analysis", *Proc. Lond. Math. Soc.* **2**, 75–115.

IS A GRAVITON DETECTABLE?*

1. Introduction

I am enormously grateful to Dr. K. K. Phua, and to everyone else who had a hand in organizing this conference, for inviting me to visit Singapore. I am also grateful to my old and new friends who came to Singapore to help me celebrate my birthday. As a former Brit, I am delighted to see this sparkling new country, which has prospered by giving free play to Chinese enterprise while still driving on the left side of the road.

Now I come to the technical substance of my talk. It is generally agreed that a gravitational field exists, satifying Einstein's equations of general relativity, and that gravitational waves traveling at the speed of light also exist. The observed orbital shrinkage of the double pulsar [Weissberg and Taylor, 2005] provides direct evidence that the pulsar is emitting gravitational waves at the rate predicted by the theory. The LIGO experiment now in operation is designed to detect kilohertz gravitational waves from astronomical sources. Nobody doubts that gravitational waves are in principle detectable.

This talk is concerned with a different question, whether it is in principle possible to detect individual gravitons, or in other words, whether it is possible to detect the quantization of the gravitational field. The words "in principle" are ambiguous. The meaning of "in principle" depends on the rules of the game that we are playing. If we assert that detection of a graviton is in principle impossible, this may have three meanings. Meaning (a): We can prove a theorem asserting that detection of a graviton would contradict the laws of physics. Meaning (b): We have examined a class of possible graviton detectors and demonstrated that they cannot work. Meaning (c): We have examined a class of graviton detectors and demonstrated

---

*Talk given at Nanyang Technological University, Singapore, August 26, 2013.

that they cannot work in the environment provided by the real universe. We do not claim to have answered the question of "in principle" detectability according to meaning (a). In Sec. 3 we look at detectors with the LIGO design, detecting gravitational waves by measuring their effects on the geometry of spacetime, and conclude that they cannot detect gravitons according to meaning (b). In Secs. 4 and 5 we look at a different class of detectors, observing the interactions of gravitons with individual atoms, and conclude that they cannot detect gravitons according to meaning (c). In Secs. 6 and 7 we look at a third class of detectors, observing the coherent transitions between graviton and photon states induced by an extended classical magnetic field, and find that they also fail according to meaning (c).

In Sec. 2 we look at a historic argument used by Niels Bohr and Leon Rosenfeld to demonstrate the quantum behavior of the electromagnetic field, and explain why this argument does not apply to the gravitational field. In Sec. 8 we briefly examine the possibility of observing primordial gravitons at the beginning of the universe by measuring the polarization of the cosmic background radiation today. No definite conclusions are reached. This talk is a report of work in progress, not a finished product. It raises the question of the observability of gravitons but does not answer it. There is much work still to do.

2. The Bohr–Rosenfeld Argument

Before looking in detail at graviton detectors, I want to discuss a general theoretical question. In 1933 a famous paper by Niels Bohr and Leon Rosenfeld [1933], was published in the proceedings of the Danish Academy of Sciences with the title, "On the question of the measurability of the electromagnetic field strengths". An English translation by Bryce de Witt, dated 1960, is in the Institute library in Princeton, bound in an elegant hard cover. This paper was a historic display of Bohr's way of thinking, expounded in long and convoluted German sentences. Rosenfeld was almost driven crazy, writing and rewriting fourteen drafts before Bohr was finally satisfied with it. The paper demonstrates, by a careful and detailed study of imaginary experiments, that the electric and magnetic fields must be quantum fields with the commutation relations dictated by the theory of quantum electrodynamics. The field-strengths are assumed to be measured by observing the motion of massive objects carrying charges and currents with which the fields interact. The massive objects are subject to the rules of ordinary quantum mechanics which set limits to the accuracy of simultaneous measurement of positions and velocities of the objects. Bohr and Rosenfeld show that the quantum-mechanical limitation of measurement of the motion of the masses implies precisely the limitation of measurement of the field-strengths imposed by quantum electrodynamics. In other words, it is mathematically inconsistent to have a classical electromagnetic field interacting with a quantum-mechanical measuring apparatus.

A typical result of the Bohr–Rosenfeld analysis is their Eq. (58),

$$\Delta E_x(1)\Delta E_x(2) \sim \hbar|A(1,2) - A(2,1)|. \tag{1}$$

Here the left side is the product of the uncertainties of measurement of two averages of the x-component of the electric field, averaged over two space-time regions (1) and (2). On the right side, $A(1,2)$ is the double average over regions (1) and (2) of the retarded electric field produced in (2) by a unit dipole charge in (1). They deduced (1) from the standard Heisenberg uncertainty relation obeyed by the measuring apparatus. The result (1) is precisely the uncertainty relation implied by the commutation rules of quantum electrodynamics. Similar results are found for other components of the electric and magnetic fields.

The question that I am asking is whether the argument of Bohr and Rosenfeld applies also to the gravitational field. If the same argument applies, then the gravitational field must be a quantum field and its quantum nature is in principle observable. However, a close inspection of the Bohr–Rosenfeld argument reveals a crucial feature of their measurement apparatus that makes it inapplicable to gravitational fields. In the last paragraph of Sec. 3 of the Bohr–Rosenfeld paper, they write: "In order to disturb the electromagnetic field to be measured as little as possible during the presence of the test body system, we shall imagine placed beside each electric or magnetic component particle another exactly oppositely charged neutralizing particle". The neutralizing particles have the following function. Suppose we have a mass carrying a charge or current J whose movement is observed in order to measure the local electric or magnetic field. The movement of the charge or current J produces an additional electromagnetic field that interferes with the field that we are trying to measure. So we must compensate the additional field by adding a second mass, carrying the charge or current $-J$ and occupying the same volume as the first mass. The second mass is constrained by a system of mechanical linkages and springs to follow the movement of the first mass and cancels the fields generated by the first mass. This cancellation is an essential part of the Bohr–Rosenfeld strategy. It is then immediately obvious that the strategy fails for measurement of the gravitational field. The test-objects for measuring the gravitational field are masses rather than charges, and there exist no negative masses that could compensate the fields produced by positive masses.

The conclusion of this argument is that the Bohr–Rosenfeld analysis does not apply to the gravitational field. This does not mean that the gravitational field cannot be quantized. It means only that the quantization of the gravitational field is not a logical consequence of the quantum behavior of the measuring apparatus. The fact that the electromagnetic field must be quantized does not imply that the gravitational field must be quantized.

3. Can LIGO Detect a Graviton?

In the LIGO experiment, if it is successful, we shall detect a classical gravitational wave, not an individual quantum of gravity. A classical wave may be considered to be a coherent superposition of a large number of gravitons. LIGO is supposed to detect a wave with a strain amplitude f of the order of 10^{-21}. According to [Landau and Lifshitz, 1975, p. 370] the energy density of this wave is

$$E = (c^2/32\pi G)\omega^2 f^2, \tag{2}$$

where G is Newton's constant of gravitation and ω is the angular frequency. For a wave with angular frequency 1 Kilohertz and amplitude 10^{-21}, Eq. (2) gives an energy density of roughly 10^{-10} ergs per cubic centimeter. A single graviton of a given angular frequency ω cannot be confined within a region with linear dimension smaller than the reduced wavelength (c/ω). Therefore, the energy density of a single graviton of this frequency is at most equal to the energy of the graviton divided by the cube of its reduced wave-length, namely

$$E_s = (\hbar\omega^4/c^3). \tag{3}$$

For an angular frequency of 1 Kilohertz, the single graviton energy density is at most 3.10^{-47} ergs per cubic centimeter. So any gravitational wave detectable by LIGO must contain at least 3.10^{37} gravitons. This wave would be barely detectable by the existing LIGO. For a LIGO apparatus to detect a single graviton, its sensitivity would have to be improved by a factor of the order of 3.10^{37}. Even this vast improvement of sensitivity would probably not be sufficient, because the detection of weak signals is usually limited not only by the sensitivity of the apparatus but also by the presence of background noise. But to see whether detection of single gravitons is possible in principle, we disregard the problem of background noise and analyze the structure and operation of a super-sensitive LIGO detector.

For a rough estimate of the sensitivity of a LIGO apparatus required to detect a single graviton, we equate (2) with (3). This gives the strain f to be detected by the apparatus,

$$f = (32\pi)^{1/2}(L_p\omega/c), \tag{4}$$

where L_p is the Planck length

$$L_p = (G\hbar/c^3)^{1/2} = 1.4 \times 10^{-33}\,\text{cm}. \tag{5}$$

The strain is derived from a measurement of the variation of distance between two mirrors separated by a distance D. The variation of the measured distance is equal to fD, so long as D does not exceed the reduced wave-length (c/ω) of the graviton. For optimum detectability we take D equal to (c/ω). Then the variation of distance is by (4)

$$\delta = (32\pi)^{1/2} L_p. \tag{6}$$

Up to a factor of order unity, the required precision of measurement of the separation between the two mirrors is equal to the Planck length, and is independent of the frequency of the graviton.

Is it possible in principle for a LIGO apparatus to measure distances between macroscopic objects to Planck-length accuracy? The following simple arguments give a negative answer to this question. First consider the case in which the objects are floating freely in space. The Heisenberg uncertainty relation between position and momentum of freely floating objects gives the lower bound

$$M\delta^2 \geq \hbar T, \tag{7}$$

for the variation of distance δ, where M is the mass of each object and T is the duration of the measurement. Now T must be greater than the time (D/c) required to communicate between the two mirrors. If δ is equal to the Planck length, (5) and (7) imply

$$D \leq (GM/c^2). \tag{8}$$

So the separation between the two mirrors is less than the Schwarzschild radius of each of them, the negative gravitational potential pulling them together is greater than Mc^2, and they are bound to collapse into a black hole before the measurement can be completed.

We next consider the situation that arises when the two mirrors are clamped in position by a rigid structure. In this case the precision of measurement of the distance between the two mirrors is limited by quantum fluctuations of the rigid structure. We use a simple dimensional argument to estimate the magnitude of the fluctuations. Let s be the velocity of sound in the structure, let D be the separation between the mirrors, and let M be the mass of the structure. There will be at least one mode of sound-vibration of the structure which gives a displacement affecting the measurement of D. The mean-square quantum fluctuation amplitude of the displacement in this mode will then be, up to a factor of order unity, at least as large as the zero-point fluctuation,

$$\delta^2 \geq (\hbar D/Ms). \tag{9}$$

The duration of the measurement must be of the order of (D/c), the time it takes the graviton to travel through the apparatus. This duration is shorter than the period (D/s) of the sound-vibration, since s cannot exceed c. Therefore the uncertainty of the measurement is at least equal to the instantaneous vibration-amplitude δ. If the uncertainty is as small as the Planck length (5), then (9) implies

$$(GM/c^2) \geq (c/s)D > D. \tag{10}$$

Again we see that the separation between the two mirrors is smaller than the Schwarzschild radius of the apparatus, so that the negative gravitational potential of the apparatus is greater than Mc^2 and it will again collapse into a black hole. It appears that Nature conspires to forbid any measurement of distance with error smaller than the Planck length. And this prohibition implies that detection of single gravitons with an apparatus resembling LIGO is impossible.

It is clear from Eq. (3) that we have a better chance of detecting a single graviton if we raise the frequency into the optical range and use a different kind of detector. When the frequency is of the order of 10^{15} Hertz or higher, a single graviton can kick an electron out of an atom, and the electron can be detected by standard methods of atomic or particle physics. We are then dealing with the gravito-electric effect, the gravitational analog of the photo-electric effect which Einstein used in 1905 [Einstein, 1905], to infer the existence of quanta of the electromagnetic field, the quanta which were later called photons. The possibility of detecting individual gravitons in this way depends on two quantities, (a) the cross-section for interaction of a graviton with an atom, and (b) the intensity of possible natural or artificial sources of high-frequency gravitons. Most of this talk will be concerned with estimating these two quantities.

4. Graviton Detectors

The simplest kind of graviton detector is an electron in an atom, which we may approximate by considering the electron to be bound in a fixed potential. We choose coordinate axes so that the z-axis is the direction of propagation of a graviton. There are then two orthogonal modes of linear polarization for the graviton, one with the wave-amplitude proportional to xy, and the other with the amplitude proportional to $(x^2 - y^2)$. We choose the x- and y-axes so that they make angles of forty-five degrees to the plane of polarization of the graviton. Then the matrix element for the electron to absorb the graviton and move from its ground state a to another state b is proportional to the mass-quadrupole component,

$$D_{ab} = m \int \psi_b^* xy \psi_a d\tau, \tag{11}$$

where m is the electron mass. Equation (11) is the quadrupole approximation, which is valid so long as the wave-length of the graviton is large compared with the size of the atom. The total cross-section for absorption of the graviton by the electron is

$$\sigma(\omega) = (4\pi^2 G \omega^3 / c^3) \sum_b |D_{ab}|^2 \delta(E_b - E_a - h\omega), \tag{12}$$

where E_a and E_b are the energies of the initial and final states. It is convenient to consider a logarithmic average of the cross-section over all frequencies ω,

$$S_a = \int \sigma(\omega) d\omega / \omega. \tag{13}$$

Integration of (12) gives the sum-rule

$$S_a = 4\pi^2 L_p^2 Q, \tag{14}$$

where the Planck length L_p is given by (4), and

$$Q = \int |(x\partial/\partial y + y\partial/\partial x)\psi_a|^2 d\tau \tag{15}$$

is a numerical factor of order unity. It is remarkable that the average cross-section (14) is independent of the electron mass and of the nuclear charge. The same formula (14) holds for the absorption of a graviton by a neutron or proton bound in a nuclear potential.

For simplicity we assume that the electron is in a state with zero component of angular momentum about the z-axis, with a wave-function $f(s, z)$, where s is the distance from the z-axis. Then (15) becomes

$$Q = \left(\int s^3 [f']^2 ds\, dz \right) \Big/ \left(2 \int s[f]^2 ds\, dz \right), \tag{15}$$

where f' means the partial derivative of f with respect to s. The inequality

$$\int s^3 [f' - (f/s)]^2 ds\, dz > 0 \tag{16}$$

implies that for any $f(s, z)$

$$Q > 1/2. \tag{17}$$

On the other hand, if the electron is in an s-state

$$f(r) = r^{-n} \exp(-r/R), \tag{18}$$

where r is distance from the origin, then

$$Q = (4/5)[1 - (n/6)]. \tag{19}$$

From (17) and (19) it appears that for any tightly-bound s-state Q will be close to unity. The cross-section for absorption of a graviton by any kind of particle will be of the same magnitude

$$4\pi^2 L_p^2 = 4\pi^2 G\hbar/c^3 = 8 \times 10^{-65}\,\text{cm}^2, \tag{20}$$

spread over a range of graviton energies extending from the binding-energy of the particle to a few times the binding-energy. For any macroscopic detector composed of ordinary matter, the absorption cross-section will be of the order of 10^{-41} square centimeters per gram.

5. Thermal Graviton Generators

We have a splendid natural generator of thermal gravitons with energies in the kilovolt range, producing far more gravitons than any artificial source. It is called the sun. Stephen Weinberg long ago calculated [Weinberg, 1965] the graviton luminosity of the sun, caused by gravitational bremsstrahlung in collisions of electrons and ions in the sun's core. A later calculation [Gould, 1985] corrected a mistake in Weinberg's paper but does not substantially change the result. For an electron-ion collision with energy E, the differential cross-section $p(\omega)$ for producing a graviton of energy $\hbar\omega$ is divergent at low energies, so that the total cross-section has no meaning. The physically meaningful quantity is the integral of the differential cross-section multiplied by the energy of the graviton,

$$\int p(\omega)\hbar\omega\,d\omega = (320/9)Z^2\alpha^2 L_p^2 E, \tag{21}$$

where α is the electromagnetic fine-structure constant and Z is the charge of the ion. Including a similar contribution from electron–electron collisions, (21) gives a total graviton luminosity of the sun

$$L_g = 79\,\text{Megawatts}, \tag{22}$$

or about 10^{24} gravitons per second with energy in the kilovolt range. This gives a flux at the earth of

$$F_g = 4 \times 10^{-4}\,\text{gravitons per cm}^2\text{ per second.} \tag{23}$$

If we imagine the whole mass of the earth to be available as raw material for the manufacture of graviton detectors, with the cross-section (20) per electron and the flux (23), the counting-rate is 2.4×10^{-17} per second. If the experiment continues for the lifetime of the sun, which is five billion years, the expected total number of gravitons detected will be four. The experiment barely succeeds, but in principle it can detect gravitons.

According to [Gould, 1985], there exist in the universe sources of thermal gravitons which are stronger than the sun, namely hot white dwarfs at the beginning of their lives, and hot neutron stars. Gould estimates the graviton luminosities of a typical white dwarf and a typical neutron star to be respectively 10^4 and 10^{10} times solar. Their luminosities are roughly proportional to their central densities. But the lifetimes during which the stars remain hot are shorter than the lifetime of the sun, being of the order of tens of millions of years for the white dwarf and tens of thousands of years for the neutron star. The lifetime output of gravitons will therefore be respectively 100 and 10^5 times solar. To stretch the theoretical possibilities of detection to the limit, we may suppose the detector to have mass equal to the sun and to be orbiting around the source of gravitons at a distance of 0.01 astronomical unit with an orbital period of eight hours. Then the expected number of gravitons detected will be of the order of 10^{13} for the white dwarf and 10^{16} for the neutron star. The detection rate is roughly one per minute for the white dwarf and 3×10^4 per second for the neutron star. The conclusion of this calculation is that graviton detection is in principle possible, if we disregard the problem of discriminating the graviton signal from background noise.

The most important source of background noise is probably the neutrinos emitted by the sun or the white dwarf or the neutron star as the case may be. These neutrinos can mimic graviton absorption events by ejecting electrons from atoms as a result of neutrino-electron scattering. The neutrinos have higher energy than the gravitons, but only a small fraction of the neutrino energy may be transferred to the electron. From the sun, about 10^{14} neutrinos are emitted for each graviton, and the cross-section for neutrino-electron scattering is about 10^{20} times the cross-section for graviton absorption, (see [Fukugita and Yanagida, 2003]). Therefore there will be about 10^{34} neutrino background events for each graviton absorption event.

For white-dwarfs and neutron-stars the ratio of background to signal is even larger, since neutrino production and scattering cross-sections increase with temperature more rapidly than graviton production and absorption cross-sections. Without performing detailed calculations, we can assert that for all thermal sources of gravitons the ratio of neutrino background to graviton signal will be of the order of 10^{34} or greater. In all cases, the total number of detected graviton events is vastly smaller than the square-root of the number of background events. The graviton signal will be swamped by the statistical scatter of the background noise.

Before jumping to conclusions about the detectability of gravitons, we must explore possible ways in which the neutrino background events might be excluded. The first possible way is to surround the detector with a shield thick enough to stop neutrinos but let gravitons pass. If the shield is made of matter of ordinary density, its thickness must be of the order 10^{10} kilometers, and its mass is so large that it will collapse into a black hole. The second possible way is to surround the graviton detector with neutrino detectors in anti-coincidence, to catch the outgoing neutrino after each scattering event. This way fails for the same reason as the shield. The neutrino detectors would need to be at least as massive as the shield. The third possible way is to build a shield or a set of anti-coincidence detectors out of some mythical material with super-high density. The known laws of physics give us no clue as to how this might be done. We conclude that, if we are using known materials and known physical processes in a noisy universe, detection of thermal gravitons appears to be impossible.

6. Non-thermal Gravitons

It is possible to imagine various ways in which energetic objects such as pulsars may emit non-thermal gravitons of high energy. One such way is a process first identified by [Gertsenshtein, 1961], the coherent mixing of photon and graviton states in the presence of an extended classical magnetic field. The graviton emission from various celestial objects resulting from the Gertsenshtein process was calculated by [Papini and Valluri, 1989]. Some interestingly high graviton luminosities were predicted.

The Gertsenshtein process results from the interaction energy

$$(8\pi G/c^4)h_{ij}T_{ij}, \tag{24}$$

between the gravitational field h_{ij} and the energy-momentum tensor T_{ij} of the electromagnetic field. This interaction expresses the fact that electromagnetic fields have weight, just like other forms of energy. Now suppose that h_{ij} is the field of a graviton traveling in the z direction and

$$T_{ij} = (1/4\pi)(B_i + b_i)(B_j + b_j), \tag{25}$$

is the energy-momentum of the photon magnetic field b_i superimposed on a fixed classical magnetic field B_i. Then the interaction (24) contains the term

$$I = (4G/c^4)h_{xy}B_x b_y, \tag{26}$$

bilinear in the graviton and photon fields. The effect of this bilinear term is to mix the photon and graviton fields, so that a particle that is created as a photon may be transformed into a graviton and vice versa. There is an oscillation between graviton and photon states, just like the oscillation between neutrino states that causes neutrinos to change their flavors while traveling between the sun and the earth. If a photon travels a distance D though a uniform transverse magnetic field B, it will emerge as a graviton with probability

$$P = \sin^2\left(G^{1/2}BD/2c^2\right) = \sin^2\left(D/L\right), \tag{27}$$

with the mixing-length

$$L = (2c^2/G^{1/2}B) \qquad (28)$$

independent of wave-length. In all practical situations, D will be small compared with L, so that

$$P = (GB^2D^2/4c^4). \qquad (29)$$

The quadratic dependence of P on D makes this process interesting as a possible astrophysical source of gravitons. The numerical value of L according to (28) is roughly

$$L = (10^{25}/B), \qquad (30)$$

when L is measured in centimeters and B in Gauss.

We may also consider the Gertsenshtein process as the basis of a graviton detector consisting of a hollow pipe of length D filled with a transverse magnetic field B. The tube must be accurately pointed at a putative source of gravitons in the sky. At the far end of the tube is a shield to block incident photons, and at the near end is a detector of photons resulting from the conversion of gravitons on their way through the tube. If D is one astronomical unit (10^{13} cm), then (29) gives

$$P = 10^{-24}B^2. \qquad (31)$$

The field B must be very strong to obtain a reasonable rate of conversion of gravitons to photons. A detector with the same design has been used in a real experiment to detect axions that might be created by thermal processes in the core of the sun [Zioutas *et al.*, 2005]. The axion field is supposed to interact with the electromagnetic field with an interaction energy similar to (26), but with a much larger coupling constant. The experimenters at CERN in Switzerland are using a surplus magnet from the Large Hadron Collider project as an axion-detector, pointing it at the sun and looking for kilovolt photons resulting from conversion of axions into photons. The length of the magnet is 9 meters and the magnetic field is 9×10^4 Gauss. They have not yet detected any axions.

The Gertsenshtein process does not require the classical magnetic field to be uniform. For a non-uniform field, the conversion of photons to gravitons still occurs with probability given by (27), if we replace the product BD by the integral of the transverse component of B along the trajectory of the photons. Likewise, the conversion will not be disturbed by a background gravitational field, even when the field is strong enough to curve the photon trajectory, because the gravitational field acts in the same way on photons and gravitons. In a curved spacetime, the photons and the gravitons follow the same geodesic paths, and the photon and graviton waves remain coherent.

7. Nonlinear Electrodynamics

However, there is an important disturbing factor which was neglected in previous discussions of the Gertsenshtein process. The disturbing factor is the nonlinearity of the electromagnetic field caused by quantum fluctuations of electron-positron pairs in the vacuum [Euler and Heisenberg, 1936; Wentzel, 1943]. The fourth-order term in the electromagnetic field energy density is [Wentzel, 1943, p. 190],

$$(\alpha/360\pi^2 H_c^2)[(E^2 - H^2)^2 + 7(E.H)^2], \tag{32}$$

where α is the fine-structure constant and

$$H_c = (m^2 c^3/e\hbar) = 5.10^{13} \text{Gauss} \tag{33}$$

is the critical magnetic field at which electron-positron pair fluctuations become noticeable.

When the field in (32) is divided into classical and photon components as in (25), there is a term quadratic in both the classical and photon fields,

$$(\alpha/360\pi^2 H_c^2)(4(B.b)^2 + 7(B.e)^2), \tag{34}$$

where b and e are the magnetic and electric fields of the photon. From (34) it follows that the photon velocity v is not equal to c but is reduced by a fraction

$$g = 1 - (v/c) = (k\alpha B^2/360\pi^2 H_c^2). \tag{35}$$

The coefficient k is equal to four or seven for a photon polarized with its magnetic field or its electric field parallel to B. We consider the case $k = 4$, since that case is more favorable to the Gertsenshtein process. Since the graviton field is not affected by the nonlinear electromagnetic interaction (32), the graviton velocity is precisely c, and the photon and graviton waves will lose coherence after traveling for a distance

$$L_c = (c/g\omega) = (90\pi^2 c H_c^2/\alpha B^2 \omega) = (10^{43}/B^2 \omega). \tag{36}$$

If the propagation distance D is larger than L_c, the Gertsenshtein process fails and the formula (29) for the photon-graviton conversion probability is incorrect. A necessary condition for the Gertsenshtein process to operate is

$$DB^2 \omega \leq 10^{43}. \tag{37}$$

Furthermore, even when the Gertsenshtein process is operating, the probability of photon-graviton conversion according to (29) and (37) is

$$P \leq (10^{36}/B^2 \omega^2). \tag{38}$$

We are interested in detecting astrophysical sources of gravitons with energies up to 100 kilovolts, which means frequencies up to 10^{20}. With $\omega = 10^{20}$, (37) and

(38) become

$$D \leq (10^{23}/B^2), \quad P \leq (10^{-4}/B^2). \tag{39}$$

We consider two situations in which (39) has important consequences. First, with typical values for the magnetic field and linear dimension of a pulsar, $B = 10^{12}$ and $D = 10^6$, (39) shows that the Gertsenshtein process fails by a wide margin. The calculations of the graviton luminosity of pulsars in [Papini and Valluri, 1989] assume that the Gertsenshtein process is producing high-energy gravitons. These calculations, and the high luminosities that they predict, are therefore incorrect. Second, in the hollow pipe graviton detector which we considered earlier, (39) shows that the Gertsenshtein process can operate with a modest field, $B = 10^5$ Gauss, and a pipe length $D = 10^{13}$ cm, but the probability of detection of each graviton traveling through the pipe is only 10^{-14}. If the field is made stronger, the length of the pipe must be shorter according to (39), and the probability of detecting a graviton becomes even smaller. The detector can work in principle, but fails for practical reasons in the real universe.

8. Conclusions

We have examined three possible kinds of graviton detector with increasingly uncertain results. First, the LIGO detector for low-energy gravitons, which we prove ineffective as a consequence of the laws of physics. Second, the gravito-electric detector for kilovolt gravitons, which we prove ineffective as a consequence of the background noise caused by neutrino processes in the real universe. Third, the coherent graviton-conversion detector for high-energy gravitons, is ineffective only because of practical limits to the size of magnetic detectors. In addition to these three kinds of detector, there is a fourth kind which actually exists, the Planck space telescope, detecting polarization of the microwave background radiation. According to Alan Guth [1997], the polarization of the background radiation in an inflationary universe could provide direct evidence of the existence of single gravitons in the primordial universe before inflation. The results of the Planck polarization measurements are not yet published, and it remains to be seen whether the observations are able to distinguish between primordial gravitons and other gravitational effects of primordial matter. The question, whether a detector of present-day microwave radiation is in principle able to detect primordial gravitons, remains open.

Many papers have been published, for example [Eppley and Hannah, 1977; Page and Geilker, 1981], claiming to demonstrate that the gravitational field must be quantized. What these papers demonstrate is that a particular theory with a classical gravitational field interacting with quantum-mechanical matter is inconsistent.

Page and Geilker assume that the classical gravitational field is generated by the expectation value of the energy-momentum tensor of the matter in whichever quantum state the matter happens to be. They performed an ingenious experiment to verify that this assumption gives the wrong answer for a measurement of the gravitational field in a real situation.

In this talk I am not advocating any particular theory of a classical gravitational field existing in an otherwise quantum-mechanical world. I am raising three separate questions. I am asking whether either one of three theoretical hypotheses may be experimentally testable. One hypothesis is that gravity is a quantum field and gravitons exist as free particles. A second hypothesis is that gravity is a quantum field but gravitons exist only as confined particles, like quarks, hidden inside composite structures which we observe as classical gravitational fields. The third hypothesis is that gravity is a statistical concept like entropy or temperature, only defined for gravitational effects of matter in bulk and not for effects of individual elementary particles. If the third hypothesis is true, then the gravitational field is not a local field like the electromagnetic field. The third hypothesis implies that the gravitational field at a point in spacetime does not exist, either as a classical or as a quantum field.

I conclude that the first hypothesis may be experimentally testable, but the second and third may not. Analysis of the properties of graviton-detectors, following the methods of this paper, cannot distinguish between the second and third hypotheses. Three outcomes are logically possible. If a graviton detector is possible and succeeds in detecting gravitons, then the first hypothesis is true. If graviton detectors are possible and fail to detect gravitons, then the first hypothesis is false and the second and third are open. If a graviton detector is in principle impossible, then all three hypotheses remain open. Even if their existence is not experimentally testable, gravitons may still exist.

The conclusion of the analysis is that we are still a long way from settling the question whether gravitons exist. But the question whether gravitons are in principle detectable is also interesting and may be easier to decide.

In conclusion, I wish to thank Tony Rothman and Steven Boughn [2006], for helpful conversations and for sharing their thoughts with me before their paper was published.

References

N. Bohr and L. Rosenfeld [1933] *Kgl. Danske Vidensk. Selsk., Mat.-Fys. Med.* **12**, 8; The English translation by B. DeWitt, "On the question of the measurability of the electromagnetic field strengths", preserves the Germanic style of the original.

H. Euler and W. Heisenberg [1936] "Folgerungen aus der diracschen theorie der positrons", *Zeits. für Phys.* **98**, 714–732.

M. E. Gertsenshtein, "Wave resonance of light and gravitational waves", *JETP* **41**, 113–114; English translation [1962] in *Soviet Physics JETP* **14**, 84–85.

A. Einstein [1905] "Über einen die erzeugung und verwandlung des lichtes betreffenden heuristischen gesichtspunkt", *Ann. der Phys.* **17**, 132–148.

K. Eppley and E. Hannah [1977] "On the logical necessity of quantizing the gravitational field", *Found. Phys.* **7**, 51–68.

M. Fukugita and T. Yanagida [2003] *Physics of Neutrinos and Applications to Astrophysics* (Springer Verlag, Berlin).

R. J. Gould [1985] "The graviton luminosity of the sun and other stars", *Astrophys. J.* **288**, 789–794.

A. H. Guth [1997] *The Inflationary Universe: The Quest for a New Theory of Cosmic Origins* (Helix Books, Addison-Wesley, Reading, Massachusetts).

L. D. Landau and E. M. Lifshitz [1975] *The Classical Theory of Fields, Course of Theoretical Physics, Volume 2*, fourth revised English edition, translated by Morton Hamermesh (Elsevier, Oxford).

D. N. Page and C. D. Geilker [1981] "Indirect evidence for quantum gravity", *Phys. Rev. Lett.* **47**, 979–982.

G. Papini and S.-R. Valluri [1989] "Photoproduction of high-frequency gravitational radiation by galactic and extragalactic sources", *Astron. Astrophys.* **208**, 345–350, and earlier papers there cited.

T. Rothman and S. Boughn [2006] "Can gravitons be detected?", *Found. Phys.* **36**, 1801–1825.

S. Weinberg [1965] "Infrared photons and gravitons", *Phys. Rev. B* **140**, 516–524.

G. Wentzel [1943] *Einführung in die Quantentheorie der Wellenfelder*, [Wien, Franz Deuticke]; Reprinted [1946] (Edwards Brothers, Ann Arbor).

K. Zioutas *et al.* [2005] "First results from the CERN Axion Solar Telescope (CAST)", preprint hep-ex/0411033v2. For this reference I am indebted to E. Witten.

INDEX